# The Iconography of Water

# The Iconography of Water

## Well and Tank Forms of the Indian Subcontinent

Fredrick W. Bunce

DK PRINTWORLD
Publishers of Indian Traditions

**Cataloging in Publication Data — DK**

[Courtesy: D.K. Agencies (P) Ltd. <docinfo@dkagencies.com>]

**Bunce, Fredrick W.**

The iconography of water : well and tank forms of the Indian subcontinent / Fredrick W. Bunce.

p. cm.

Includes bibliographical references (p. ).

**ISBN** 13: 9788124606971

1. Stepwells – India – History. 2. Stepwells – India – Designs and plans. 3. Tanks – India – History. 4. Tanks – India – Designs and plans. 5. Swimming pools – India – History. 6. Swimming pools – India – Designs and plans. 7. Hindu temples – India – History. 8. Hindu temples – India – Designs and plans. 9. Monuments – India. 10. Architecture, Ancient – India. I. Title.

DDC 725.740954 23

**ISBN** 13: 978-81-246-0697-1 **ISBN** 10: 81-246-0697-8

First published in India, 2013



*Printed and published by:*

D.K. Printworld (P) Ltd.

*Regd. office : "Vedaśrī,"* F-395, Sudarshan Park

(Metro Station: Ramesh Nagar)

New Delhi - 110 015

*Phones* : (011) 2545 3975, 2546 6019; *Fax* : (011) 2546 5926

*e-mail* : indology@dkprintworld.com

*Web* : www.dkprintworld.com

# Contents

# Contents (Continued)

# List of Plates

## List of Plates (Continued)



### Appendix C

# List of Plates (Continued)

**Plate** **Page**



## Addendum

# List of Figures

# List of Figures (Continued)



## Addendum

## Abbreviations

**AK** Arnar Nath Khanna, *Archaeology of India*.
**CT** Christopher Tadgell, *The History of Architecture in India*.
**HC** Henry Cousens, *The Chalukyan Architecture of the Kanarese Districts*.
**HS I** Henri Stierlin, *India*.
**HS II** Henri Stierlin, *Islamic India*.
**HS III** Henri Stierlin, *Hindu India from Khajuraho to the Temple City of Madurai*.
**HSa** H. Sarkar, *An Architectural Survey of the Temples of Kerala*.
**FB I** Fredrick W. Bunce, *An Encyclopaedia of Buddhist Deities, Demigods, Godlings, Saints and Demons: With Special Focus on Iconographic Attributes*.
**FB II** Fredrick W. Bunce, *A Dictionary of Buddhist and Hindu Iconography — Illustrated — Objects, Devices, Concepts, Rites and Related Terms*.
**FB III** Fredrick W. Bunce, *The Iconography of Plans: A Study of the Influence of Buddhism and Hinduism on Architectural Plans of South and Southeast Asia*.
**FB IV** Fredrick W. Bunce, *Numbers Found in Buddhist and Hindu Practices: An Iconographic Consideration*.
**FB V** Fredrick W. Bunce, *Islamic Tombs in India: The Iconography and Genesis of Their Design*.
**FB VI** Fredrick W. Bunce, *The Sacred Dichotomy: Thoughts and Comments on the Duality of Female and Male Iconography in South Asia and the Mediterranean*.
**FB VII** Fredrick W. Bunce, *Royal Palaces, Residences and Pavilions of India: 13th through 18th Centuries — An Iconographic Consideration*.
**JN** Jutta Jain-Neubauer, *The Stepwells of Gujarat: In Art–Historical Perspective*.
**KD** Krishna Deva, *Temples of India*.
**ML** Morna Livingston, *Steps to Water: The Ancient Stepwells of India*.
**SK** Stella Kramrisch, *The Hindu Temple*.

# Preface

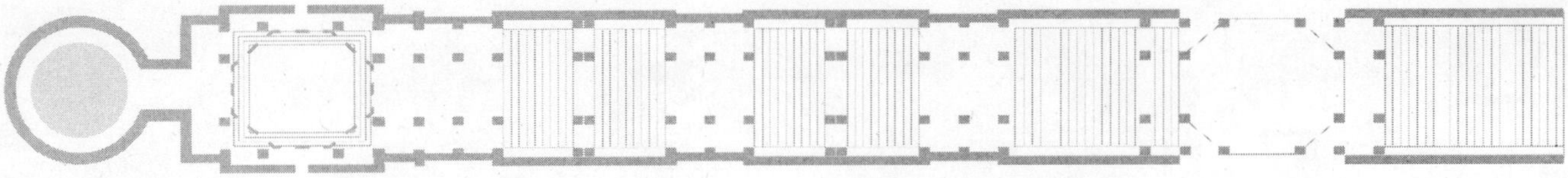

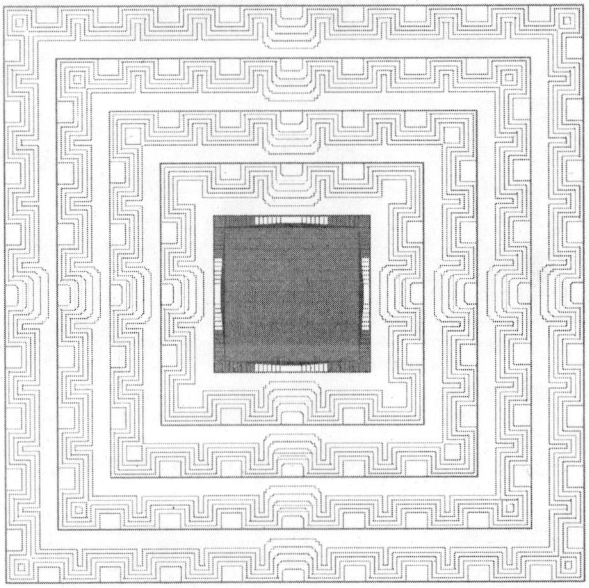

# Preface

GENERATIONS of historians and art/architectural historians have lauded and dissected many of the "major" monuments of the Indian subcontinent. Who in the modern, technological world has not heard of the Taj Mahal? Who of those who have been educated in art and architectural history have not heard of the excavated temples of Ellorā and Ajantā, the monuments of Sāñcī, Mahābalipuram, Khajurāho, Ranakpur; of the great palaces of the Rājpūt at Jodhpur, Jaipur and Datiā; of the magnificent Mughal mosques; Imperial palaces of Fatehpur Sīkrī or Shāhjahānābād; the Tomb of I'timād-ud-Daulāh; and the temple cities of Madurai, Tiruchirapalli and Tiruvannamalai?

Yet, few are familiar or have even heard of the splendid the magnificent Sūrya Temple Tank in Moḍherā (**Plate 8**), the vertiginous Cānda Bāolī at Ābhānerī (**Plate 10**), the incomparable Rānī-kī Wāv of Pāṭan (**Plate 11**), the beautiful Kalayni Tank of Hulikere (**Plate 12**), or the impressive Rudabāī-nī Wāv at Adalaj (**Plate 19**). One is at a loss as to why these monuments as a genre have been so generally neglected in the oeuvres of art/architectural history, particularly those that deal with the Indian subcontinent.[1] Is it because they do not rise above the surrounding areas in which they are found? Is it because they do not exhibit many of the structural/engineering qualities apparent in other soaring structures? Or, is it because their raison d'être was less spiritual and more practical? Certainly it is not because they are so few! In fact, there are literally thousands of these structures throughout India.[2] But, they have been generally neglected and one wonders why![3]

Therefore, this brief study will be devoted to a number of these monuments, both tanks (*kuṇḍa*) and step wells (*wāv, bāolī*). The emphasis will be on the plan of the various structures and other salient information. This is not meant to be an exhaustive study, rather it is undertaken with the hope that it will offer a number of keys by which others may use in unlocking the mysteries and beauties of these neglected monuments.

**Fredrick W. Bunce**

1 Of course, the exception is the recent, thought provoking work of Morna Livingston, *Steps to Water: The Ancient Stepwells of India*, Princeton: 2002; and the earlier study of Jutta Jain-Neubauer, *The Stepwells of Gujarat: In Art – Historical Perspective*, Bangalore: 1981.

2 **Appendix A** provides but a brief list of the step wells and tanks found throughout the subcontinent.

3 Every researcher has his or her preference, but to allow these to seep into their research is highly problematical. Krishna Deva in his important study of the *Temples of India* (1995) merely mentions the major sites of the south. None of the great temple cities is considered in any length — e.g. Śrī Raṅganāthaswāmī, Śrī Jambukeśvara, Aruṇācaleśvara, Cidambaram Naṭarājar Mīnākṣī Sundareśvara temples. This preference seems to apply to monuments as well. Virtually every general study of Indian architectural history will provide photos and often plans of the temple city of Mīnākṣī Sundareśvara at Madurai. But, is it the exemplar of the southern temple cities? What about the elaborate Śrī Raṅganāthaswāmī Temple Complex? One begins to view this state of things as proceeding from some conspiracy. Whether Vaiṣṇava or Śaiva, whether individual preference is for Draviḍa, Nāgara or Vesara ought not to enter into ones consideration, unless that is the focus of the undertaking.

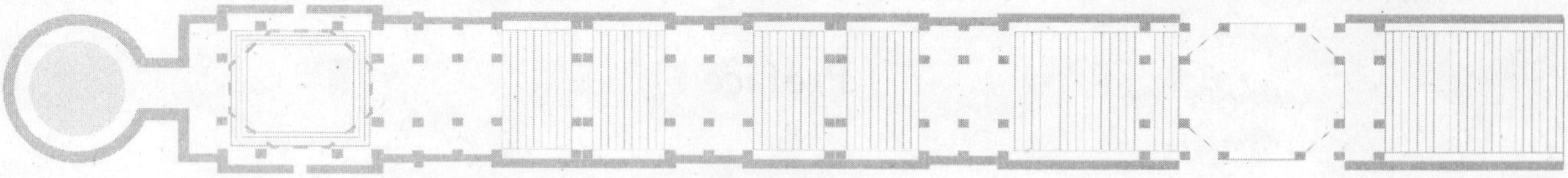

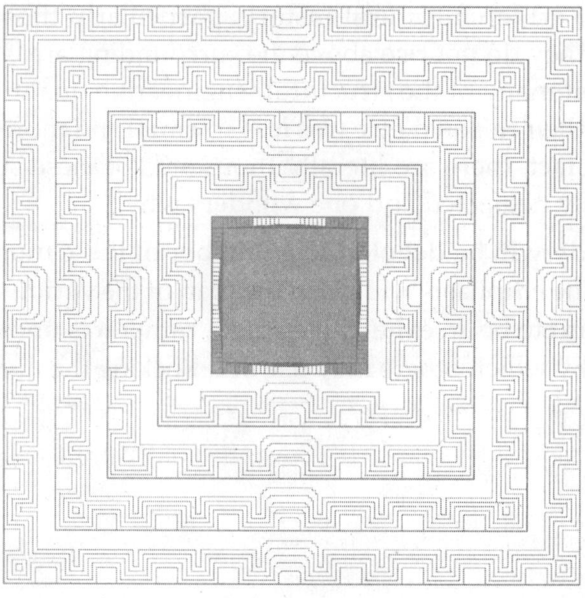

# Introduction

# Introduction

"Water, water everywhere, and not a drop to drink."
— *The Rime of the Ancient Mariner*, Samuel Taylor Coleridge

HUMANKIND can do without many things that we in our modern world have learned to expect or tol that we need — e.g. automobiles, telephones, televisions, radios, airplanes and paved streets. But, suitable, potable drinking water is not one of them! It, like the air we breathe, is an absolute requirement, a categorical necessity to sustain human life.

There are many regions on this planet between the Arctic and Antarctic circles where the lack of sufficient rainfall simply does not support human life. Areas which do not have sources of suitable drinking water without modern and costly technological applications are sprinkled across the globe — e.g. the Sahara Desert, the Namib Desert, the Gobi Desert, the Arabian Desert, the Great Victoria Desert and the Atacama Desert. If these desert areas are to support human existence, then life-giving water must be brought in and frequently at enormous cost. On the opposite side, there are other areas which seem quite beneficent to the sustenance of human and animal life. The northern Great Plains of North America, particularly the state of Minnesota — i.e. the "Land of Ten-thousand Lakes" — sees a constant and abundant supply of life-giving water via rain, lakes, rivers and easily accessible aquifers. There the water table, in many areas, lies but a few feet below the surface. There are many other regions of the world which are able to easily and regularly supply this basic need of humankind.

**Water as a Cultural Imperative**

An abundant supply of water was and is an absolute requirement for the establishment and sustenance of a civilization. The Nile gave birth to the brilliant Egyptian civilization; the Tigris and Euphrates fostered the powerful Sumarians, the Babylonians, the Assyrians and the Persians; and the Indus sustained Mohen̊jo-Dāṛo, Dholāvīra and Haṛappā. However, abuse of the environment can create catastrophic effects that may affect the availability of water. Archaeologists and ecological historians posit that the sudden collapse of the Anasazi civilization in south-west United States and the Maya civilization in Central America were due, in part, to the ravaging of the local ecology.

In Asia, the Irrawaddy supported the brilliant Pagan period. The city grew in size, prosperity and reputation. Scores, Nay! hundreds of small and large, brick temples were constructed in and adjacent to that lustrous city. The eventual effect of this building frenzy was the razing, the denuding of the forests around Pagan and along the Irrawaddy. These stands of timber were felled to feed the insatiable brick producing kilns. This ultimately created an ecological backlash and the region suddenly suffered a series of severe, catastrophic droughts. Pagan was abandoned! It became virtually uninhabitable! Akbar the Great, raised the fabled Fatehpur Sīkrī and even provided it with the Anupa Tālāo or Cāra Camana for his own personal, imperial enjoyment. It was an indication, a symbol, an icon of his power and wealth. This small tank served no other purpose than to supplement the aesthetic and personal enjoyment of the Emperor. However, even the Emperor, the Great Akbar could not command the rains that fed the small river that flowed near the city's northern borders nor the rains that were needed to fill the great artificial lake (*tālāo*) of his new capital. There was simply not enough water to sustain Fatehpur Sīkrī's ballooning population. Fatehpur Sīkrī was abandoned! It became virtually uninhabitable!

India has witnessed the rise of some of the earliest civilizations on earth — e.g. Mohen̊jo-Dāṛo,

Dholāvīra and Haṛappā mentioned above. This great subcontinent[1] has evidences the efflorescence of human attainment in religion, philosophy and architecture. India has seen the growth and sustenance of its population to where it now teeters at the pinnacle of sustainable human population. Like with the genesis of other fledgling civilizations, a river seemed to have been the agar. In India's case it was the Indus River, then the great Mother Gaṅgā (the Ganges) and the Brahmaputra.

From the thick forests of the Terai along the northern borders of the Gangetic plain, to the impenetrable tangle of the Sunderban at the head of the Bay of Bengal, to the lush woodlands of the south; from the burning desert of western Rajasthan to the frigid wastes of the high mountains of Jammu-Kashmir and Himachal Pradesh, India provides a wide range of geologic features. From the Rann of Kutch and the Thar Desert in the far west where any rainfall at all is seen as a divine gift to be savoured, to Assam and Arunchal Pradesh in the north-east where some of the heaviest rainfalls in history are recorded, India provides a wide range of climatic conditions.

The vast subcontinent extends into the Indian Ocean. South of the Tropic of Cancer it is surrounded by water. But, this never-ending source of water, the Indian Ocean, does not give up its liquid riches readily except during the monsoons. The rocky Himalayas and their foothills provides a steady supply of glacial runoff for those who live near the rushing streams. But, the Eastern Ghāṭs and the Western Ghāṭs, the Arāvalī range, the Vindhyas, the Satpuṛās and the great Deccan Plateau do not allow for the retention of much of the water that is deposited during the yearly monsoons. Only the vast Gangetic plain and delta, regions along the Coromandel Coast and small areas along the western coast of the subcontinent are relatively flat and where the water table is relatively close to the surface providing a readily accessible supply of this life-giving liquid.

Water has always been a concern in the arid and semi-arid regions of India, particularly Gujarat and western Rajasthan. During the three-month wet season, the western monsoons generally deposit bountiful amounts of rain upon the land. But the monsoons can be capricious! They may come early and stay late flooding the region with an overabundance of rain. Or, it may be delayed and deposit only a fraction of the average, a fraction of what is required.

Then the dry season follows the season of plenty, a sere season with a duration that averages nine months. The earth which had drunk inches of monsoonal rain slowly dries out, gives up that gift of nature and the green, shallow-root vegetation soon thereafter turns colour and withers.

Water is not only needed for human consumption, but is also vital for domestic animals as well. In addition, the population of India has for millennia required two crops per year to sustain it. The wet season easily provides for one, generally a large crop. Early in the subcontinent's history, farmers realized that they had to construct large catchment basins (Ind.: *khadin*, *dhora*) to impound and retain monsoonal runoff in order to irrigate a second, less abundant, crop. This they did. Through necessity and ingenuity, they developed forms to capture and retain the life-giving rain — i.e. wells and tanks.

### Water as a Religious Imperative

Water, for the major religions of the world, has had ritualistic and dogmatic importance. It is required in a number of sacred rites. The use of water is universally connected with iconic purification and

[1] The term "subcontinent" is herein employed to include the countries of India, Pakistan and Bangladesh.

spiritual cleansing in every major faith. In Christianity, baptism via sprinkling or immersion is seen as an elemental ritual, an iconic application connected with initiation into the faith. Without it, some feel that the soul is lost. It is also used in rites of blessing — holy water — and physical regeneration — the water of Lourdes. A Muslim must appear in prayer before Allāh clean, physically and spiritually. Birka (Arab.: *aka fisqiya, haud, saqaya, sihrij*) or ritual ablution of the arms, hands, feet, face and ears is a requirement of every Muslim before prayer. All mosques have a pond (Arab.: *haud*) or other source of water so that this ritual may be practised. Ritual bathing of the corpse is also required before burial in Islam. For the Theravāda Buddhists of Thailand, it is considered desirable for every temple to face a water source — whether a tank, pond or river. Water is an integral element in the rituals for the dead in Buddhism. Even in Judaism, water plays an important role in rites for the dead.

However, for the Hindu and the Buddhists alike, water (Ind.: *ap*) is considered elemental. It is one of the *pañca-tanmātras*, the five essential elements; the other four being: earth (Ind.: *kṣiti*), ether (Ind.: *ākāśa*), wind (Ind.: *marut*) and fire (Ind.: *tejas*). For the Hindu, water is seen as being not only a purifying elemental source, it also has regenerative qualities. There are numerous religious and/or ritual occasions in which bathing is an absolute requirement. Normally one bathes before prayer at a temple.[2] *Tarpaṇa, śodhana* and *abhiṣekam* are rituals for Hindus in which water is a necessary element. Virtually every major Hindu temple will have associated with it a tank or other source of water, an iconic necessity. In these tanks the devotees ritually bathe in preparation for *pūjā*. Even within individual homes, devout Hindus bathe or cleanse themselves before their morning prayers at the family shrine. Bathing in the sacred Gaṅgā at least once in one's lifetime is the wish of every Hindu. The Kumbha Melā and the Mahā Kumbha Melā are sacred rites attached to bathing which are attended by millions of Hindus. There are a number of other rivers in India that are considered sacred and bathing in their waters is spiritually seen as beneficial — the Godāvarī, Kāverī, Narmadā, Sarasvatī, Sindhu and Yamunā rivers.

**Water as a Source of Power**

Water may be seen as a source of wealth and/or power, an emblem of authority and domination. Those who commanded the source of water, commanded the people. The puissant Khmer rulers understood this even though their capital, Angkor was constructed near a huge natural lake — the Tonle Sap. They required that varying amounts of water be brought to Angkor as part of the yearly tribute from their many feudatories. These offerings of water were icons of submission. The Sukhothai Kingdom announced its separation from the Angkorian yoke by purposely failing to supply their yearly quota of water. In India, virtually every fort and/or fortified city was strategically constructed where there was a source of water, hopefully abundant. This was a physical necessity. If a flowing

[2] Of the major world religions, only Christianity does not require ritual bathing before entering a church for worship. This may have been the reason that the Christians, as they visited other lands, were considered, "dirty". Required ablution for the Hindu and the Muslim before prayer was a natural extention of their everyday life. Generally, the peoples of eastern and south-eastern Asia bathed regularly. This is a practice that is centuries old. In Europe, particularly after the fall of the Roman Empire, bathing fell into disuse. Physical cleanliness was never a rampant concern in Europe. When European explorers and adventurers ventured into the Orient, particularly to the Courts of Japan, China and India they were universally viewed with unabashed and repulsed horror! They reeked of body odor, their clothes were filthy and their physical bodies were covered with vermin. In the sixteenth century, Diane de Poitiers, mistress of Henri II Valois, King of France, was considered highly unusual at the time. This was because she bathed every morning, and in cold water! It is said that Louis XIV Bourbon, The Sun King of France, took but four baths in his adult life.

source was not present, deep wells, great cisterns, vast *tālāo*s were constructed of such a size as to sustain the population of the citadel in case of a siege and stagger the imagination. The impregnable Meherāngaṛh of Jodhpur, resting on its high, rocky outcrop, possessed a number of deep wells and cisterns for their use and protection. It stood and stands in isolated, sustainable splendour, an symbol of power and wealth.

### Access to Water Sources

Within India, as with many other regions of the world there are five naturally available sources of water for human consumption and use — rain (monsoonal), aquifers, glacial runoff, natural lakes and rivers. India developed a number of forms for access, catchment and the retention of water — simple domestic wells, step wells (*bāolī*), tanks or small ponds, artificial lakes and channels. Simple wells (Ind.: *kūpa*) were an early answer to the dry months and parched throats. These wells were sunk where the water table, the aquifer was relatively close to the surface of the land, generally no more than 6 or 7 m. They later became elaborated, expanded and deepened in the form of a step well to provide not only access to the life-giving fluid, but also to furnish a commodious place for relaxation, a locale that offered respite from the scorching heat of the dry season. "Designed to bring the people and the gods together, these wells attempted to entice the gods to leave their abodes for a cool drink of water — the elixir of life."[3] This enticement, this step-well was in the form of a trench dug into the

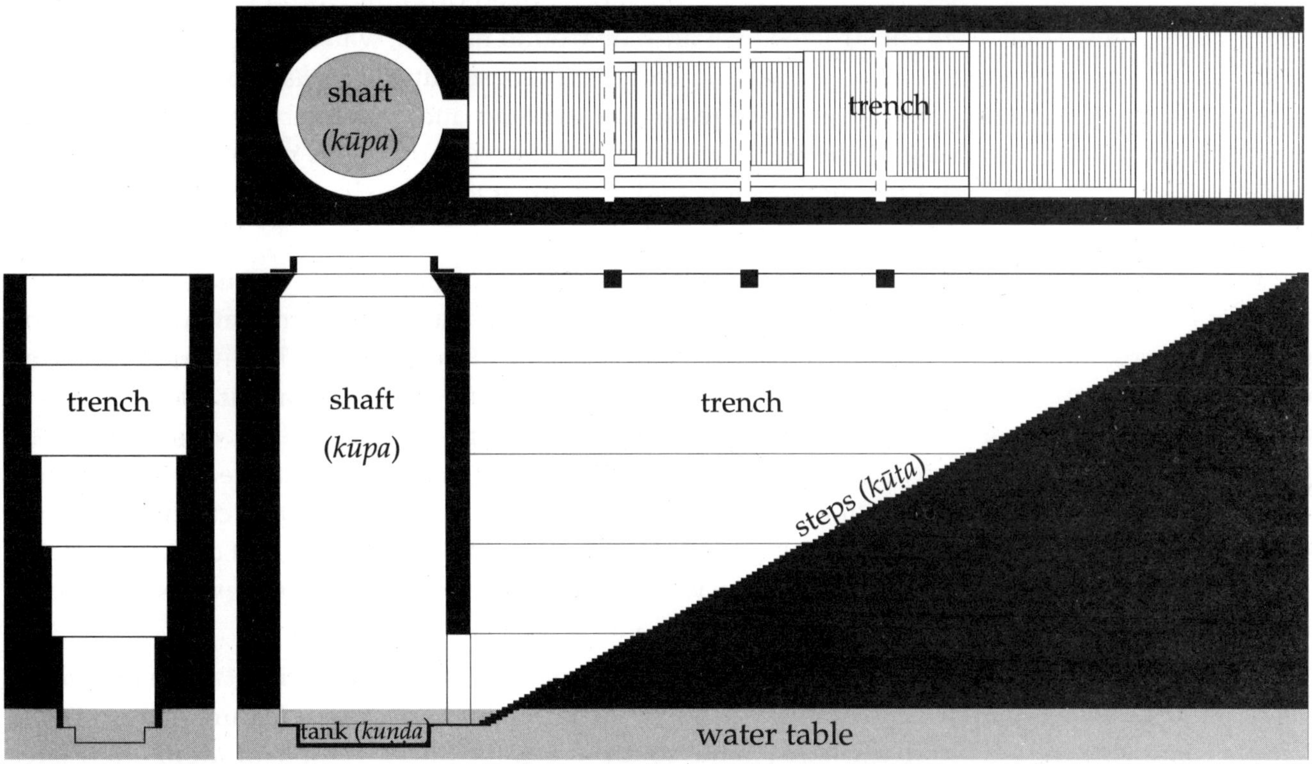


**Figure 1:** East Well (*bāolī*), Rohtās, Pakistan, plan (top) and sections (bottom) indicating the general construction and geology of a step well.

[3] Harpreet Kaur, *Vāvs, the Stepwells of Gujarat*.

earth and angled downwards until the water table or aquifer was reached. In addition, a shaft (Ind.: *kūpa*) was constructed which allowed the drawing of water from ground level (**Figure 1**). The sides of these trenches were lined with great slabs of stone, suitable flights of steps were laid and frequently there were colonnaded levels where people could gather out of the glare of the sun to relax and cool off in the shade. This was the main feature of the stepped well or step-well in northwestern India. In Rajasthan they were known as *bāvaḍīs* (Ind.: *bāolīs, bāoṛīs, bāvs, bāvḍīs, bauḍīs, bavṛī*) or *wāv* (*vāv, vāvaḍī*) in Gujarat or *baurīs* in Bījāpur. As mentioned, the goal of the stepped wells was to dig a hole or trench deep enough to reach the subterranean water table. In some areas this meant the excavation of a mere 20 or 30 ft. In other regions — e.g. areas of Gujarat, particularly around Ahmedabad and Pāṭan — the water table was 60 or 70 ft below ground level. In these deep step wells the water level remained relatively constant throughout the year and provided suitable drinking water. This water had been naturally purified by percolating down through the layer upon layer of soil providing ample drinkable water. Only extreme drought or unusually heavy rainfall would effect the level within these wells to any great extent.

Generally, step wells, tanks (*bāolī*) possessed a common architectural feature — i.e. there are a flight of stairs leading down to or surrounding the surface of the water. *Bāolīs* have one further feature which is a shaft, as mentioned above, by which water may be drawn up without descending the steps. In most cases the steps of a tank would continue downward in an even descent — even below the surface of the water. Only in the driest of seasons or during a prolonged drought would the lowermost step be revealed. A few *bāolīs* were constructed over and/or around natural, water filled clefts in the rock. Steps would lead down to the natural drop and end there. These natural *bāolīs* could be dangerous, particularly if one was not familiar with its geological makeup and was unable to swim. Then, as now, signs are posted warning of the potential danger. However, this does not always work. The Niẓāmuddīn Bāolī (**Plate 17**) is infamous as it is said to claim at least one life a year. Otherwise *bāolīs* were relatively safe, even during high-water periods.

From the earliest of times in the subcontinent, tanks and ponds had been constructed to catch the abundant rainfall, to contain natural runoff of the monsoon and save it not only for human consumption but also for irrigation during the long, hot dry season. The natural runoff would be diverted via raised embankments (Ind.: *khadin, dhora*) and channelled into catchment areas with a suitable depth as to mitigate excessive evaporation before the water was utilized. Large ponds or reservoirs (Ind. *taḍāka*) were also excavated for the catchment and retention of water that may later be utilized in irrigation in many regions of the subcontinent. In Karnataka these are known as *kere*. In Ladakh catchment ponds (Lad.: *zings*) were often constructed on solid bedrock and utilized to collect the runoff of glacial waters. Raised on a high plateau, the builders of the Shādiābad Mandira also included two extensive tanks in their plan — i.e. the Muñja Tālāo and the Kaphur Tālāo (**Plate 36**).

Smaller tanks, sacred tanks — variously known as *devara kalyāṇī, kalyāṇī, kovil kuḷam, kuḷam, kuṇḍa, puṣkariṇī, tīrtha-kuḷams, tīrtham* and *sarovara* — for ritual bathing also became a feature of most of the temple complexes in India from the north to the south — e.g. the Sūrya Temple of Moḍherā (**Plates 7-8**), Virūpākṣa-Pampī Temple Complex at Hampī (**Plates 30-31**), the Viṭṭhalapura also at Hampī (**Plates 32-33**), or the Mīnākṣī Sundareśvara Temple, Madurai (**Plates 56-57**). These temple tanks were generally within the precincts, proper of the temples and consisted of steps surrounding the whole

**Figure 2:** Typical temple tank section

and leading down to the water level (**Figure 2**). Frequently the water in these tanks was supplied by rainfall, however, most of the sacred tanks did possess outside sources such as streams and/or rivers that flowed through channels to the tank to sustain their water level throughout the year. These sacred tanks served four purposes.[4] They provided water for *abhiṣeka* of the deity, provided aquatic flowers for offerings, allowed for irrigation within the temple precincts and furnished water for ritual ablution.

**Basic Design Features**

One detail must be mentioned here. By modern standards, the angle of descent and/or ascent into or up from a majority of the step wells and/or tanks appears unduly steep. The rise of a step in relation to the depth of the tread is approximately 1:1 in most of the step wells and tanks of India (**Figure 3A**). Ergonomics was not a science that was fully appreciated or known or practised in India or the rest of the world prior to the twentieth century. In modern residences and other structures which employ stairs the ratio tends to be 1:1.5 (**Figure 3B**). This allows for greater ease in ascent and particularly descent. A stairway which has an angle of 45° (1:1) presents certain problems, particularly in descending a long flight. There is a tendency to stumble downward unless one is conscious of and focused on every step. Furthermore, the angle (45°) is visually uncomfortable, even disturbing for many people when descending. On the other hand a flight of stairs with a 1:1.5 ratio is far easier and comfortable to descend. It is ergonomically preferable. It is seen that in nearly all the step wells the flight of stairs does not descend uninterrupted, rather there are landings suitably placed along the descent. These landings give physical and visual pause to the person descending or climbing. This angle of descent or the rise of a step in relation to the depth of the tread may well have been employed for purely economic reasons. A 1:1 ratio requires less stone than a 1:1.5 ratio. In a 1:1 ratio with 50 steps would require a third less stone than if a 1:1.5 ration was employed.

A B C

**Figure 3A-C:** Proportions of rise to tread: (**A**) *bāolī*/tank, (**B**) modern residential, and (**C**) Angkor/Khmer

[4] K.V. Krishnamurthy, "Sacred Tanks: Glimpses from Epigraphy", in K.R. Srinivasan et al., *Sacred Tanks of South India*, Chennai: 2002, pp. 39-44.

On the other hand, the staircases at the temples of Angkor — the Bayon and Angkor Wat — are far steeper — i.e. 1.5:1 or the rise is one and a half times higher than the tread (see **Figure 3-C**). In order to ascend to the Khmer temples, one was and is literally forced to do so with not only the feet, but with the hands as well. Literally, one had to crawl up to the sanctuary, a proper attitude when approaching the abode of a deity.

**Engineering Constraints and Requirements**

*Ṭaṅkas, kere, zing*s and *tālāo*s all had one basic feature, regardless of their size. They had to provide a volume of space that retained the water without allowing it to seep with unrestrained rapidity into the surrounding earth. In Ladakh, the *zing*s were frequently constructed upon solid table-rock, as noted above, with suitable water-proof walls to retain the runoff. The table-rock was a natural barrier. The *zing*s were supplemental to the normal and plentiful glacial runoff, feeding the many mountain streams along the southern slopes of the Himalaya. The area surrounding Bikaner, in the Great Indian Desert, did not have this luxury. There households built small reservoirs or tanks (Ind.: *ṭaṅka*s) below ground level and generally within the confines of the house for the drinking water of the individual families. These small, residential *ṭaṅka*s would have a layer of local clay applied to the inner surface. This resisted permeation. These *ṭaṅka*s may be further lined with tile, fired brick or even stucco to keep the water as clear as possible (**Figure 4**).

The *tālāo*s, *kere*s and other catchment areas similarly needed to be excavated to table-rock or lined with suitable, impermeable material. In many cases, such as the Mūñja Tālāo at Māṇḍu, the construction of these reservoirs, these *tālāo*s was an immense, monumental task. But the necessity for unrestricted access to water far outweighed the cost of construction (**Figure 5**). The man-hours required to excavate these *tālāo*s, a back-breaking task with simple tools, were hardly a consideration, and by today's standards, a monumental task.

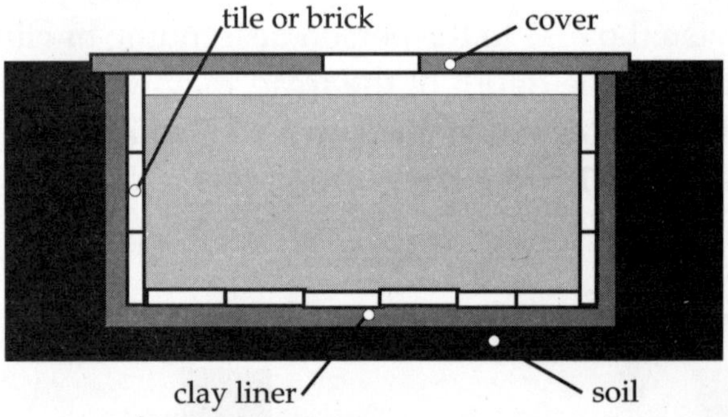


**Figure 4:** Residential *ṭaṅka*, general construction

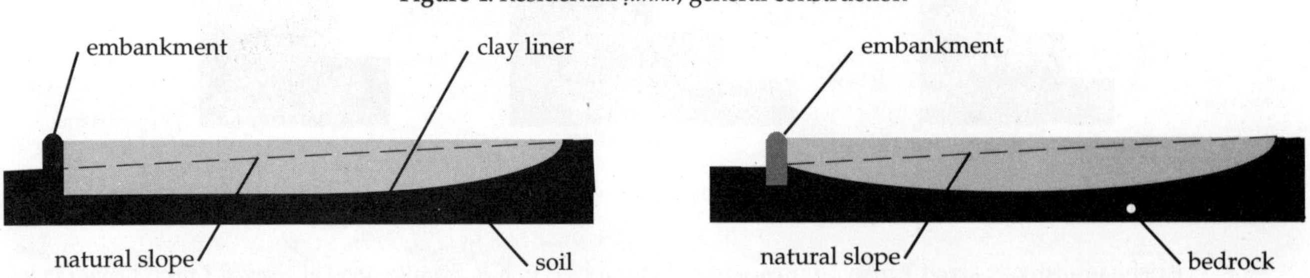


**Figure 5:** Indicating the general construction and geology of a tank, *khadin*, *zing* or *tālāo*

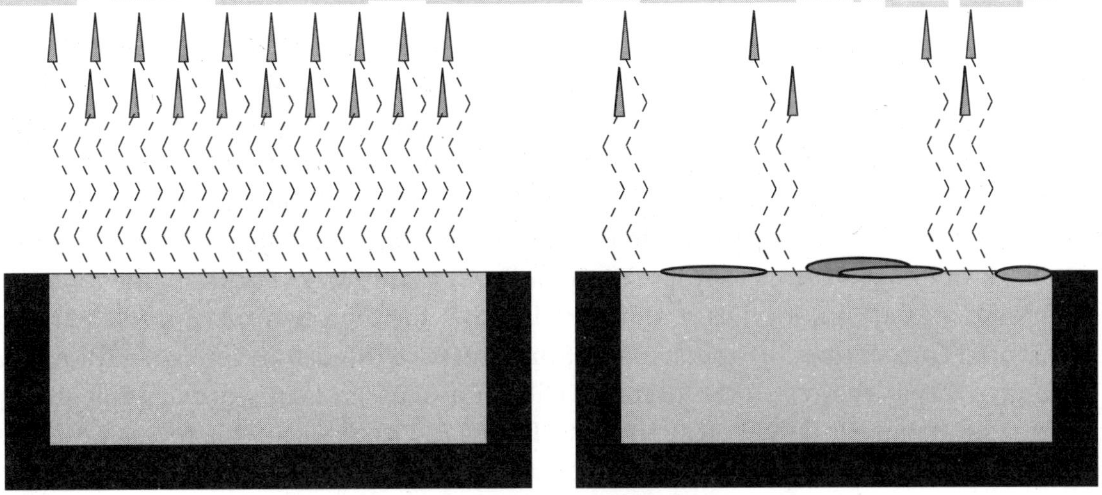

**Figure 6:** Evaporation without and with water plants

Once the problem of permeation or loss of water through the surrounding earth was solved, the next greatest problem was evaporation. Unfortunately, there was no feasible solution to this natural loss of water in tanks (*tālāo*s) that was technically available at the time. It is suggested that water plants were cultivated in the sacred tanks to lessen the water surfaces exposure and the inevitable evaporation.[5] This would have cut the amount of exposed surface of the tank and thereby lessened the rate of evaporation, somewhat (**Figure 6**).

On the other hand, the residential *ṭaṅka*s were frequently supplied with a cover. Initially this prohibited unwanted material from falling into the water. The covers also inhibited evaporation to some extent. The water level in all the open tanks — *tālāo*s and *kere*s — would recede during the long dry period under the burning sun. This evaporation would reveal one step after another as the level fell and the bottom was reached (**Figure 7**). Frequently, in many tanks the last step would be somewhat deeper, particularly in a bathing tank. There was a relationship between the surface area of a tank and its depth that mitigated excessive loss of the volume. Simply put: a 1 cu m volume of water with an exposed surface of 4 sq m would be expected to evaporate four times faster than a 1 cu m volume of water with an exposed surface of but 1 sq m. Therefore, surface area was never a solution, depth was. However, where there was a readily accessible water supply and the power of the overlord was substantial, the proliferation of tanks, regardless of size seemed to appear — e.g. Vijayanagara and Bījāpur.

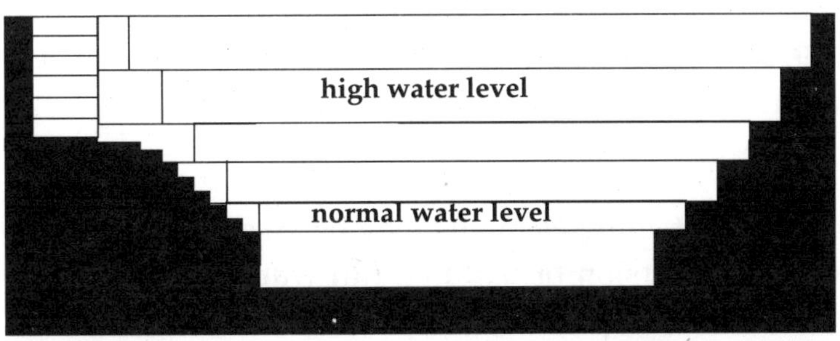


**Figure 7:** Residential bathing tank: section indicating high and normal high water level

[5] K. Ganesan, "Sacred Tanks", in K.R. Srinivasan, et al., *Sacred Tanks of South India*, Chennai: 2002, pp. 45-51.

Not only is the angle of descent steep, but as the step wells descend into the earth the side walls are stepped in at regular intervals (**Figure 1**). When standing at the topmost level and gazing towards the bottommost level, this diminution of width creates an optical illusion of increased depth — i.e. more than what is actually present. This principle, in reverse is incorporated in the temple-pyramids of Angkor. There, each ascending level is lower than the previous and the stairs leading from the lower levels to the upper terraces are narrower, making the whole appear taller or higher than it actually is. The temple of Ta Keo is a prime example (**Appendix C, Plate 71**). In addition, the towering *gopura* that are a hallmark of the temple cities of southern India — the temples of Aruṇācala (Annamalaiar), Śrī Jambukeśvāra (Thiruchirapalli) and Mīnākṣī Sundareśvara (Madurai) — also employ this optical illusion as each of the successive stories becomes shorter as it ascends upwards giving an impression of a height greater than it actually is (**Appendix C, Plates 72-73**). *En plan*, this recession as seen in the step wells is not apparent — e.g. the Cānda Bāolī plan (**Plate 10**). However, a sectional view indicates the level of descent (**Figure 10**). Yet, the full impact is only realized in person and at the site.

In addition, this diminution of space between the side walls is one solution in dealing with the lateral force exerted by the surrounding earth within a trench. Regardless of the composition of the surrounding material — laterite, for example — the weight of the earth exerts a downward thrust that is compounded by depth. If a trench is dug, then this force assumes a lateral thrust, at times, of immense proportions. In a number of cases the earth into which the trench was dug was not stable. If the walls were not diminished as the *bāolī* deepens, the likelihood of collapse is a real and present danger. Even with the stepping in of the side walls, the lateral force has caused collapse over the years and/or displacement of the walls as is seen in a number of step wells that employ this method. Several step wells of Gujarat — the Rānī-ki Wāv (**Plate 11**) of Pāṭan, the Rudabāī-nī Wāv (**Plate 19**) at Adalaj and the Dādā Harīr Wāv (**Plate 20**) of Asarwa — employed another solution. In these *bāolī*s the walls were not stepped in as the depth was reached (**Appendix C, Plate 74**). Rather, bracing in the forms of stacked columns with substantial lintels and solid-slab floor elements was employed (**Appendix C, Plate 75**). This solution ameliorated the lateral force of the surrounding earth. One further solution was employed, although not extensively — an octagonal shaft with a trench. The Octagonal Well of Malpannaguḍi (**Plate 27**) was constructed in an octagonal form that did not require an exceptional depth to reach the ground water. The octagon or circle could conceivably be employed for a well of great depth as the octagon and/or circle's walls are self supporting and diffuses the lateral force of the surrounding earth (**Appendix C, Plate 76**). This method could have been employed for a deep well and circular stairs could have been utilized to gain access to the water — circular stairs as seen in the Rudabāī-nī Wāv (**Plate 19**) and the Dādā Harīr Wāv (**Plate 20**). However, such a design would have allowed only the smallest amount of space for relaxation.

### Well/Tank Types

Seven general types of water containing elements may be assembled into four major divisions — (i) drinking water, (ii) water of immersion or bathing, (iii) water of irrigation and stock, (iv) and the ornamental use of water. The seven types may be seen as:[6]

[6] It must be noted that the terms for the various forms of water catchment and holding vary considerably, used in one area to denote one form and in another region for a completely different form and are interchanged throughout the vast subcontinent.

1. *Ṭaṅka*s, the most elemental form
2. *Kuṇḍa*s or simple wells dug into the earth
3. Stepped wells or *bāolī*s, which may consist of:
    (a) a trench-shaft,
    (b) a shaft-shaft,
    (c) a trench-shaft-shaft,
    (d) a square form, and
    (e) a natural defile,

and variously identified as:

(a) *bāolī,*
(b) *bārādarā,*
(c) *bauḍī,*
(d) *baulī,*
(e) *bauṛī,*
(f) *bāv,*
(g) *bav,*
(h) *bāvadī,*
(i) *bavaḍī,*
(j) *bāw,*
(k) *bawḍī,*
(l) *bawrī,*
(m) *boulī,*
(n) *wāv* and
(o) *wāvaḍī*

4. Sacred tanks or pools which are variously identified as:
    (a) *chera*s,
    (b) *kalayni,*
    (c) *karai* (Mysore),
    (d) *kohil,*
    (e) *kuḷam,*
    (f) *kuṇḍa,*
    (g) *puṣkariṇī,*
    (h) *sāgara,*
    (i) *sarobara,*
    (j) *sarovara,*
    (k) *taṅka,* and
    (l) *tīrtham*
5. Bathing tanks or pools for delight
6. Large artificial lakes which are variously identified as:

(a) *eri,*

(b) *kere,*

(c) *taḍāka,*

(d) *tālāo,*

(e) *yert* and

(f) *zing*

7. Ornamental uses.

**The Plans**

The plans contained within this brief study were carefully redrawn on a computer using the Adobe PageMaker 6.0 program. Each drawing was checked and double-checked during the lengthy process that was required. Unfortunately, very few plans of either step wells or tanks exist or are readily available to either researchers or the general public. The Livingston work is an exception. Two or three are illustrated in the more comprehensive studies — the Rudabāī-nī Wāv and Dādā Harīr Wāv Gujarat — but others are not. There are myriad illustrations or photographs of step wells and tanks in both printed material and the Internet, but most are not comprehensive in their focus or frame. A number of the sources presented unusual problems.[7] A number of the plans contained, herein, were carefully and meticulously reconstructed utilizing a number of photographs gleaned from the Internet. Every detail was assiduously analysed, compared and rechecked with the photographs in order to present, as clearly as possible, the dimensions, proportions and relationships of each of the various elements of the structures. Those which were thus drawn from photographs are not purported to be scale drawings, although many may closely conform to the individual monuments.

Although there are numerous stepped wells, tanks and pools scattered and/or concentrated across the vast subcontinent, thousands by conservative estimates, this study will concentrate on but a few out of necessity in this consideration. Finally, the monuments are arranged chronologically.

[7] K.M. Suresh, *Temples of Hampi: and Its Environs,* Delhi: 2003. This volume presented a number of plans — thirty-nine in all — and attending descriptions. The descriptions were thorough and were greatly appreciated as the section displaying the various plans — pp. 135-74 — presented a number of plans which simply did not follow the descriptions or vice versa. For example, the description of the Durgā Temple, p. 55, states "The *garbhagṛha* is square . . . ." However the plan of that temple, found on p. 139, indicates a clearly rectangular *garbhagṛha* and columns of the *sabhāmaṇḍapa* that are also rectangular. Immediately, I realized that the plans and elevations that Suresh had provided the publisher had been "widened", obviously to Conform more readily to the format of the page. This strange malformation appears on nearly everyone of these thirty-nine pages. One particularly odd, even amusing plan is found on p. 150, in which the "Bāla Kṛṣṇa Temple Complex" is "squashed" down to two-thirds of its normal east–west configuration. For such an important book, this is sad, inexcusable!

# Stepped Wells, Tanks and Pools

# Mañjuśrī Wāv, Dhank, Gujarat (*c.* CE 650)

The Mañjuśrī Wāv[1] was constructed during the Maitraka Dynasty of the Kingdom of Vallabhi, probably during the reign(s) of Dhruvasena II (r. *c.* 640-44) or Dharasena IV (r. *c.* 644-51) or Dhruvasena III (r. *c.* 651-56) or Kharagraha II (r. *c.* 656-62) (**Appendix D**). It is considered to be one of the earliest manifestations of a step well (*wāv*) in Gujarat. It indicates a direction that later *wāv*s or *bāolī*s would take.

Unlike the later step wells, the Mañjuśrī Wāv is "L" shaped. Consisting of a *kūpa* and trench with two relief shrines at the turning point of the trench, the *wāv* is essentially quite simple. The shaft is circular, but rather than being cylindrical, it is cone-like — the diameter of the shaft lessens as it descends into the earth (**Figure 8**). Similarly, the width of the stairs lessens after the 90th turn, down to the water's edge. As the stairs become less wide, the walls of this part of the trench also angles in as the steps go deeper. This feature not only creates less material to be excavated, it also creates a forced perspective — the depth appears more than it actually is (**Appendix C, Plates 71-72**). Lintels are placed at four points, ostensibly to support the walls and act as counterthrusts. However, this may not have been necessary as a few meters below the surface, porous rock is encountered which is able to support the excavated walls.

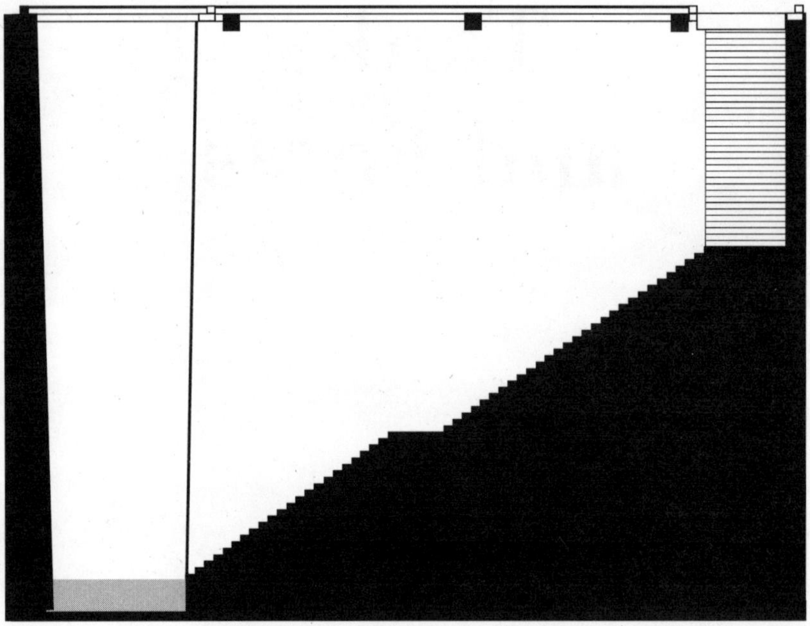

**Figure 8:** Mañjuśrī Wāv, sectional view

1 The plan is redrawn using Adobe PageMaker 6.0, from: Morna Livingston (ML), p. 24.

# Plate 1: Mañjuśrī Wāv plan

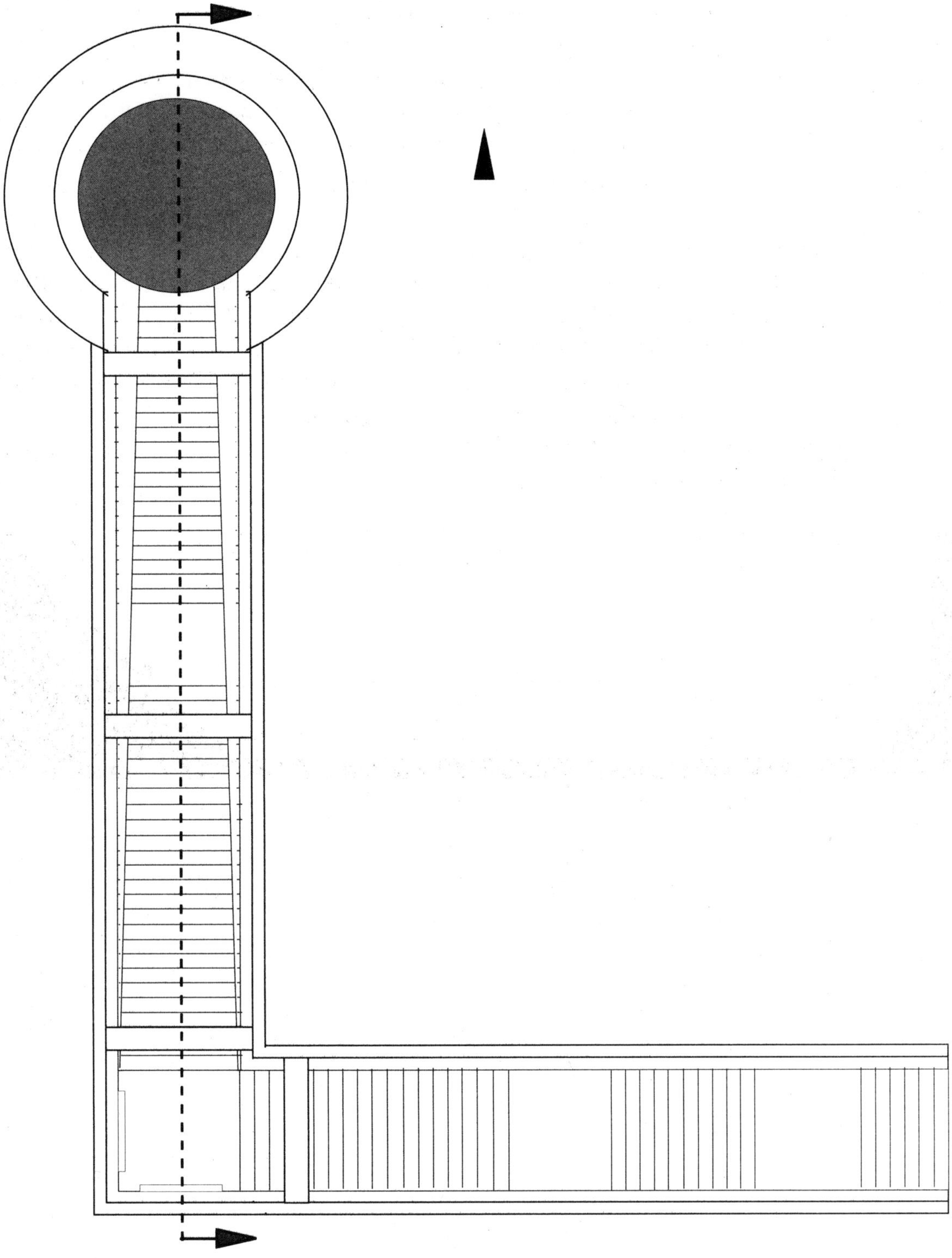

## Viṣṇu-puṣkariṇī, Mahākūṭeśvara Temple, Bādāmī, Karnataka (*c.* CE 696)

A short distance to the west of Bādāmī is the complex of Mahākūṭeśvara Temple. The various temples and shrines are not among the singular temples built by the Cāḷukyas. Constructed during the reigns of Vinayāditya I (r. CE 681-96) and Vijayāditya III (r. CE 696–733) (**Appendix D**) and dedicated to a form of the Lord Śiva, the main temples are oriented towards the east. In the compound is a large tank, the Viṣṇu-puṣkariṇī[2] (Lotus Pool of the Lord Viṣṇu), an interesting nomenclature in a Śaiva complex.

The Viṣṇu-puṣkariṇī, the sacred tank of the complex, is rectangular, but not orthogonal. This tank is fed by a stream from a spring providing it with a constant supply of water. The runoff from this tank supplies a smaller ablution tank, the Pāpavināśana Tīrtha. Two sets of steps lead down to the water level, one on the north side and one on the south of this large tank. The tank proper is initially set deep within the surrounding soil. Narrow ledges or steps surrounding the tank's interior lead downward. However, these are rather narrow and would not lend themselves to vast numbers of pilgrims (**Figure 9**). Essentially, the Viṣṇu-puṣkariṇī is designed to serve a limited number of devotees commensurate with the importance of the Mahākūṭeśvara Temple. To the north, near the northern steps is a small *nirazhi maṇḍapa* floating on the surface. This small shrine is composed of four columns supporting an elaborate roof which shelters a *pañcamukha liṅga*.

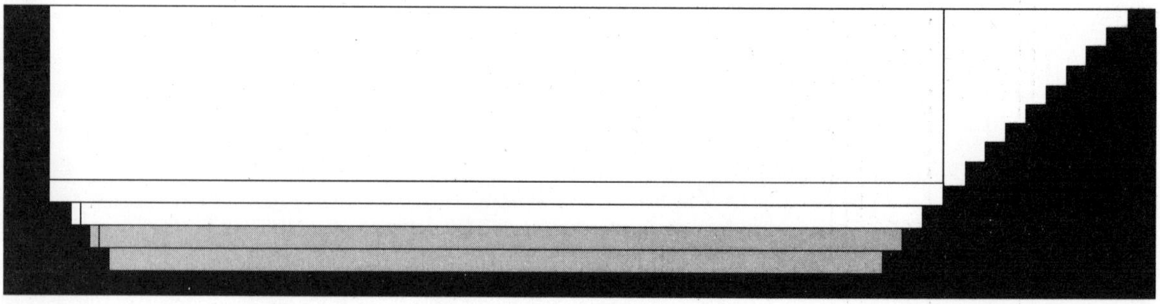

**Figure 9:** *Viṣṇu-puṣkariṇī, Mahākūṭeśvara Temple,* sectional view

[2] The plan is redrawn using Adobe PageMaker 6.0, from Henry Cousen (HC), Plate XXXVII.

## Plate 2: Viṣṇu-puṣkariṇī plan

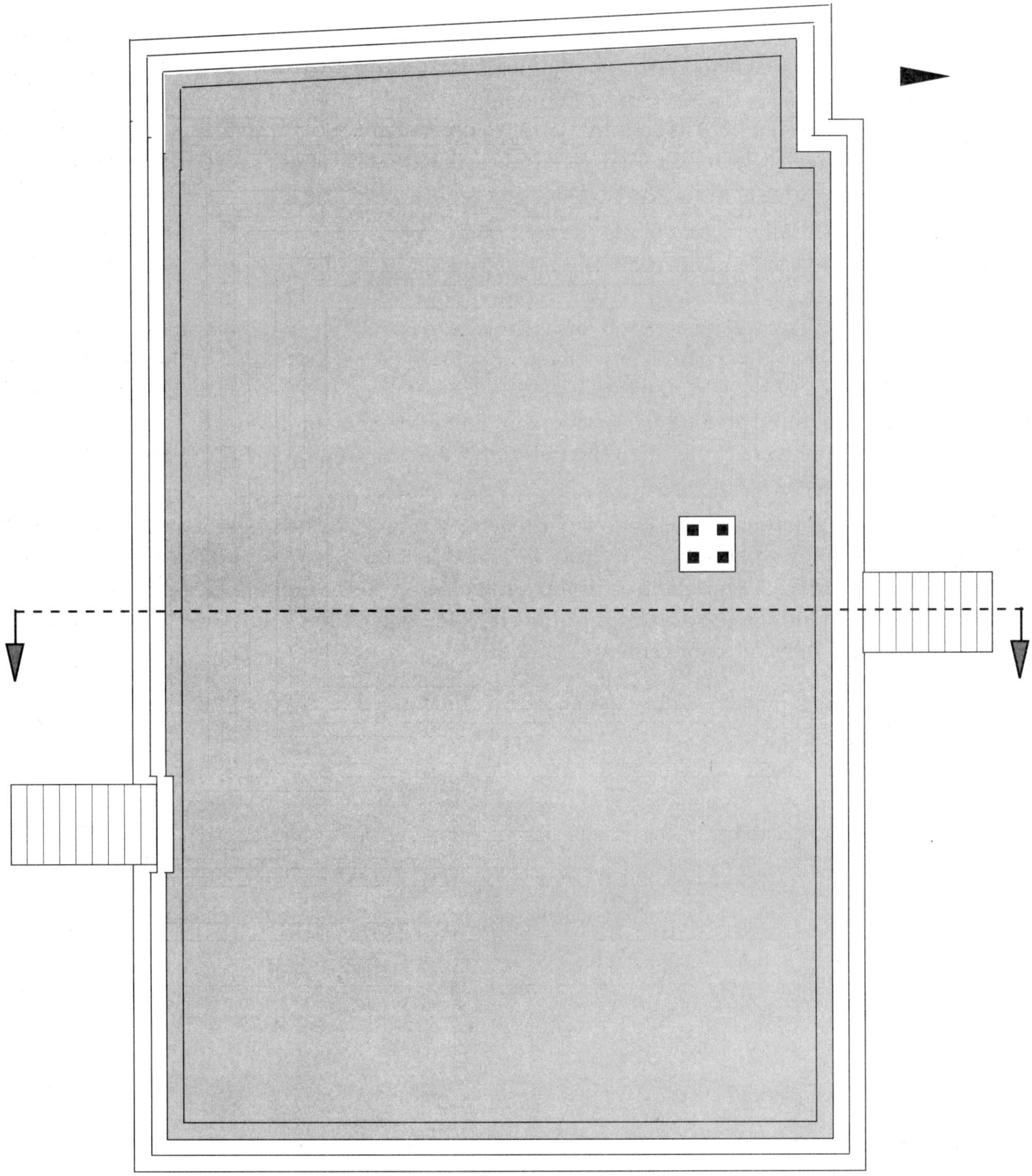

# Roḍā Kuṇḍa, Roḍā, Gujarat (a. CE 700)

The Roḍā Kuṇḍa[3] in Gujarat is a unique structure for a number of reasons. It is an example of an early *kuṇḍa* with regularly spaced step elements that lead, progressively, down to the level of the water. As with a number of tanks, it possesses certain religious implications and/or character[4] as there are four small chapels placed within its confines — two aligned south, the third facing east and the fourth oriented towards the west. Its rectangular plan is unusual as most *kuṇḍa*s have a square plan. Finally, unlike so many structures in India, the Roḍā Kuṇḍa is not symmetrical.

The *kuṇḍa* is essentially aligned to the cardinal points of the compass — i.e., its length is basically east–west with the entrance in the east. The northern part of the Roḍā Kuṇḍa is in a state of ruin, but it is believed that, like a number of other contemporary tanks, it was merely a flat wall which facilitated the drawing of water directly. If this is the case, it is unlikely that the water would have been consumed, rather it would have been employed for *śodhana* (ritual bathing). On the southern side of the tank are a number of regularly placed five-step elements leading up to ground level (**Figure 10**). Similar elements are found on the eastern side, but their arrangement differs due to the width of the tank. However, on the west, rather than the five-step elements seen on the south and east, there the elements rise nine steps or two levels.

The asymmetrical nature of the Roḍā Kuṇḍa is most unusual. The placement and alignment of the four chapels are interesting but not unusual. This orientation is seen in the later axial arranged chapels of the Lakṣmaṇa Temple at Khajurāho. However, even though the chapels hint at an axial solution, the placement of the five- and/or nine-step elements of the south, east and west are such that any type of symmetrical arrangement is voided.

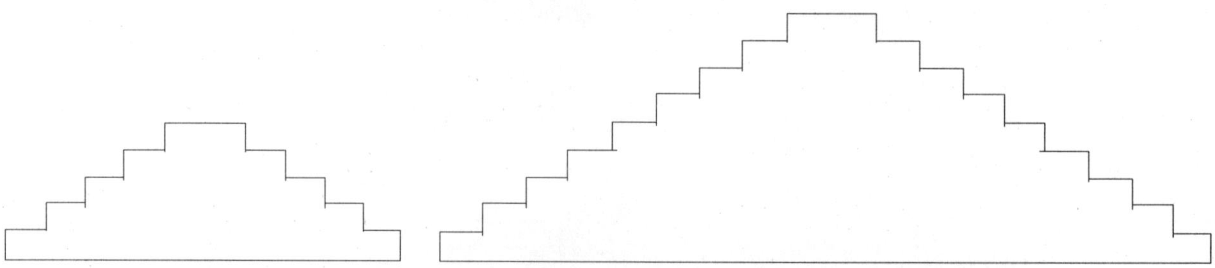

**Figure 10:** Roḍā Kuṇḍa, southern and eastern five-step elements (left) and western nine-step elements

[3] The plan is redrawn using Adobe PageMaker 6.0, from ML, p. 33.

[4] ML, pp. 47-55, as well as a number of other references through her work.

## **Plate 3:** Roḍā Kuṇḍa plan

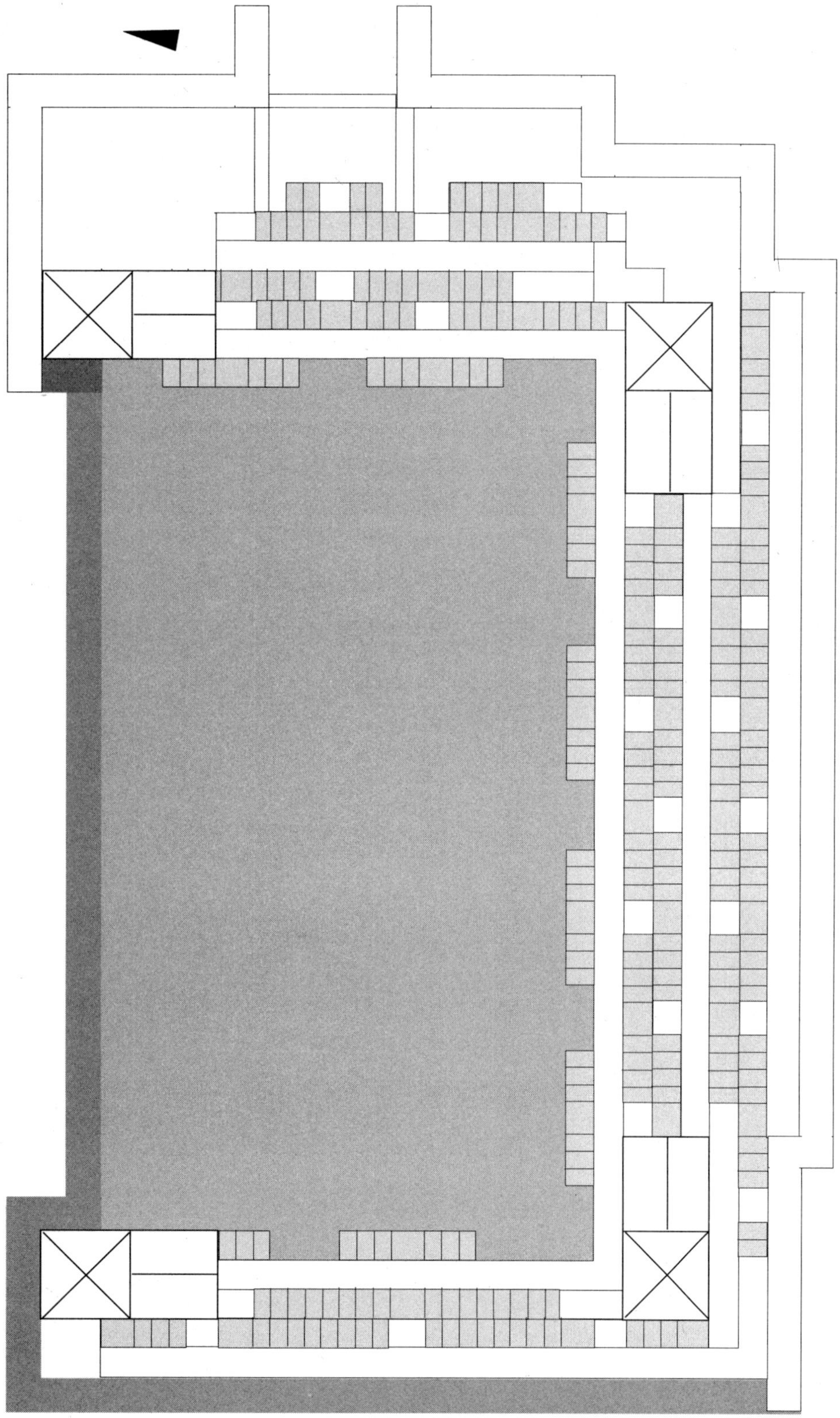

# Naṭarājar Temple, Cidambaram, Tamil Nadu (ninth-seventeenth century CE)

One of the most ancient and revered temples in India is the Naṭarājar Temple[5] at Cidambaram. It is one of the major pilgrimage temples in the subcontinent which draws tens of thousands annually. The Naṭarājar Temple is of the Bhūta Liṅgam category and is associated with the elemental, ether. This venerable temple was adorned by the Coḷa rulers Āditya-Rāja I (*c.* CE 871–907) and Parāntaka I (CE 907-55) (**Appendix D**). Its sacredness and fame were such that it was endowed and added on to by virtually every subsequent ruler and dynasty of the region. The holy of holies of the temple is dedicated to Śiva-Naṭarāja (**Plate 4, A**), but subsequent Śaiva temples and shrines were added — e.g.the Pāṇḍyas Nāyakam, Śivakāmi Amman Shrine and the Nandī Shrine — creating an elaborate and eccentric assemblage of structures.

The temple ensemble consists of four enclosures (*prākāra*) (**Plate 4, I-IV**). To the north of the inner sacred ensemble's *prākāram* is the Śivagaṅgā Tank (**Plate 4, N**). The tank is aligned on a north–south axis similar with the *cit sabhā* with its images of Śiva (Naṭarāja) and Pārvatī (Śivakāmī) and *kanaka sabhā.* These two most sacred structures of the whole complex nestle within the inner *prākāra* which is, in turn, surrounded by the *mūlasthāna sabhā, deva sabhā* and *nṛtta sabhā* and the third *prākāra*. The Śivagaṅgā Tank is just to the north of this grouping and would seem to have been constructed, initially, after the third *prākāra* or, at least, contemporaneously with it.

Next to the substantial *rāja sabhā,* the Śivagaṅgā Tank is the largest element, *en plan,* of the whole ensemble. Its placement and size attest to the importance of this tank. It had to be large enough to comfortably accommodate the hundreds, sometimes thousands of devotees who came to the Naṭarājar Temple on high feast days. It was entered from the north via an imposing pillared gate/*śāla* as well as from the south-east and south-west corners. Within the precinct of the Śivagaṅgā Tank, the devotees could perform their ritual ablution as required.

**A** *Cit sabhā* and *kanaka sabhā*
**B** Govindarāja shrine
**C** *Nṛtta sabhā*
**D** *Deva sabhā*
**E** *Mūlasthāna sabhā*
**F** East *gopuram*
**G** South *gopuram*
**H** West *gopuram*
**J** North *gopuram*
**K** Pāṇḍyas Nāyakam
**L** *Navagraha* shrine
**M** Śivakāmi Amman Shrine
**N** Śivagaṅgā Tank
**O** *Rāja sabhā*
**P** Hundred-pillared hall
**Q** *Sthala vṛkṣa*
**R** *Yoga sabhā*
**S** Gaṇeśa Shrine
**T** Nandī Shrine
**I.** Inner *prākāra* wall
**II.** Third *prākāra* wall
**III.** Second *prākāra* wall
**IV.** First *prākāra* wall

[5] The plan is redrawn using Adobe PageMaker 6.0 from Christopher Tadgell (CT), p. 100 and from the Internet.

## **Plate 4:** Naṭarājar Temple plan

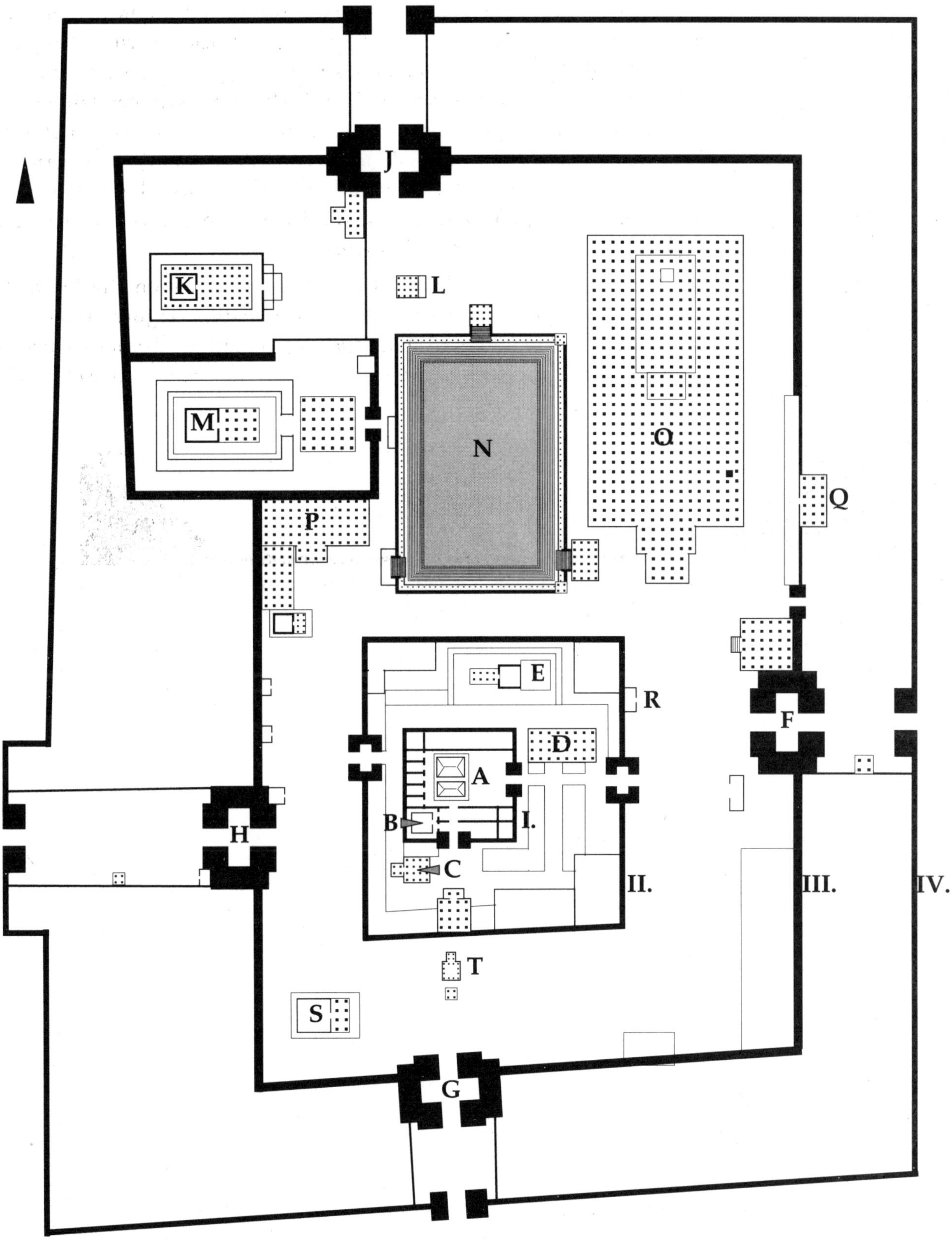

# Naṭarājar Temple, Śivagaṅgā Tank

The sacred temple tank, the Śivagaṅgā Tank[6] is oriented towards the north and opens onto a commodious plaza with the *navagraha* shrine to the west. It is a sacred ablution tank associated with a temple. An impressive pillared *śālā* or *sabhāmaṇḍapa* announces the tank's main entrance and leads to a broad flight of steps while to the south are entrances to the east and west. These steps descend down to the upper, broad terrace of the tank. In the south-west and south-east quadrants are two other entrances to the tank. That on the south-east is via an imposing *sabhāmaṇḍapa,* larger than the northern, main entrance. All four sides of the tank are surrounded by a covered colonnade. This colonnaded area not only provides suitable shelter from the noonday sun but also furnishes the pilgrims with a generous area to rest and relax before and after their ablution and prayers.

Should the Śivagaṅgā Tank be near full, especially at the end of the rainy season, the depth of the water would be considerable and would pose a danger for any hapless bather (**Figure 11**). However, the surface would soon sink to a depth that would pose little danger to the bathing devotees. Generally, the quality of the water in the tank posed no health hazard to those who came to immerse themselves, but it was not suitable for drinking.

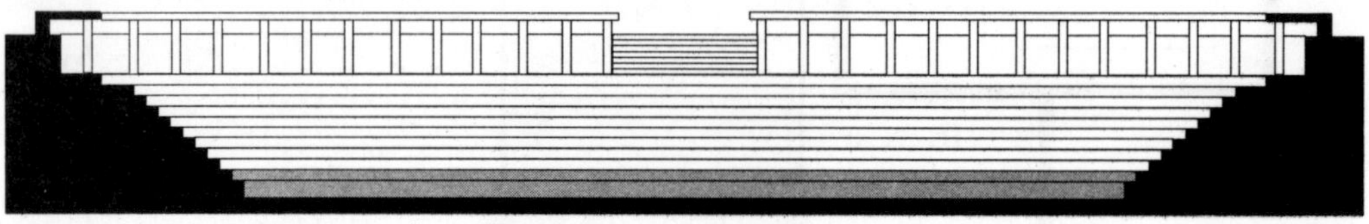

**Figure 11:** Śivagaṅgā Tank, Naṭarājar Temple, sectional view

[6] The plan is redrawn using Adobe PageMaker 6.0 from the Internet.

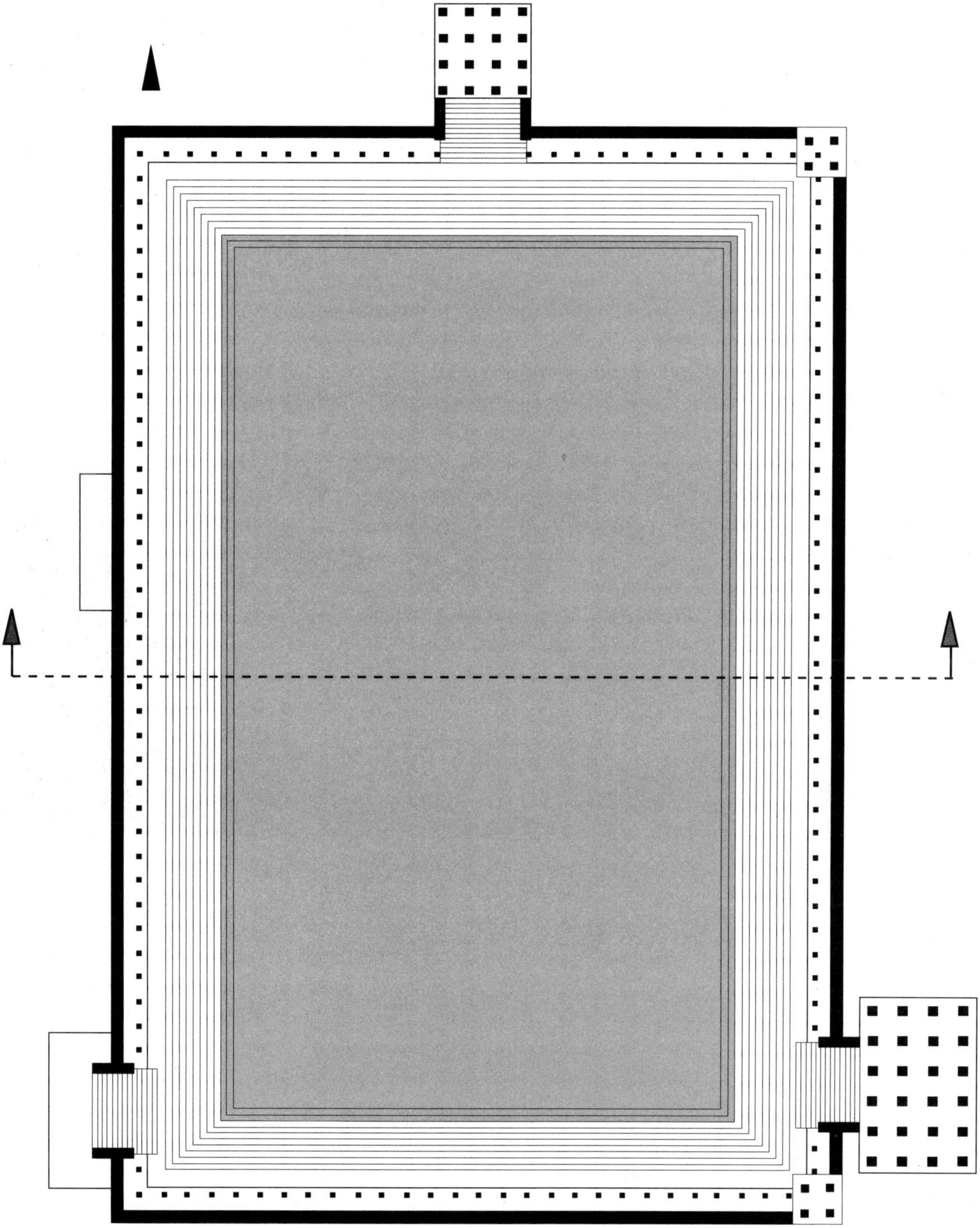

## Maṇikarṇikā Kuṇḍa (Cakra-Puṣkariṇī Kuṇḍa), Varanasi, Uttar Pradesh (b. tenth century CE)

Varanasi is reputed to be one of the oldest inhabited cities in India, certainly one of the holiest locales on the subcontinent. There many go to die, become cremated and have their ashes scattered on the sacred Ganges (Mother Gaṅgā). Of all the *ghāṭ*s at Varanasi, the Maṇikarṇikā Ghāṭ is one that is preferred by many for the site of cremation. In its environs is found the Maṇikarṇikā Kuṇḍa (Cakra-Puṣkariṇī Kuṇḍa),[7] a sacred tank independent of a temple. Until recently it had been neglected to the point that it had become completely filled with dirt and refuse over the years. Now it is restored. Adjacent to this *kuṇḍa* is found a small shrine with images of the Lords Viṣṇu and Śiva. The true origin and date of this *kuṇḍa* are not known.

Even though it is close to the Ganges, the Maṇikarṇikā Kuṇḍa (Cakra-Puṣkariṇī Kuṇḍa) depends on rain and its runoff to supply the *kuṇḍa*. The original construction was of carefully dressed blocks of stone and mortar, making perculation from the *kuṇḍa* relatively minimal. It is rectangular in shape. The steps descend steeply into the earth to the water level (**Figure 10**). The proportions of rise to tread is at 1:1 rate (**Figure 3A**). Small platforms, large enough to provide seating for devotees, are located on every third step. These platforms may be seen to mimic the cremation platforms that are found at the Maṇikarṇikā Ghāṭ and other *ghāṭ*s throughout Varanasi.

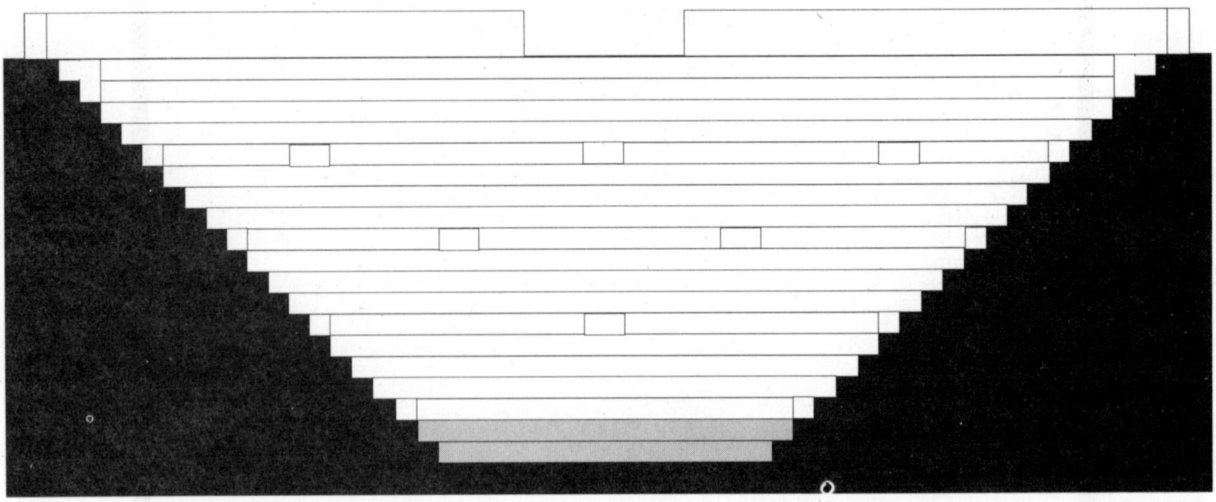

**Figure 12:** Maṇikarṇikā Kuṇḍa (Cakra-Puṣkariṇī Kuṇḍa), sectional view

[7] The plan is redrawn using Adobe PageMaker 6.0 from the Internet.

## **Plate 6:** Maṇikarṇikā Kuṇḍa (Cakra-Puṣkariṇī Kuṇḍa), plan

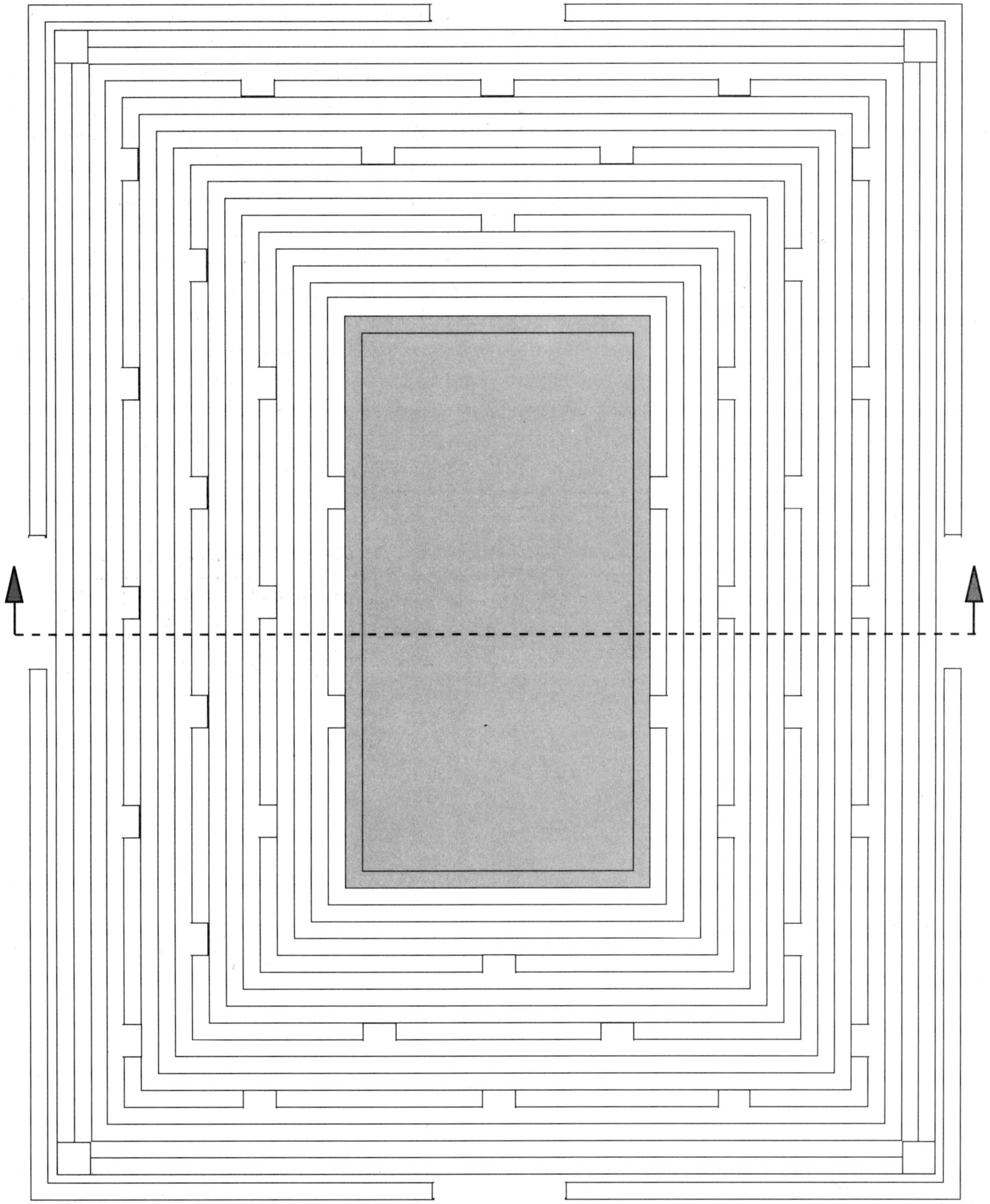

# Sūrya Temple, Moḍherā, Gujarat (CE 1026)

The Solaṅkīs of Gujarat were a powerful and important dynasty in north-western India from the tenth through the thirteenth centuries CE. The noteworthy monarch Bhīma I (r. CE 1023-65) (**Appendix D**) sat on the throne when the impressive Sūrya Temple[8] ensemble was conceived and raised at Moḍherā. This complex marks the pinnacle, the efflorescence of the impressive Solaṅkī style in this region of India. Even though it fell within the sphere of the destructive Muslim invaders — e.g. Mahmud Ghazni (CE 998–1030) — it still stands as one of the most impressive and innovative temple complexes in India.

The Sūrya Temple at Moḍherā is unusual for a number of reasons. First, it is one of the few temple complexes in India dedicated to the sun deity, Sūrya — the other major complex being the Sūrya Temple[9] Complex at Koṇārak in Orissa, raised some 200 years later. Second, the Sūrya Temple ensemble exhibits a tank of unusual size, particularly in comparison to both the *sabhāmaṇḍapa* and the *mūlaprāsāda* (*vimāna*). In many cases, a temple tank may be somewhat larger than the *vimāna*, however, this tank is over ten times larger in area than the *vimāna* of the Sūrya Temple. And, third, unlike so many other temple complexes, the tank was conceived as being a choate part of or an extension of the mirror axial layout of the *vimāna* and *sabhāmaṇḍapa* — i.e. the eastern sector of the east–west axis. Since the east is an important realm of the sun, its rising location, it would seem that the placement of this monument and elaborate tank in this alignment is of major importance. The tank is an integral part of the design unit of the whole.

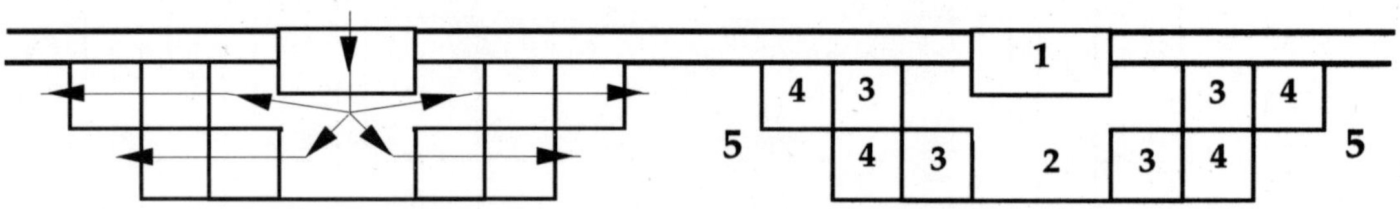


**Figure 13:** A step element found in the Sūrya Temple Tank

[8] The plan is redrawn using Adobe PageMaker 6.0 from Christopher Tadgell, *The History of Architecture in India*, London: 1990, p. 110 and Krishna Deva, *Temples of India*, New Delhi: 1995, p. 195.

[9] Others include shrines or *garbhagṛhas* in the Basavanna (Traipuruṣa) Tri-kūṭa Temple, Malghana; the Mallikārjuna Tri-kūṭa Temple, Basaralu; the Someśvara Tri-kūṭa Shrine, Kabali and the Sūrya-Nārāyaṇa Tri-kūṭa Temple, Magala.

# **Plate 7:** Sūrya Temple Ensemble plan

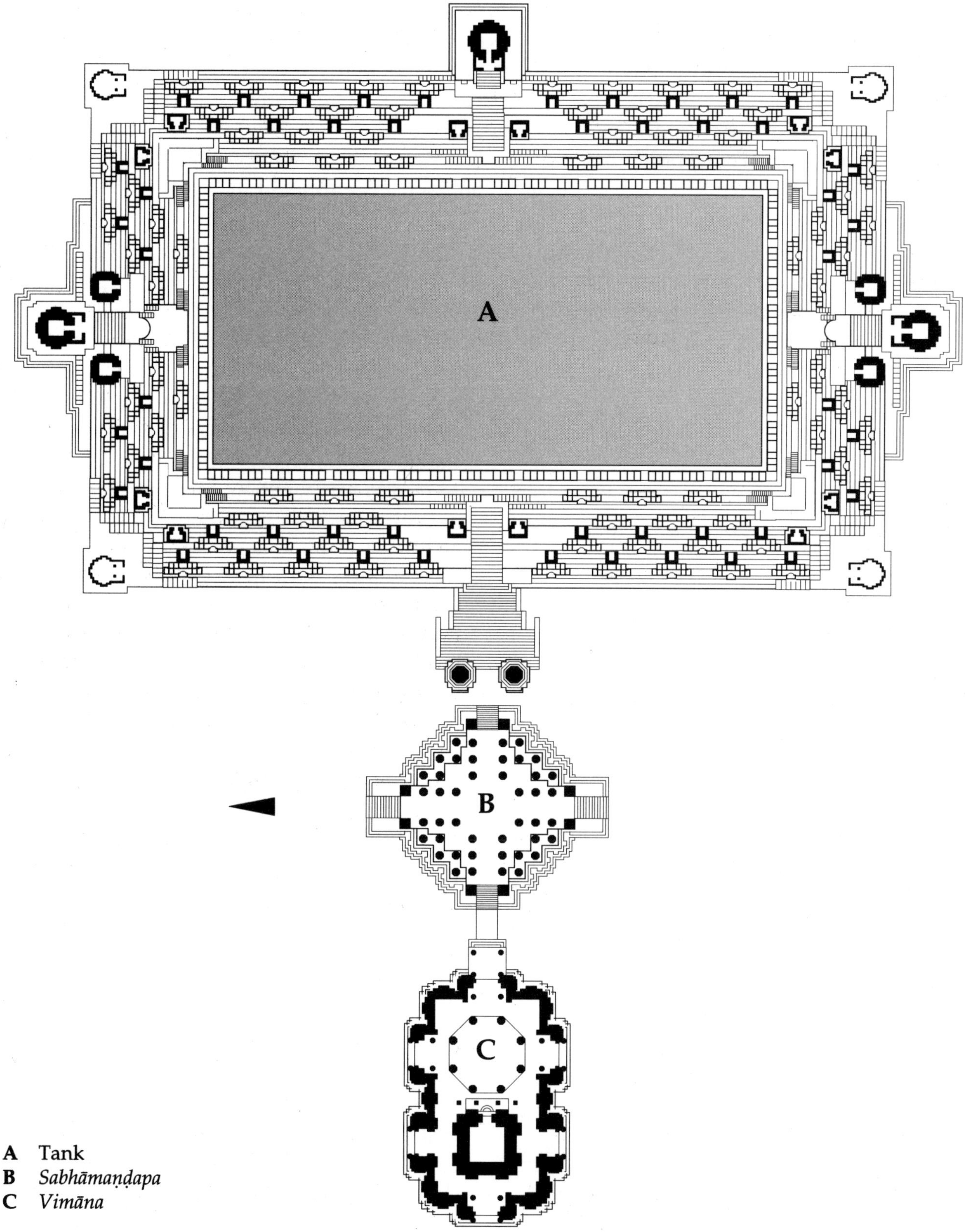


**A** Tank
**B** *Sabhāmaṇḍapa*
**C** *Vimāna*

# Sūrya Temple Tank

The Sūrya Temple Tank[10] is an extravagant example of conspicuous wealth and power, as well as a sacred tank of incomparable beauty associated with a temple. The tank is rectangular and consists of a number of planes, punctuated by four levels on which there are stepped elements (**Figure 13**). These elements descend in five steps from one terrace to the next and allow for four different ways of descent. Ninety-six of these elements are duplicated around the tank. There are conspicuous shrines placed at three of the cardinal points of the compass as well as the magnificent staircase leading from the *sabhāmaṇḍapa* to the tank at the west. In addition, there are shrines at the intercardinal points of the tank — i.e. the corners.

It is of some interest to note that there are seventy-two shrines, minus the three major elevated shrines at the cardinal points (**Figure 14**). Seventy-two is a number of some importance, particularly amongst the Buddhists and Hindus. It is one-fifth of the circle's circumference or degrees and is therefore related to the sacred numeral, five. It also possesses links with the numbers six, eight, nine and twelve; all of which are most auspicious. Within the various world religions, seventy-two is prominent. The Chinese count seventy-two saints; there are seventy-two bishops recognized by the Cathars, seventy-two martyrs were counted in the battle of Karbalā; Islam attests to seventy-two sects; and Adam fathered seventy-two children. Seventy-two also is recognized in numerous architectural monuments. There are seventy-two *caṇḍī perwara* in the inner court surrounding the central ensemble of Chandai Sewu, Prambanan, Java; seventy-two Buddha images on the fourth level of Borobudur, Java; seventy-two subsidiary *stūpas* (Indo.: *dagaba*) on levels six through eight of Borobudur, Java; and, seventy-two *chedi* surrounding the *ubosot* (Thai.) complex of Wat Phra Chetuphon, Bangkok.

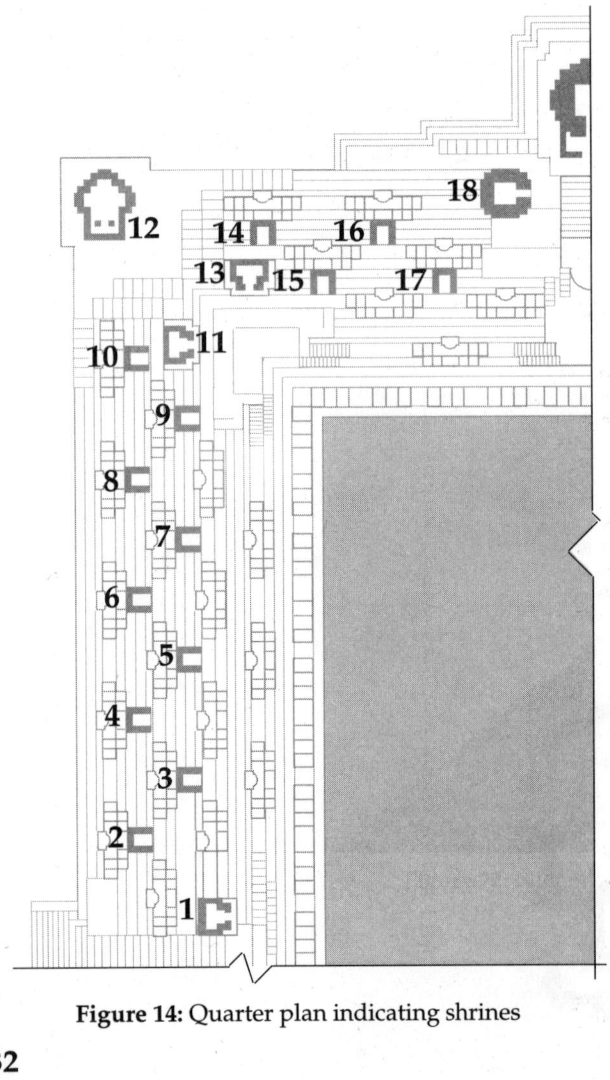

**Figure 14:** Quarter plan indicating shrines

The alignment of this elaborate tank in the east of the east–west axis of the ensemble. This places it, iconographically, in a most important position — i.e. the masculine east.

10 The plan is redrawn using Adobe PageMaker 6.0, from Christopher Tadgell, *The History of Architecture in India*, London: 1990, p. 110 and Krishna Deva, *Temples of India*, New Delhi: 1995, p. 195.

## Plate 8: Sūrya Temple Tank plan

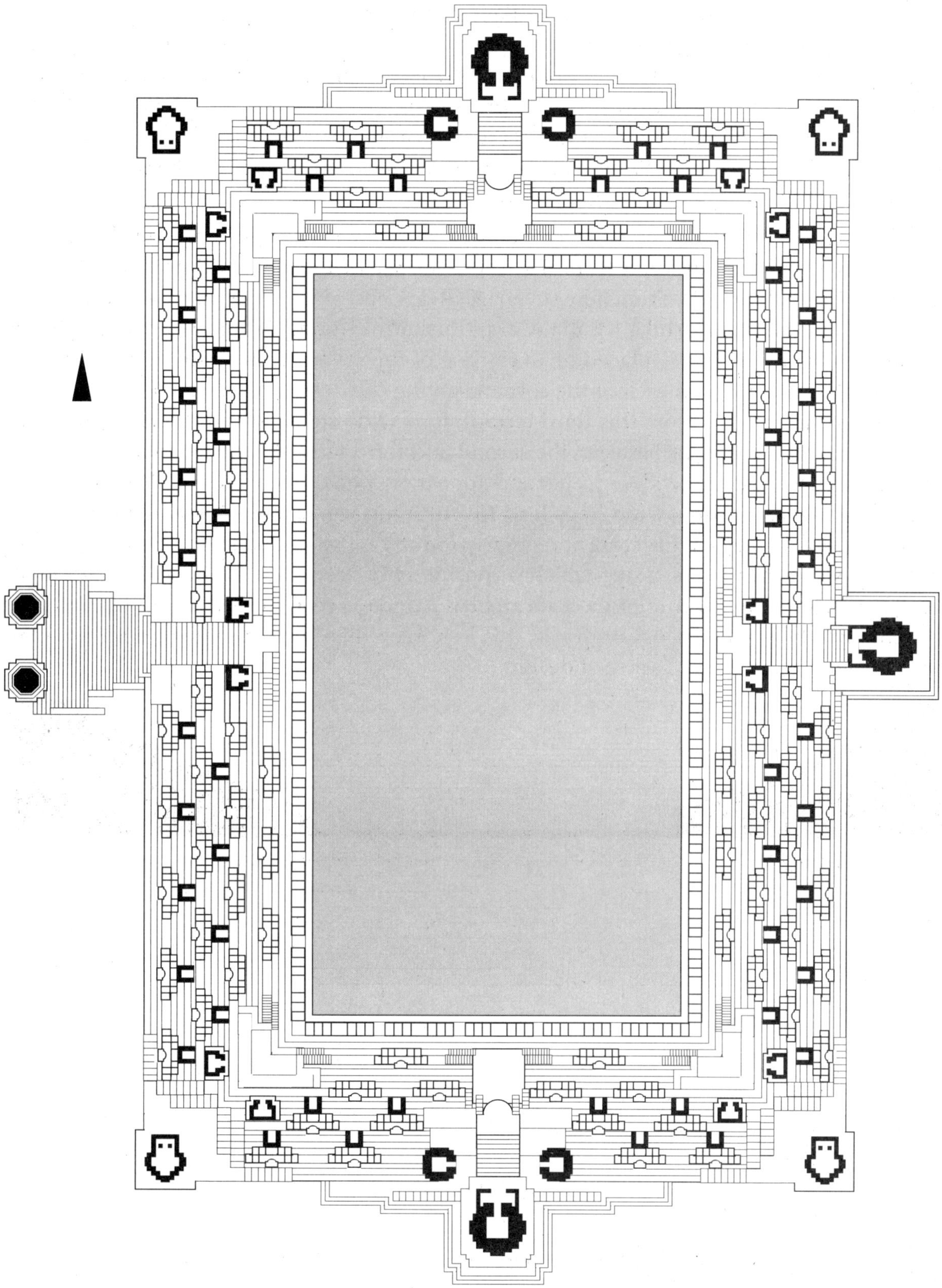

## Pañcamahal Kuṇḍa, Pañcamahal, Gujarat (a. CE 1026)

In north-eastern Gujarat is found the district of Pañcamahal and the city of the same name. There is found a *kuṇḍa* (tank) of exceptional design. The Pañcamahal Kuṇḍa[11] may be seen to find its design basis in the Solaṅkī Sūrya Temple Tank of Moḍherā (**Plates 7-8**) and the Maṇikarṇikā Kuṇḍa (Cakra-Puṣkariṇī Kuṇḍa) of Varanasi (**Plate 6**) as well as pressaging the square tanks of Rajasthan (**Plates 10, 55**). The design is such that it causes one to suspect that it was not primarily used as a *bāolī* or source of drinking water, but as a place for bathing, possibly even ritual bathing. Access to the water is achieved in every direction and that access is accomplished with ease.

Unlike many tanks, the Pañcamahal Kuṇḍa is square, *en forme.* Here, sets of double stairs of five steps each, regularly placed, descend from the upper terrace of the *kuṇḍa* (**Plate 9, I**) down to the second, wide terrace (**Plate 9, II**). From here seven steps descend to the third, narrower terrace (**Plate 9, III**). As is seen in the Maṇikarṇikā Kuṇḍa (Cakra-Puṣkariṇī Kuṇḍa) (**Plate 6**), small platforms or seats are alternately and regularly placed on every one of the seven steps. Also, from the second to the third terrace, steps appear in each of the corners, giving the design of this section of the *kuṇḍa* a redented appearance, *en plan.* From this third terrace, four wide steps descend to the fourth terrace (**Plate 9, IV**). Like the seven steps between the second and third terrace, seats or small platforms are incorporated, but, here, they are wider, deeper and appear only on the second step. The fourth terrace mimics the third in width. Three wide steps lead to a fifth terrace, usually below the water line. As with the previous set of steps, wide seats at regularly found on the second step. From here a narrow terrace (**Plate 9, V**) leads to the next step which is considerably deeper and leads to the bottom of the *kuṇḍa.* Another noteworthy feature of the Pañcamahal Kuṇḍa is seen in the rate of descent (**Figure 15**). Unlike many other tanks, *bāolīs* and pools, it is less steep than seen in many others and may be seen as a testament to the Solaṅkī sense of design.

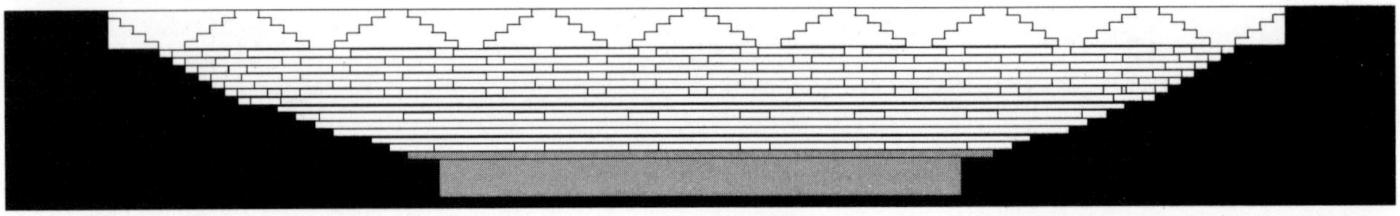

**Figure 15:** Pañcamahal Kuṇḍa, Pañcamahal, section indicating the angle of descent

[11] The plan is redrawn using Adobe PageMaker 6.0 from the Internet.

# Plate 9: Pañcamahal Kuṇḍa plan

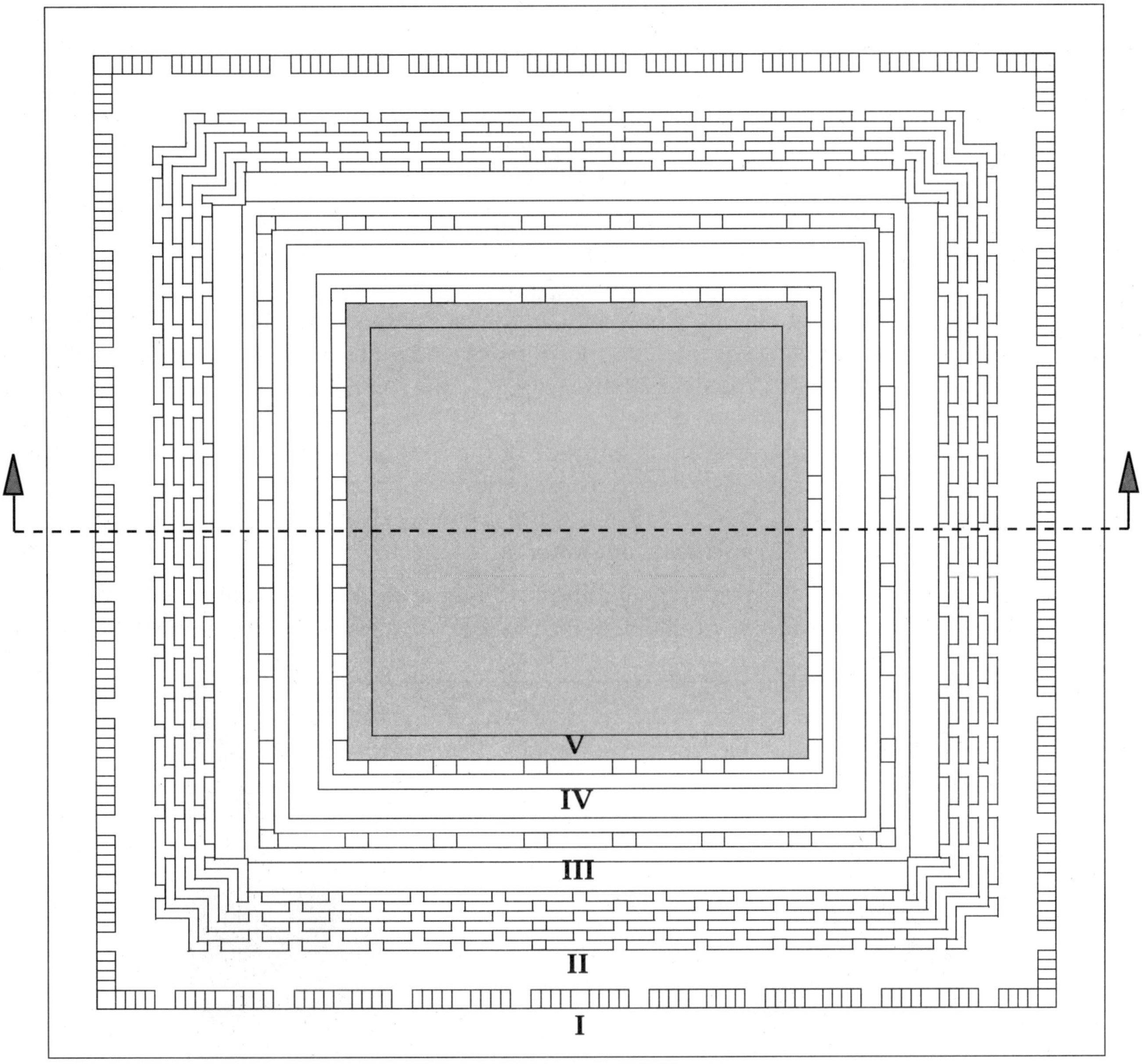

# Cānda Bāolī, Ābhānerī, Rajasthan (eleventh century)

The Cānda Bāolī[12] at Ābhānerī, Rajasthan, is one of the most spectacular step wells to be found in India. Spectacular, not because of the design of the individual step elements, which are somewhat similar to those of the Sūrya Temple Tank at Moḍherā (**Plates 7-8**) or the impressive Rānī-kī Wāv of Pāṭan (**Plate 11**), but spectacular because of the depth and the steepness of the descent in order to be able to reach the water level (**Figure 16**). Anyone with the slightest amount of vertigo or acrophobia would be paralysed at coming to the topmost level and gazing downward for nearly 30 m.

Not adjacent to the Thar Desert nor the Great Indian Desert, but in a region of some elevation and where the rainfall can be sparse, the Cānda Bāolī was built essentially for obtaining drinking water. Although it has been stated that step wells were often constructed as "an act of great generosity and benevolence", they were often built out of necessity. In the case of the Cānda Bāolī, both reasons may be applicable as there is a fair amount of decorative detail attendant to this *bāolī*. Thirteen levels with a repetition of stepped elements drop over 30 dizzying m to the relatively small, lowest level in this the largest stepped well ever constructed! Here, at the driest part of the year, water could normally be obtained. Verandas and rooms rise up on one side of the *bāolī* from which the populous can rest and relax. In this sense, it differs in basic design from the other step wells. Like the Rānī-kī Wāv to the south of the Great Indian Desert, the Cānda Bāolī was a source of drinking water rather than a tank for ritual ablution as the Sūrya Temple Tank at Moḍherā.

The thirteen levels may have iconographic significance. It is not an unlucky number as is erroneously believed in the West. Thirteen may be a referent to the number of cycles or months within a lunar year.

**Figure 16:** Cānda Bāolī, Ābhānerī, section indicating the angle of descent

[12] The plan is redrawn using Adobe PageMaker 6.0. The plan of the Cānda Bāolī has been drawn from several photographs, none of which shows the total. However, the number of levels and the disposition of the step elements, particularly at the corners, is accurate.

# **Plate 10:** Cānda Bāolī plan

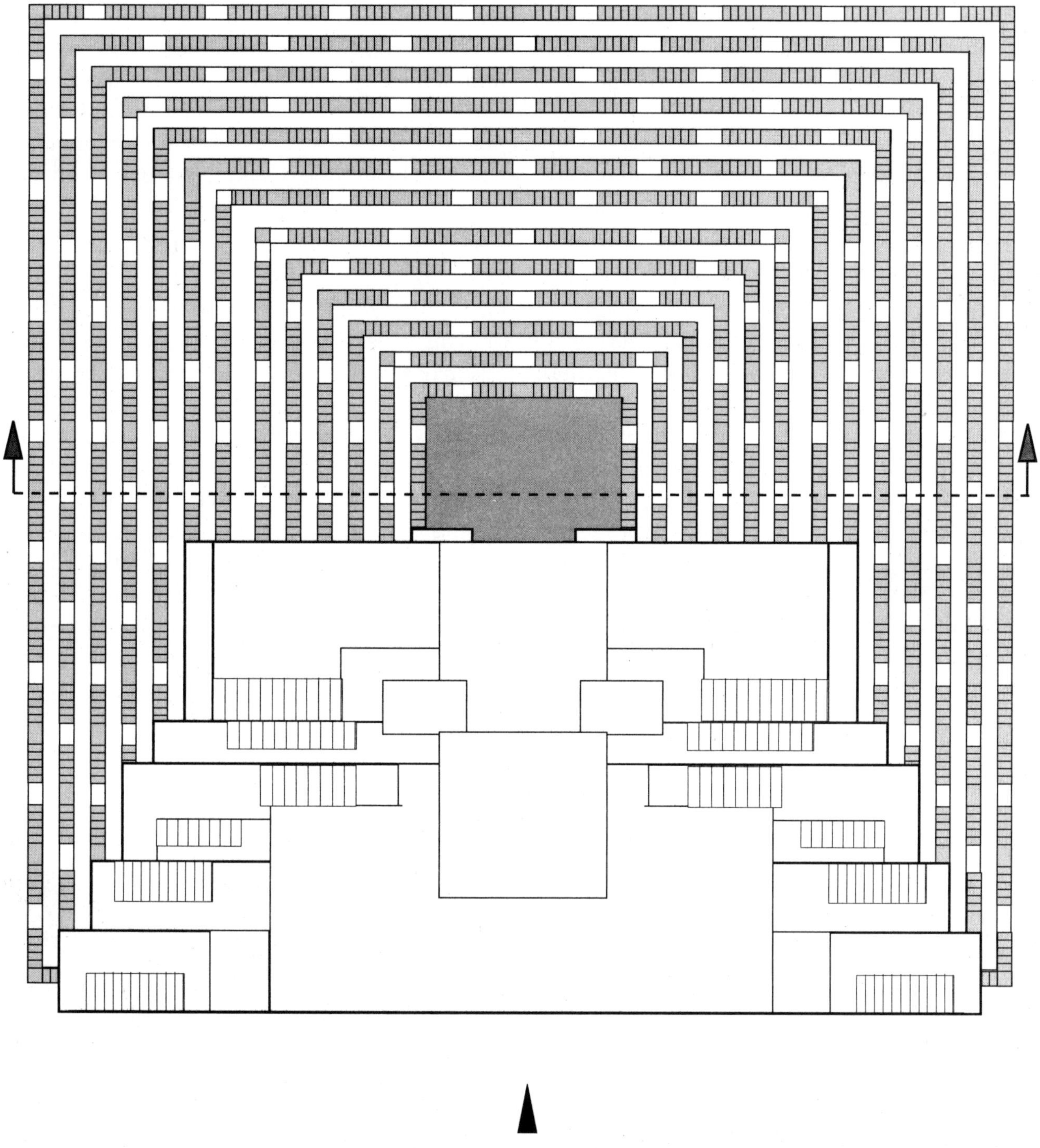

## Rānī-kī Wāv, Pāṭan, Gujarat (*c.* CE 1050)

The Rānī-kī Wāv or Ran-kī Wāv[13] of Pāṭan in Gujarat is considered by many to be one of the most elaborate and pleasing of all the *bāolīs* of Gujarat. Indeed, there is much here to deserve that approbation. As with most *bāolīs* or *wāvs*, it was commissioned by a wealthy patron. In this case, the Queen Udayamatī, consort of Bhīmdev I (r. CE 1022-63) (**Appendix D**) who was the son of Mūlarāja, the founder of the brilliant Solaṅkī Dynasty. It is stated that she constructed this *wāv* of memory of her husband. It is a well that was essentially constructed for obtaining drinking water as well as for the delight of those who use it.

In size and complexity it is rivaled by few other *bāolīs* and sets the standard for subsequent step wells, particularly in Gujarat. A set of steps leads down to a broad terrace, the initial width of the well and from there a series of step elements, similar to those found at the Sūrya Temple (**Plate 8**) and the Cānda Bāolī (**Plate 10**) drop to the next terrace. This level is covered by two ranks of columns supporting a roof/ceiling (**Plate 11, A**). Then another drop of similar stepped elements and the second covered, columned terrace (**Plate 11, B**). Rather than two ranks, this terrace displays three ranks and is two storeys high. From this level, there is a drop with stepped elements to an uncovered terrace and another drop to a second uncovered terrace (**Plate 11, C-D**). From this second, uncovered terrace, a broad flight of steps drops to the third covered terrace (**Plate 11, E**). This terrace, like the one above consists of three ranks of columns, but, four storeys high (**Figure 17**). From this level a set of stepped elements descends to the lowest level and a double rank of columns, open at the top (**Plate 11, F**). The next to the last element of the Ran-kī Wāv is a square space, supported by sixteen pillars and six storeys high (**Plate 11, G**). This is followed by the well shaft (**Plate 6, H**). It is to be noted that each storey, starting with the uppermost, is accompanied by a diminution of width similar to that seen in **Figure 1**.

The walls of each storey, except the uppermost, are decorated with architectural and sculptural elements. The design of the storeys becomes more elaborate as each successive storey is reached (**Appendix D, Plate 77**). The result of the whole is a pleasing and a complex visual delight.

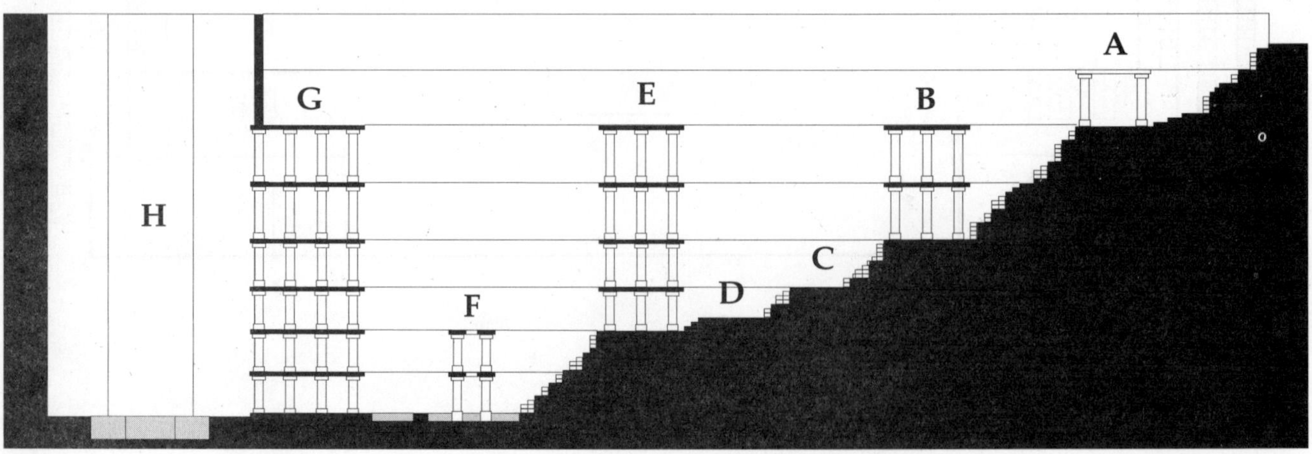


**Figure 17:** Rānī-kī Wāv, sectional view (simplified)

13 The plan is redrawn using Adobe PageMaker 6.0 from the Internet.

# **Plate 11:** Rānī-kī Wāv plan

A 1st covered terrace
B 2nd covered terrace
C 1st open terrace
D 2nd open terrace
E 3rd covered terrace
F open colonnade
G 4th covered terrace
H well shaft (*kūpa*)

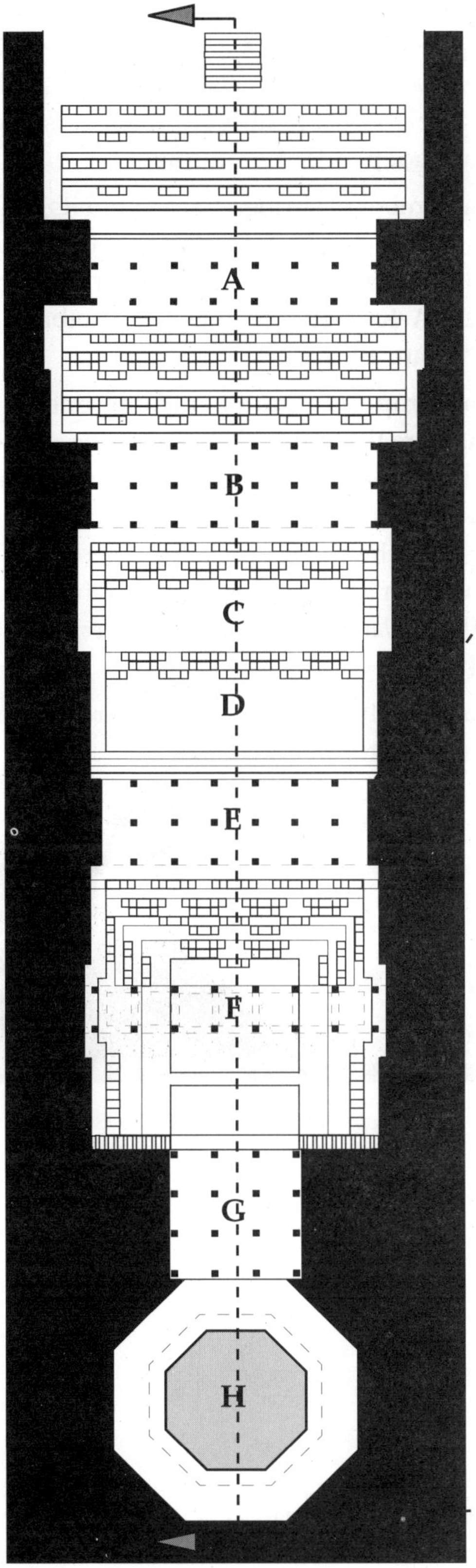


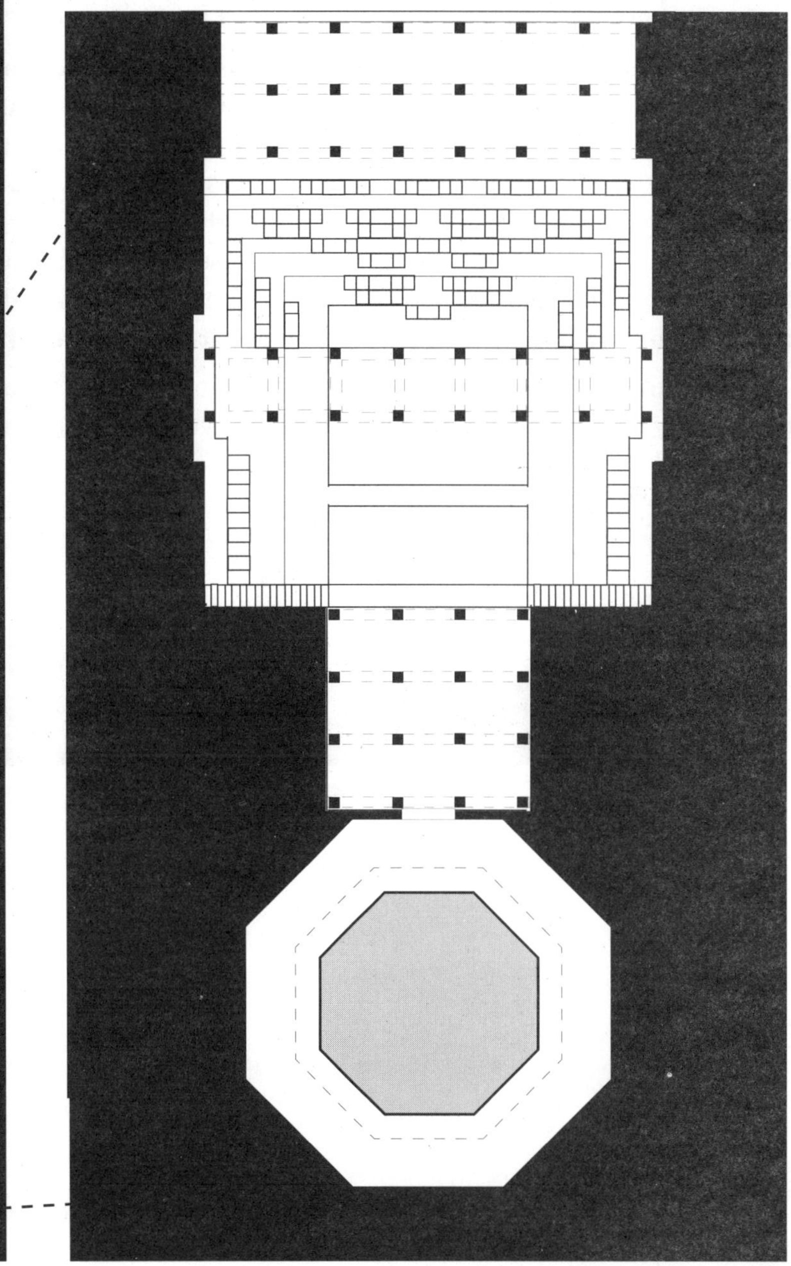

# Kalayni Tank, Hulikere, Karnataka (a. CE 1050)

Far to the south of the large and elaborate *bāolīs* of Gujarat, such as the Cānda Bāolī of Ābhānerī and the Sūrya Temple Tank at Moḍherā is found a small but impressive, Hoysaḷa architectural jewel — Kalayni Tank.[14] Nestled in the small village of Hulikere, Karnataka, rests this elaborate tank. Unlike the Sūrya Temple Tank, this tank does not seem to be part of an elaborate temple complex. It is one of a number of sacred tanks that are independent of a temple; an independent monument.

Like the Sūrya Temple Tank, the Kalayni Tank consists of a number of shrines. Set around its square shape are twenty-seven shrines, *en toto*, which are set amongst the descending stairs that surround the water deep within (**Figure 18**). Eleven major shrines with individual *śikhara*, sitting on elaborate bases rise from the third terrace of the Kalayni Tank (**Plate 12, A-K**). Between these and setback are sixteen minor shrines, hidden by the floor of the first terrace (**Plate 12, a-p**).

Why this arrangement? Twenty-seven is not a particularly important number, representing the first masculine number cubed — i.e. $3^3$. However, if one considers the implied shrine at the point of the entrance stairs (**Plate 12** [ghost image]), then there are twenty-eight shrines. Twenty-eight is a number that is auspicious within the Buddhist–Hindu world as it refers to the days in the cycle of the moon. Eleven for the Hindu is important as the eleventh day after the full moon is considered highly auspicious for prayers. Similarly, twelve is also a favourable number as it not only refers to the Zodiac, the months of the year, but also the twelve-petalled lotus of the *anāhata cakra*. Sixteen, referring to the minor shrines (**Plate 12, a-p**) is important as it is the sacred $4^2$.

None the less, the Kalayni Tank, set within the small village of Hulikere is a most interesting and beautifully designed monument. A bright jewel of Hoysaḷa architecture.

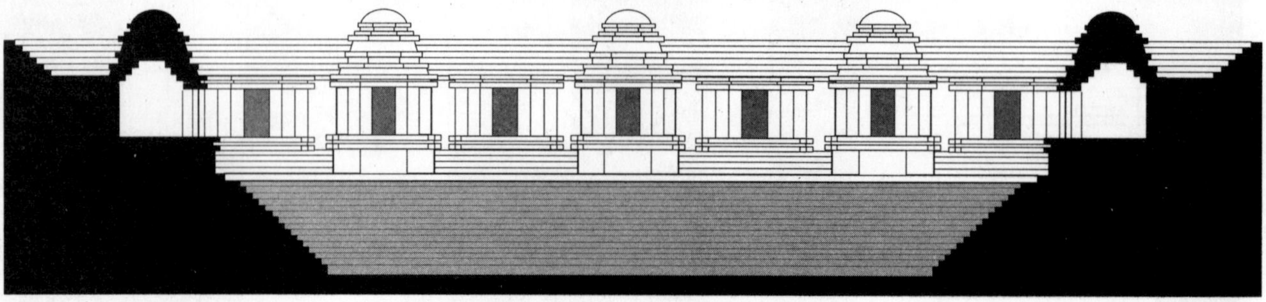

**Figure 18:** Kalayni Tank, sectional view

[14] The plan is redrawn using Adobe PageMaker 6.0, from Gerard Foekema, *A Complete Guide to Hoysala Temples*, New Delhi: 1996, p. 64 and the Internet.

# **Plate 12:** Kalayni Tank plan

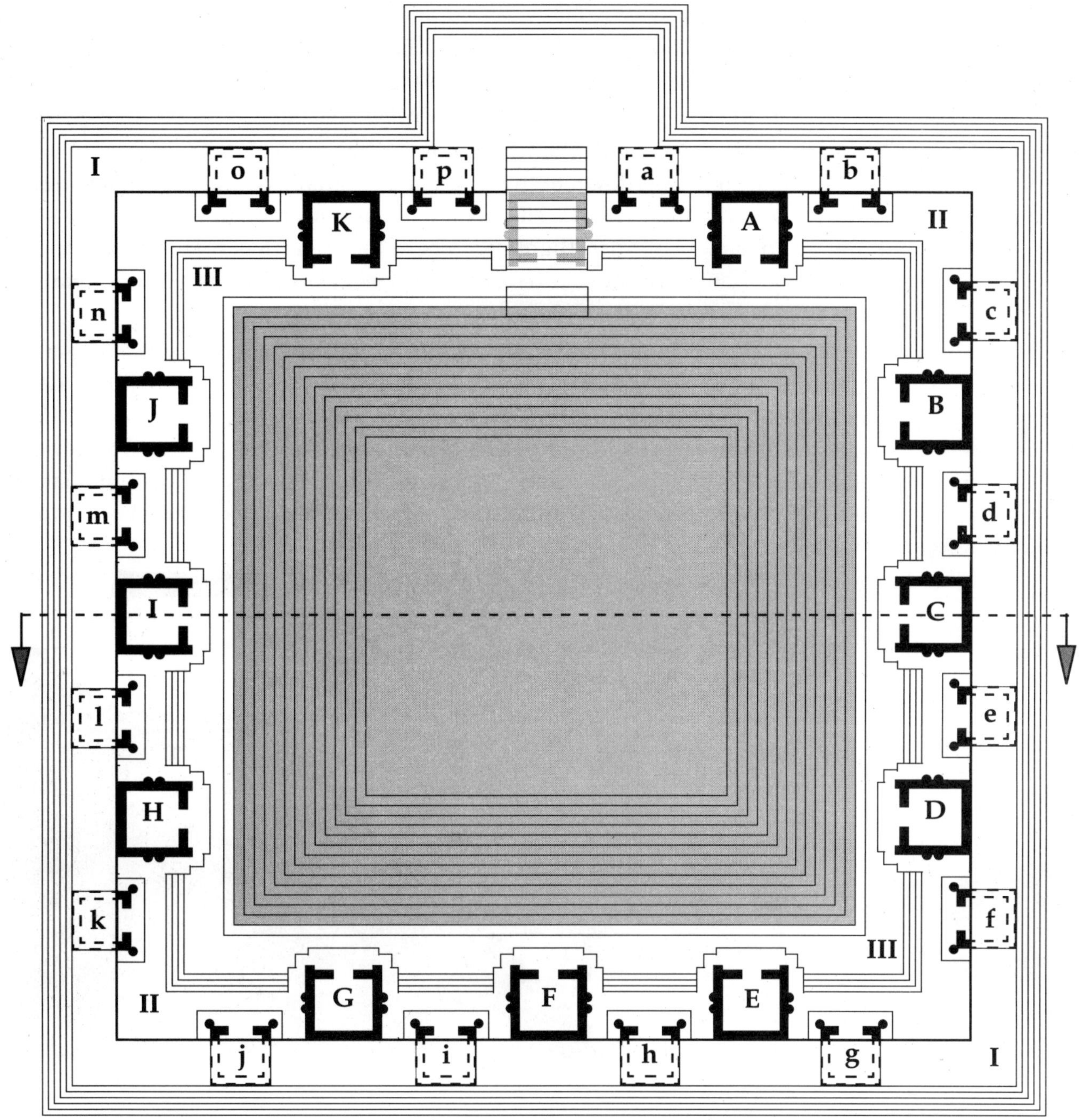


**I** First terrace
**II** Second terrace
**III** Third terrace
**A-K** Major shrines
**a-p** Minor shrines

# Cennakeśava Temple Tank, Beḷūr, Karnataka (CE 1117)

The Hoysaḷa king Viṣṇuvardhana (Bittiga) (r. 1110-52) (**Appendix D**) constructed the Cennakeśava Temple at Beḷūr as a memorial of his victory over the Coḷas at Talakad. As with most temple complexes in the south of India a sacred tank was also constructed within the precincts for ritual ablution of the devotees. Although not large by prevailing standards, the Cennakeśava Temple Tank[15] is unique, *en plan*.

With the exception of the entrance pavilion, the whole was constructed below the ground-level of the main compound. A flight of steps laid down to the entrance of a small, simple cube, the entrance pavilion. On the interior, this shrine-like entrance sits upon its own five-stepped podium. The steps of the podium continue downward in a broad flight of stairs to the water level. The tank proper is essentially rectangular with two small shrines on redented corners of the main level. This creates a tank of eccentric proportions as is seen in **Plate 13**.

The north, south and east aspects of the tank's steps are precipitously narrow, with a rise-tread ratio of less than 2:1 (**Figure 3, C**). This effectively creates only one way by which one may easily approach the water for the required ablution — from the west or from the main entrance. In addition, the depth of the tank, when it is nearly full, is quite deep (**Figure 19**). This design seems to indicate that although the Cennakeśava Temple may have been of some import locally, it was not a site of major pilgrimage like the Naṭarājar Temple at Cidambaram, otherwise the tank would have a more open and accessible design.

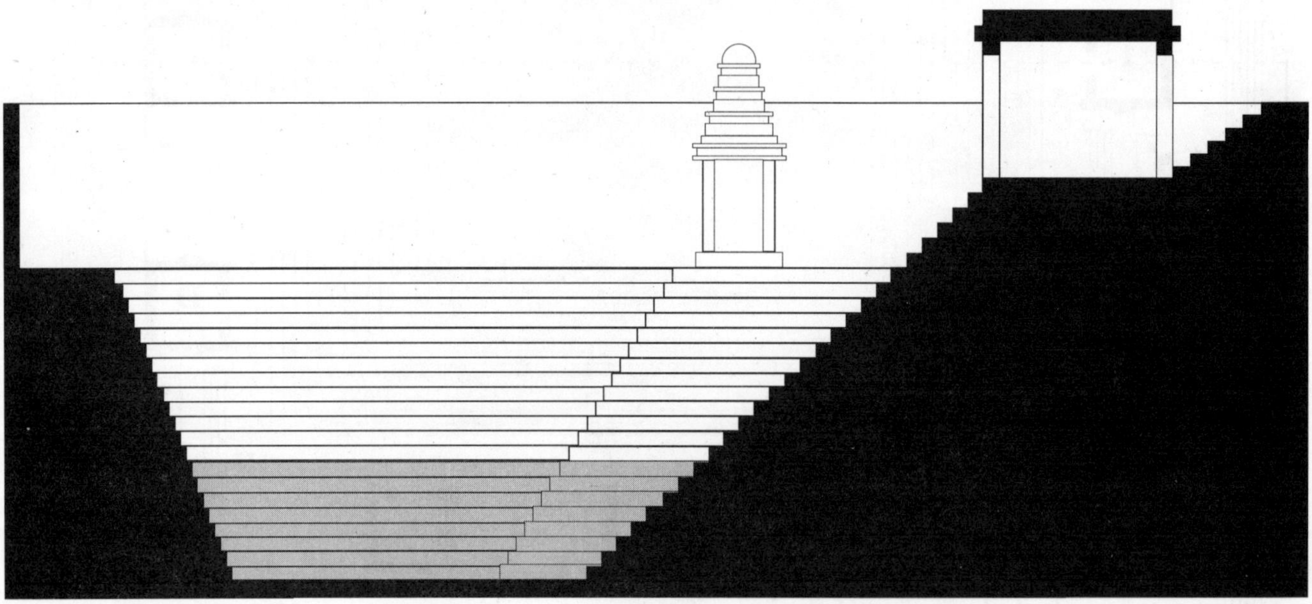

**Figure 19:** Cennakeśava Temple Tank, sectional view

[15] The plan is redrawn using Adobe PageMaker 6.0 from the Internet.

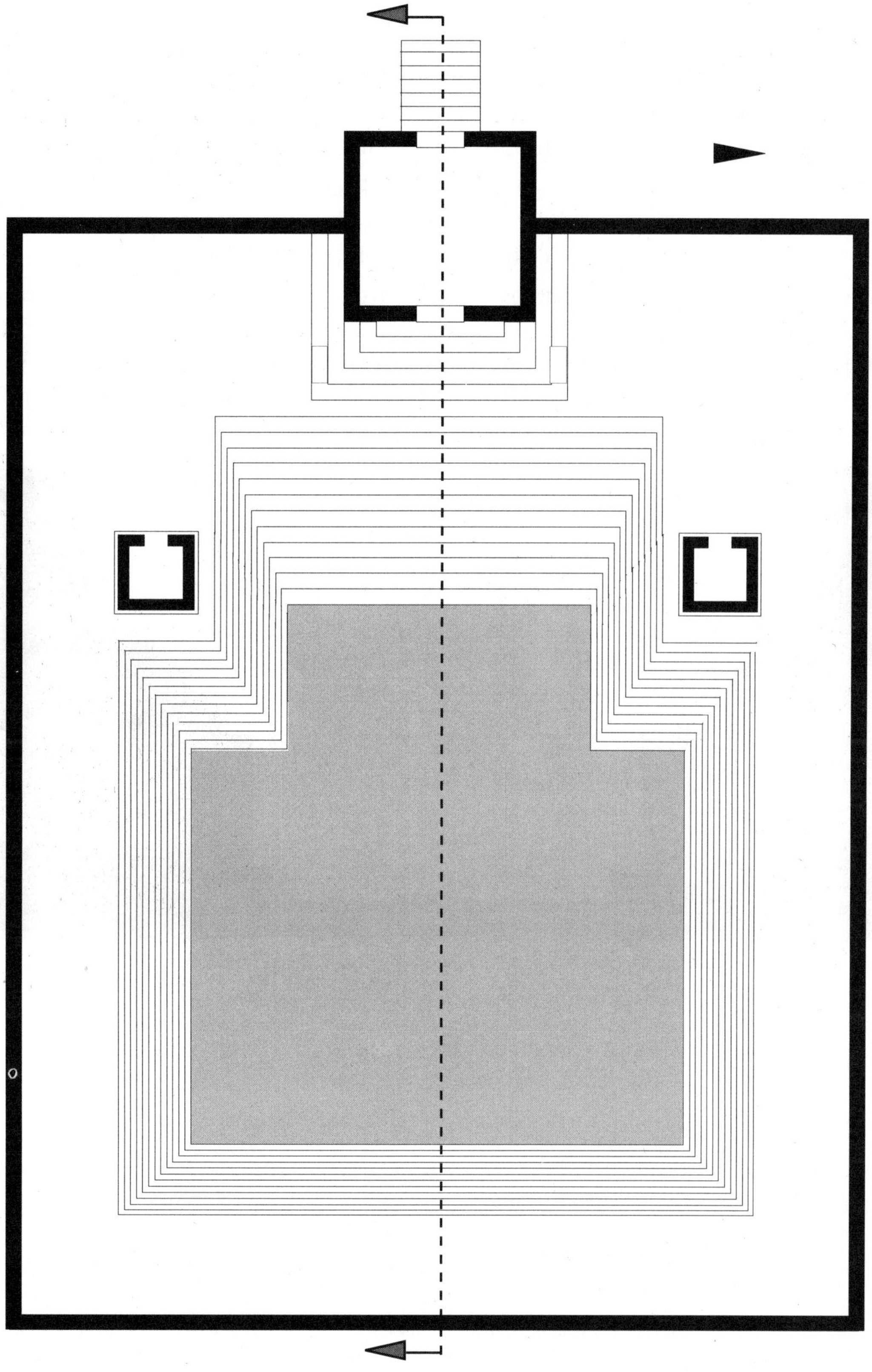

## Mahāmāyā and Navaliṅga Temples' Tank, Kukkanūr, Karnataka (*c.* CE 1150)

In the small village of Kukkanūr is a complex of several temples and the Mahāmāyā Temple and the older Navaliṅga Temple are eccentric, *en plan*. To the north and east of the Mahāmāyā Temple is the Mahāmāyā and Navaliṅga Temples' Tank.[16] Constructed by the Cāḷukyas, probably during the reigns of Jagadekamalla II (r. CE 1135-51) and Taila III (r. CE 1151-56), before the Kaḷacuri (Kalyāṇa) dominance (**Appendix D**), the tank is one of the thousands sacred tanks that are associated with a temple .

This tank depends on rainfall as well as the aquifer to keep it supplied with water. The Mahāmāyā and Navaliṅga Temples' Tank is essentially a sacred tank, the water of which is used for *abhiṣeka* (the ritual ablution) of the images rather than ritual bathing of the devotees.

Rectilinear, *en plan*, the first flight of steps descends from the south to the second terrace (**Plate 14, II**). Then two more flights descend eastward to the first terrace and thence to the water (**Plate 14, I**). Due to the depth of the aquifer, the tank required considerable excavation (**Figure 20**). The design of the plan is pleasing and somewhat unusual from other tanks of the area. The design of the Mahāmāyā and Navaliṅga Temples' Tank allows the depth of the water table to be reached with relative ease.

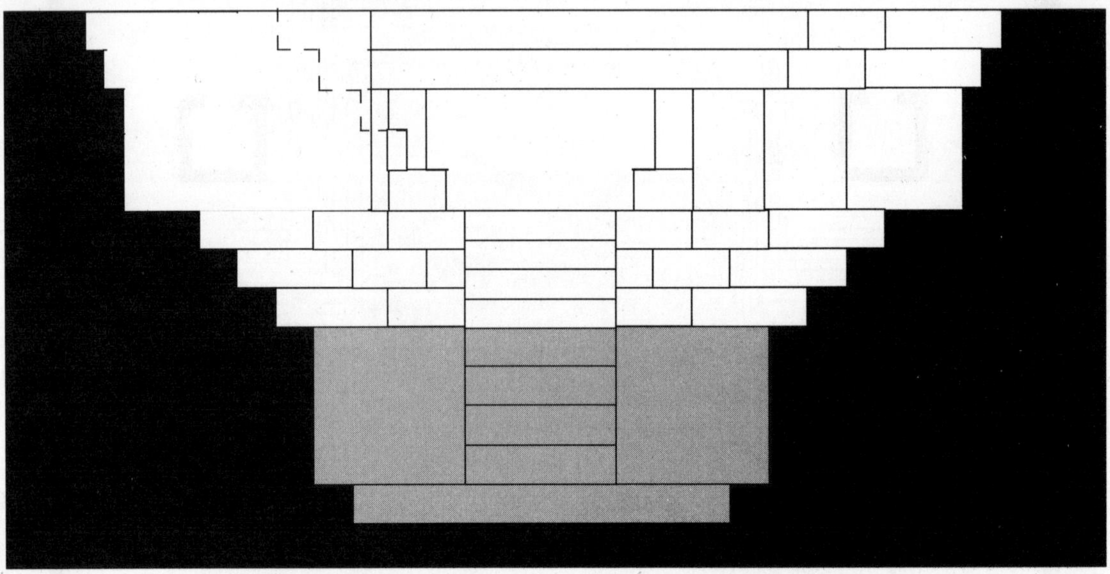

**Figure 20:** Mahāmāyā and Navaliṅga Temples' Tank, sectional view

16 The plan is redrawn using Adobe PageMaker 6.0 from Henry Cousens, *The Calukyan Architecture of the Kanarese Districts*, New Delhi: 1996, Plate LV.

## Plate 14: Mahāmāyā and Navaliṅga Temples' Tank plan

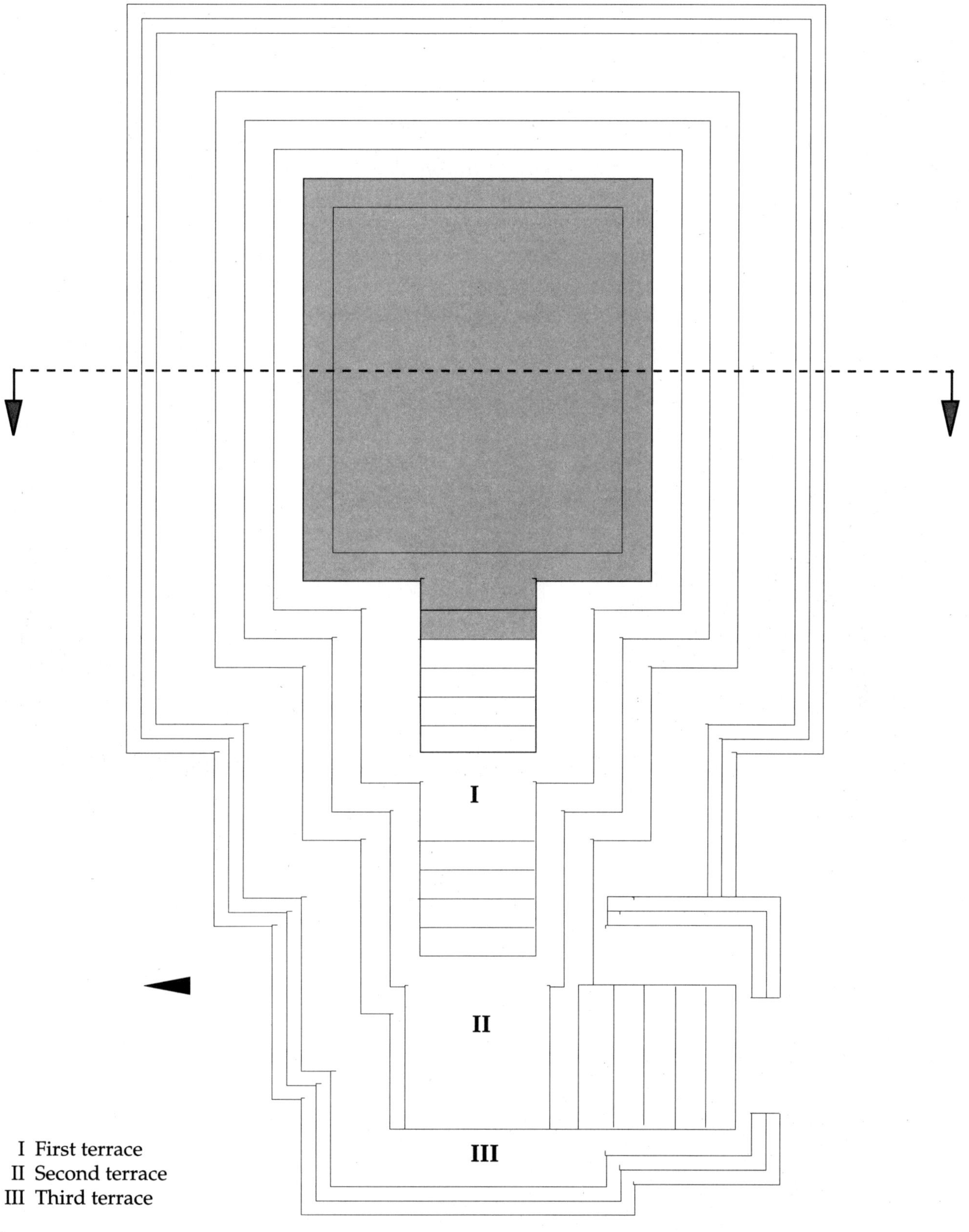


I First terrace
II Second terrace
III Third terrace

## Brahma Kuṇḍa, Kāverī Temple, Telakāverī, Karnataka (*c.* thirteenth century)

Whereas, many of the other tanks in India are filled either by rainfall and/or via an aquifer, a channel or viaduct from a nearby stream or river, the Brahma Kuṇḍa[17] is said to be fed by a spring. This spring and the *kuṇḍa* are seen as the source of the sacred Kāverī River[18] and the *kuṇḍa* is a sacred tank independent of a temple. The Kāverī flows eastward into the Bay of Bengal and around the famed Śrīraṅgam island, home of the sacred Śrī Raṅgamswāmi Temple.

The Brahma Kuṇḍa is by comparison, one of the smallest *kuṇḍa*s (tanks or step wells) in India. This does not detract it from the importance as a place of pilgrimage. Two staircases lead to the main level of the *kuṇḍa* — one broad stairway leading into the upper corner and a narrow stairway at the adjacent corner. At the base of the narrower stairs is a recessed niche. These stairs bracket a small shrine found on the main level of the *kuṇḍa*. Opposite this shrine, on the other side of the *kuṇḍa,* is an image of a kneeling bull — Nandī — projecting from the main level. Asymmetrical, *en plan*, the *kuṇḍa* is essentially square with three stairways that lead from the main level down and into the sacred water (**Figure 21**). These allow numerous pilgrims to descend to the water level for their ablution. The *kuṇḍa* is set within the side of a small hill, over a spring which, in part, accounts for its smaller size.

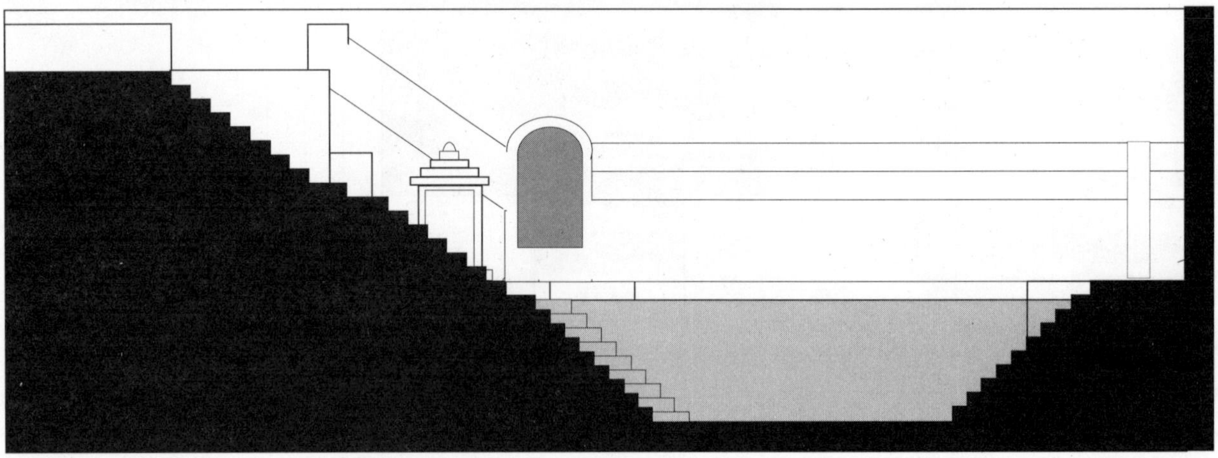

**Figure 21:** Brahma Kuṇḍa, Kāverī Temple, sectional view

17 The plan is redrawn using Adobe PageMaker 6.0 from the Internet.

18 Although not as revered as the Mother Gaṅgā, the Kāverī is, none the less, one of the most sacred rivers of India, and bathing at its source is considered quite auspicious.

**Plate 15:** Brahma Kuṇḍa plan (partial)

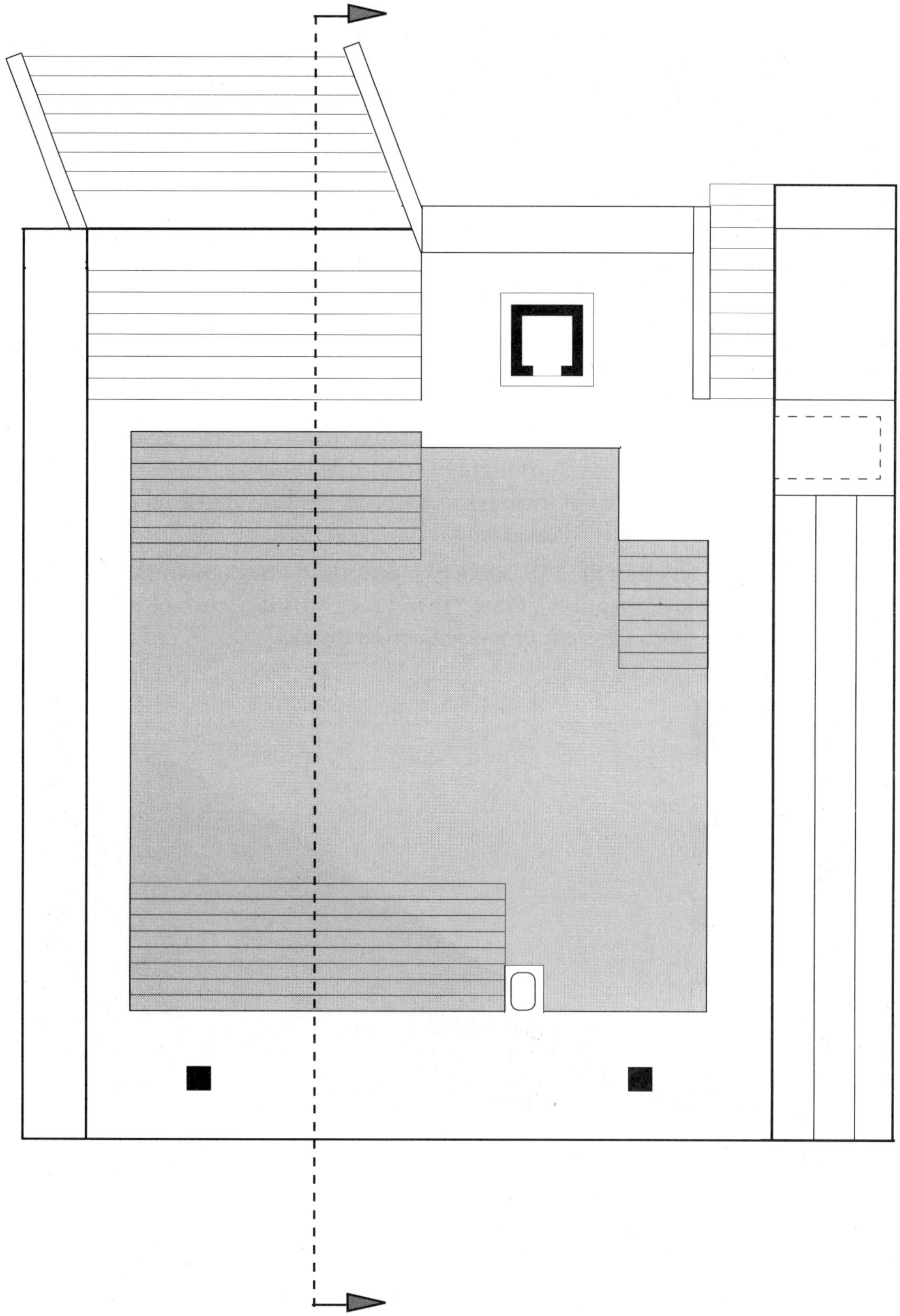

## Gandhak-kī Bāolī, Delhi (a. CE 1211)

In addition to consolidating his reign militarily soon after his ascension, the Sultan Shams-ud-din Iltutmish (r. 1211-36) (**Appendix D**) was responsible for the construction of a number of *bāolī*s in and around Delhi. The Gandhak-kī Bāolī,[19] which he built near the famed Qutb Minar, was one of them and was a well (*bāolī*) essentially for obtaining drinking water. The Qutb Minar was the paladin of Muslim rule in the north and the Sultan's construction of this source of water near it underscored his and the Muslim Sultanate legitimacy.

The Gandhak-kī Bāolī, by prevailing standards, was not only functional, it was rather simple and unadorned. The only variation from the plain interior vistas is to be seen in the upper level towards the openings to the shaft (**Appendix C, Plate 74**). Here four columns and two engaged columns punctuate the end of this level.

*En plan*, the *bāolī* proceeds from the ground level via two opposing stairways which lead a short distance to the entrance level (**Plate 16, I**). From here three broad stairways descend to the third level (**Plate 16, IV**). Three more stairways, in ever-decreasing widths, lead to the lowest level of the *bāolī* (**Plate 16, VIII**). The side walls of all the levels are, as mentioned above, unadorned. Opposite the entrance on level **II**, on both sides is a short flight of stairs that leads up to the upper shaft level (**Plate 16, X**). From here, those who were not inclined to descend to the water level could draw the much needed water directly from the shaft (**Plate 16, IX**).

The decreasing width levels of the *bāolī* serve two purposes. They ameliorate the lateral forces of the earth (See **Figure 22;** and Appendix C, **Plate 74**) and their width allows for a number on people to rest in the shade of either side as the sun processed across the sky.

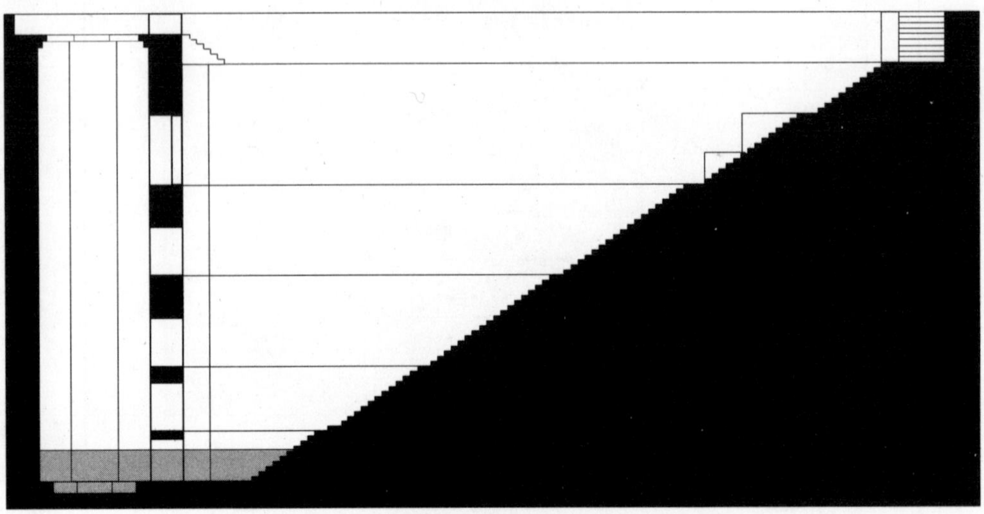

**Figure 22:** Gandhak-kī Bāolī, sectional view

[19] The plan is redrawn using Adobe PageMaker 6.0 from the Internet.

# **Plate 16:** Gandhak-kī Bāolī plan

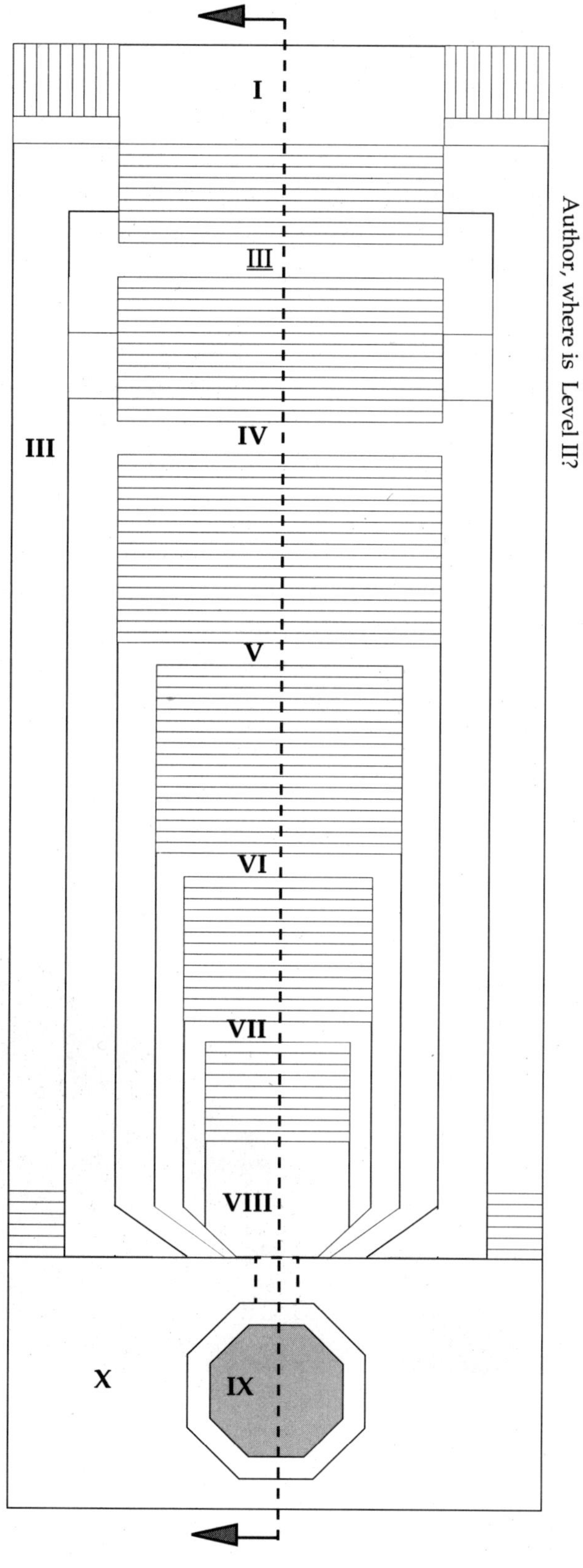


I Entrance level
II 1st level
III Intermediate level
IV 2nd level
V 3rd level
VI 4th level
VII 5th level
VIII Lowest level
IX Shaft
X Upper shaft level

# Niẓāmuddīn Bāolī, Delhi (*c.* CE 1300)

Nearly 100 years after the construction of the Gandhak-kī Bāolī the prominent Sufic *guru*/saint, Haẓrat Niẓāmuddīn Auliyā, began to build his *bāolī*. The construction of the *bāolī* was an act of merit and is a well essentially for obtaining drinking water. Unfortunately it was at the time when the Sultan Ghiyās-ud-Dīn Tughlaq Shāh I (Ghāzī Malik) (r. 1320-25) (**Appendix D**) was constructing his royal city, Tughlaqābād. The Sultan commandeered all available labourers in and around the region for his citadel, including those working on the Niẓāmuddīn Bāolī.[20] This angered Niẓāmuddīn who pronounced a curse on Tughlaqābād. His stature and fame within the Muslim community at the time was such that even the Sultān was loath to discipline the Sufic saint for his temerity. However, in time the *bāolī* was completed.

The Niẓāmuddīn Bāolī is one of the few *bāolī*s of the period in and around Delhi that is still full of water and available to the surrounding populous more for bathing than drinking water. The shaft is no longer utilized and, indeed, it is not accessible to the public as buildings now surround most of the *bāolī*. Like the Gandhak-kī Bāolī, the Niẓāmuddīn Bāolī was constructed so that people could gain access to drinking water.

The Niẓāmuddīn Bāolī is rectilinear, *en plan*, with a flight of steps and a landing as an entrance. This leads down to the main stairway. These stairs are the width of the *bāolī* and lead down to the bottommost level with a number of landings. An unusual feature of this *bāolī* is the space at the bottom step when the tank is full. The Niẓāmuddīn Bāolī is now bounded by buildings from which young boys dive to retrieve coins thrown into the exceedingly deep water (**Figure 23**). Unfortunately the *bāolī* still claims at least one life a year.

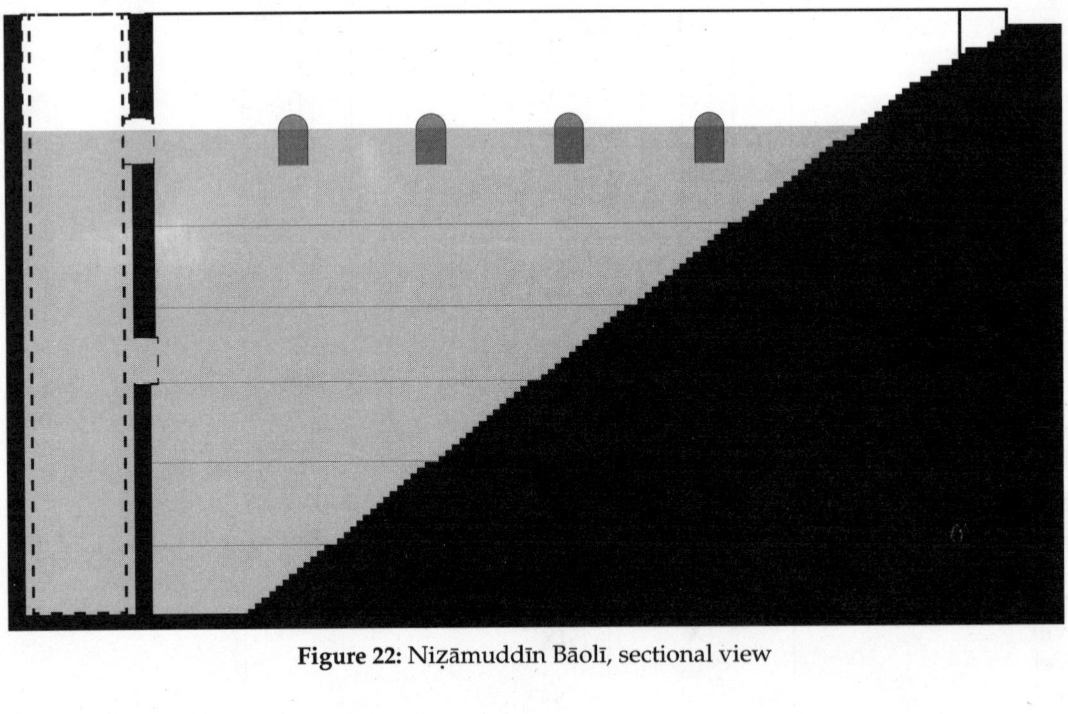

**Figure 22:** Niẓāmuddīn Bāolī, sectional view

[20] The plan is redrawn using Adobe PageMaker 6.0 from the Internet.

# Plate 17: Niẓāmuddīn Bāolī plan

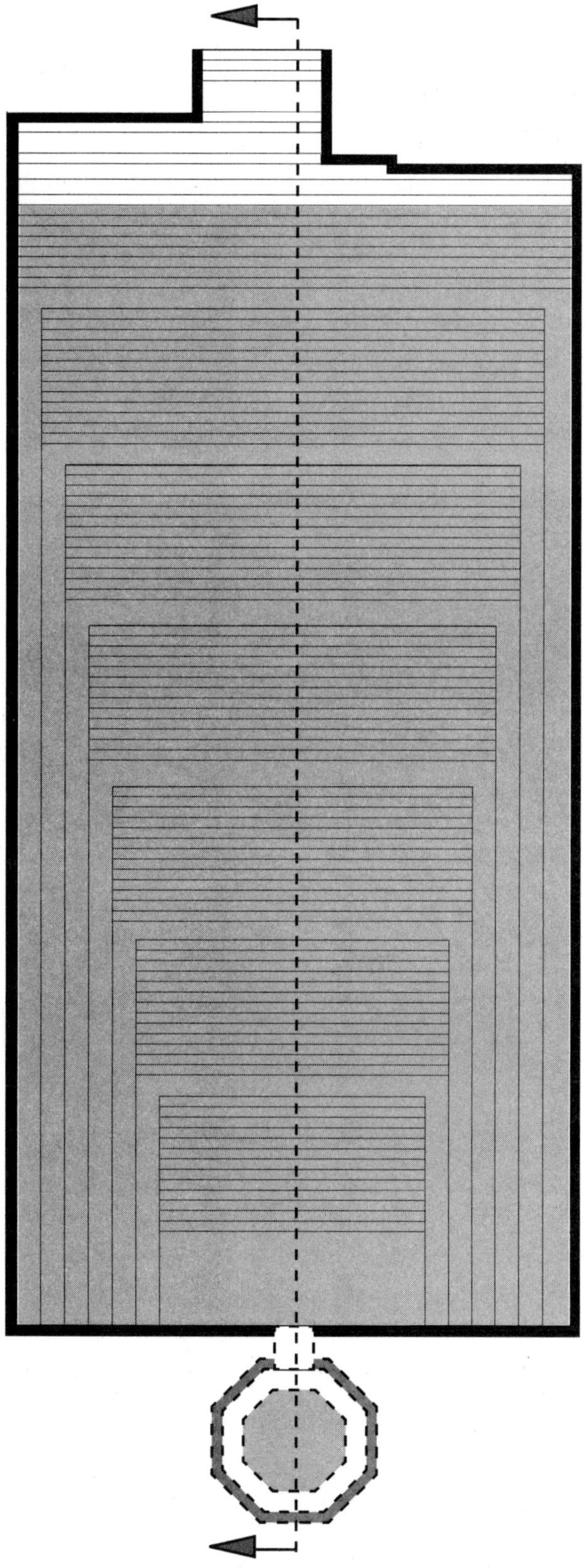

# Ugrasena-kī Bāolī, Delhi (a. CE 1300)

One of the most impressive step wells (*bāolīs*) in Delhi is the Ugrasena-kī Bāolī.[21] Legend states that it was constructed in the thirteenth century by one Rājā Ugrasena. However, the design of the elevation causes many to surmise that it was constructed during the Tughluq or Lodi Dynasties (**Appendix D**). This would place the construction of the *bāolī,* as now evidenced, after CE 1300.

The Ugrasena-kī Bāolī is a well (*bāolī*) essentially for obtaining drinking water, and consists of three levels and seven storeys. The side walls of the two storeys of the bottom level are blank, but the two storeys of the next levels that are separated by narrow walkways possess open arcades on the lower storey and a blind arcade on the upper storey. The public could process along these narrow walkways and enter the cool interior of the open arcade and gain welcome relief from the heat of the unremitting sun. At the end — the transition from the open trench to the cylindrical shaft — are openings from the trench to the shaft. A wall, the uppermost storey, extends above ground level and borders the long sides and the end towards the shaft. From the shaft, one could draw water without descending the long way down to the water level (**Figure 24**).

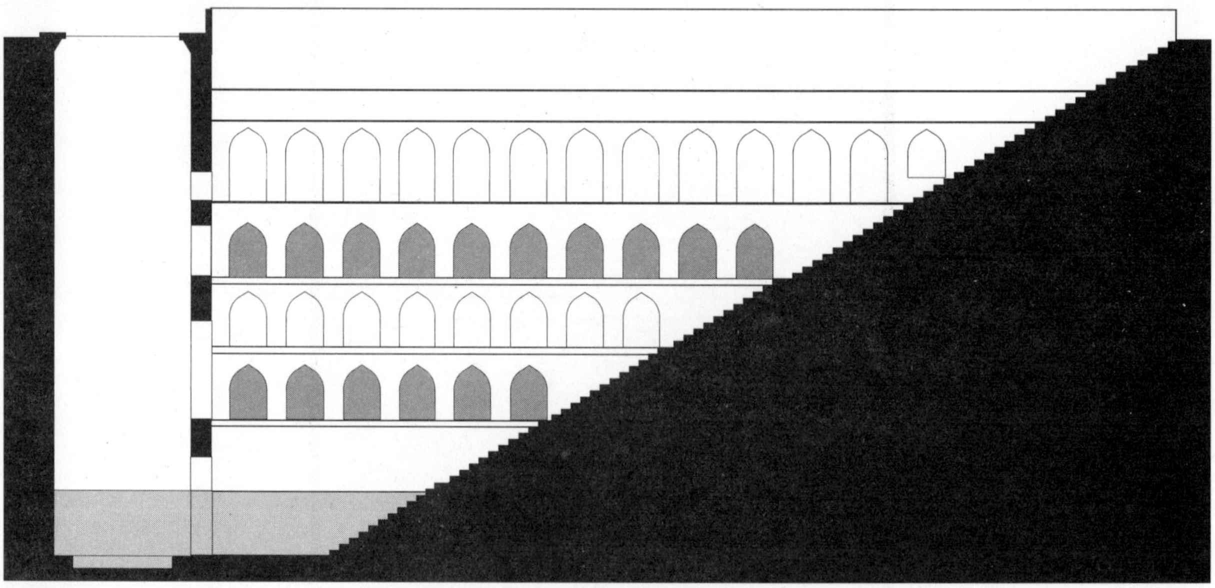

**Figure 24:** Ugrasena-kī Bāolī, sectional view

[21] The plan is redrawn using Adobe PageMaker 6.0 from the Internet.

# Plate 18: Ugrasena-kī Bāolī plan

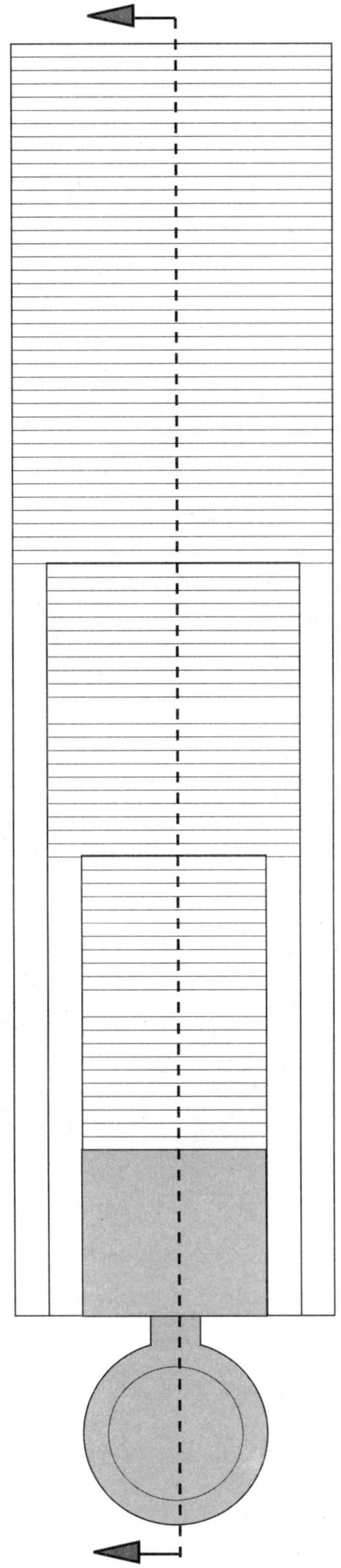

# Rudabāī-nī Wāv, Adalaj, Gujarat (CE 1499)

During the reign of Sultān Nasīr-ud-Dīn Mahmūd Shāh I Begarah (r. 1459–1511) (**Appendix D**), a *wāv* or *bāolī* was constructed at Adalaj. Sometimes known as the Adalaj Wāv, it is known in the area as the Rudabāī-nī Wāv.[22] The Rudabāī-nī Wāv is named after Ruda, the wife of a wealthy Vaghela chieftain, Vīrsinh. He was related to the Vaghela branch (CE 1293-1304) of the Cāḷukya-Solaṅkī Dynasty. 'Alā-ud-dīn Muhammad Khaljī (r. CE 1296–1316) subdued the Vaghela in CE 1304 and Gujarat was in turn taken over by the Muzaffarid Shāhi in CE 1391.

Of the numerous *wāv*s or *bāolī*s of Gujarat — e.g. the Rānī-kī Wāv, or the Dādā Harīr Wāv — the Rudabāī-nī Wāv is second to none in its sumptuous detail and plan. Oriented towards the north, the entrance to the *wāv* is threefold — i.e. a broad flight of stairs to the north leading down to a common level and a duplicate sets of stairs to the east and to the west.[23] This common level is composed of a number of columns with a large octagonal opening allowing an abundance of light to flood this commodious area (**Figure 25**). Unlike the Gandhak-kī Bāolī or the Rānī-kī Wāv, the side walls do not diminish in width as each successive level is reached. Rather, they remain the same as does the width of the descending set of steps. In place of the diminishing width, lateral bracing in the forms of substantial beams and floors supported by ranks of columns absorb and neutralize the lateral force of the surrounding earth (**Appendix C, Plate 75**). In addition, the Rudabāī-nī Wāv indicates four means by which one could gain access to the water — i.e. first from the circular well-shaft, second and third via the circular stairs — i.e., one set to the east and one to the west of the lower level — and the last by descending the broad steps from the entrance(s).

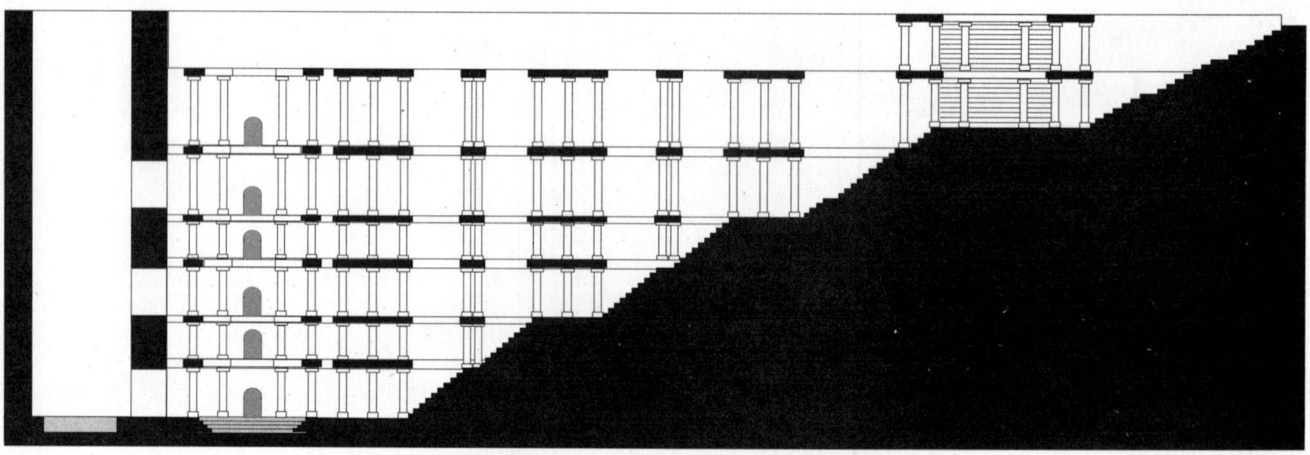

**Figure 25:** Rudabāī-nī Wāv, Adalaj, sectional view

22 The plan is redrawn using Adobe PageMaker 6.0 from Christopher Tadgell, *The History of Architecture in India*, London: 1990, p. 174 and the Internet.

23 JN, p. 26, catagorizes this type as "Type two . . . with three entrances".

## Plate 19: Rudabāī-nī Wāv plan

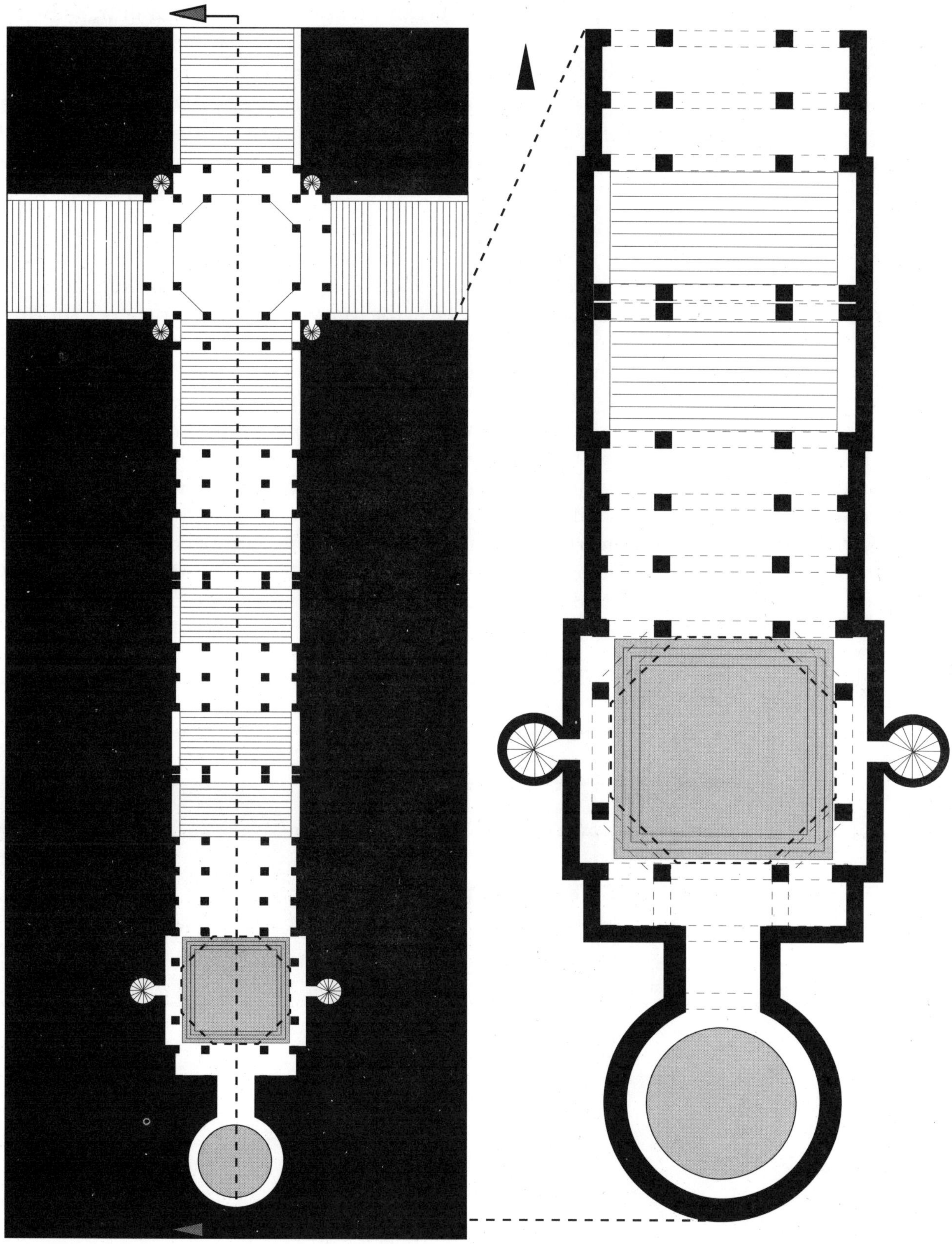

# Dādā Harīr Wāv, Asarwa, Gujarat (*c.* CE 1500)

Also constructed during the reign of Sultān Nasīr-ud-Dīn Mahmūd Shah I Begarah (r. 1459–1511) (**Appendix D**) is a step well known variously as the Bāī Hari Wāv or the Dādā Harīr Wāv, the Bāolī Bāī Hari.[24] It is a well (*bāolī*) essentially for obtaining drinking water and for the delight of those who venture there. As with most step wells in Gujarat this *wāv* was constructed as an act of merit and patronage. It is stated that a Lady of the Court of Mahmūd Shāh Begarah, one Bāī Śrī Harīr was responsible for the construction of the Dādā Harīr Wāv. It is of interest to note that a number of important step wells of Gujarat were built under the auspices of prominent women of the realm — e.g. the Rānī-kī Wāv and the Rudabāī-nī Wāv.

In many ways the Dādā Harīr Wāv is the most pleasing of the major *bāolīs* of Gujarat. It consists of an inclined trench with stairs and three shafts — a large shaft for obtaining water also known as a *kūpa* and two smaller shafts for spiral stairs. The *kūpa* is open at the top which allows for water to be drawn manually rather than descending the many flights of stairs. The trench consists of a number of flights of stairs punctuated with landings. The employment of a short flight of steps with a landing before the next short flight is not only visually pleasing, *en plan*, but it is also physically more agreeable when descending. At intervals the flights are broken by square or narrow, rectangular covered, pillared areas where one might take refuge in the cool shade (**Figure 26**). These are some time referred to as *maṇḍapas*. These pillared elements are stacked one upon the other as the depth of the *bāolī* increases. They serve a real and necessary purpose — the neutralizing of the lateral force or thrust of the surrounding earth (**Appendix D, Plate 75**). At the bottom, a larger area — five bays square with an octagonal central area that is open through the intervening four storeys — allows light to filter down to this lowest level. On each level or storey, access is gained to the spiral stairs and thence to the other levels.

**Figure 26:** Dādā Harīr Wāv, Asarwa, sectional view

[24] The plan is redrawn using Adobe PageMaker 6.0 from Satish Grover, *Islamic Architecture in India,* New Delhi: 1996, p. 70.

# **Plate 20:** Dādā Harīr Wāv plan

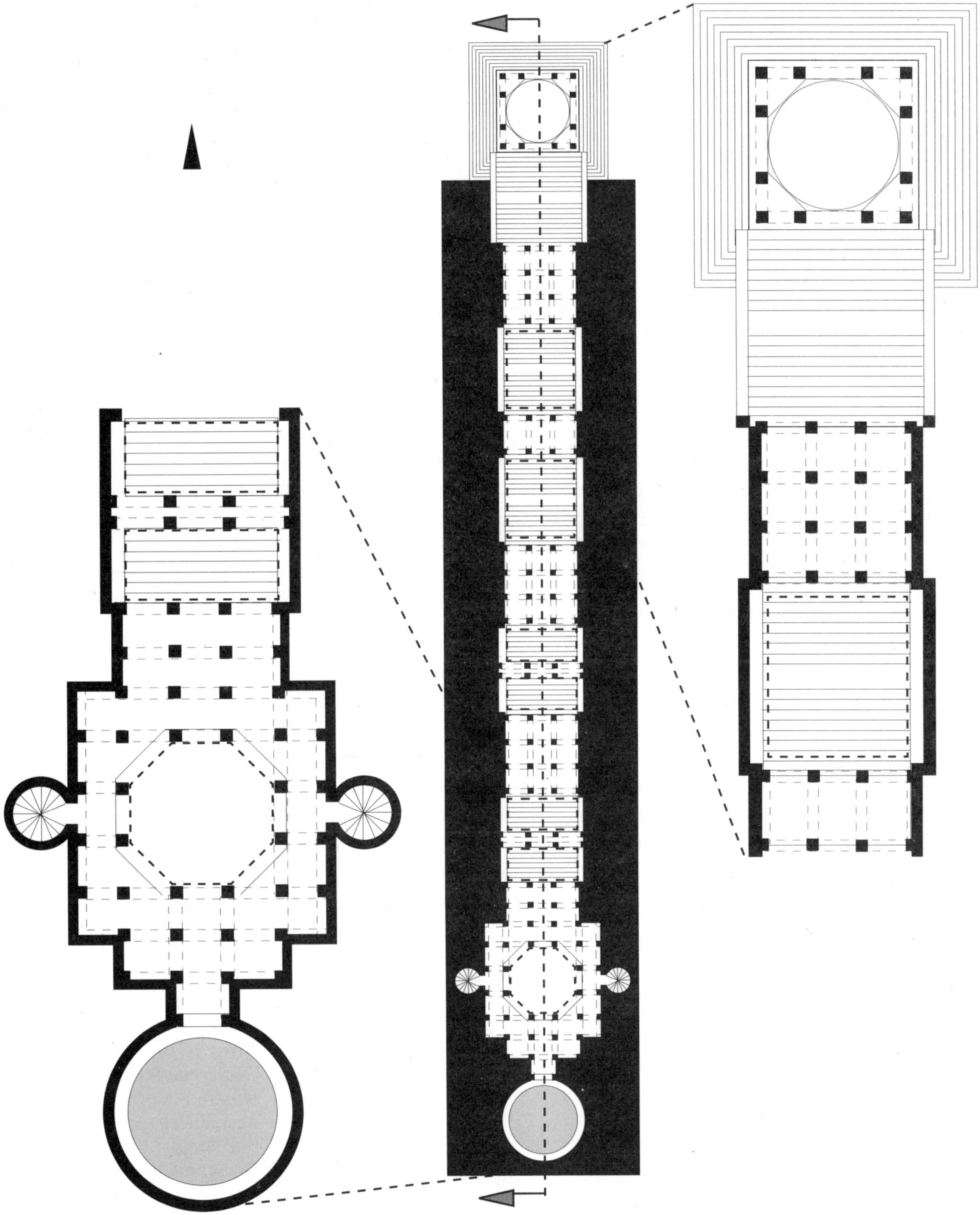

# Imperial Vijayanagara, Karnataka (CE 1346–1565)

Vijayanagara, the last of the great Hindu empires, anchored the southern third of the subcontinent while Islam was expanding in the north. Its brilliance, pomp and sumptuous detail was known as far as the Mediterranean and the Imperial Court of China.

Southern India had been looked upon with covetous eyes by the Islamic powers of the north. Khaljī Sultān of Delhi, 'Ala-ud-dīn Muhammad Khaljī (r. CE 1296–1316) became envious of the riches of the south and moved his armies into the Deccan where he attacked and plundered Devagiri (Daulatābād) in 1296. By 1309 the Kākatīya kingdom succumbed to the overwhelming onslaught of the Delhi Sultānate under the then leadership of Sultān Ghiyās-ud-Dīn Tughlaq Shāh I (r. CE 1320-25). One of the terms imposed upon the Kākatīya Hindu was that they were to accept Islam, *en masse*. This was an edict forced upon them in order for them to retain their realm as a feudatory of Delhi. In 1323 the Kākatīyas revolted, were reconquered, all former rights being removed and the kingdom ceased to exist because of their perfidy towards the Sultān.

Two of the five Kākatīya brothers or cousins — Hakka and later known as Harihara (r. CE 1336-56) and the other called Bukka (r. 1356-77) (**Appendix D**), Saṅgamas of the Kuruba/Golla clan — were among the seventy-five Nāyakas of the Kākatīya Court at Wāraṅgal who were sent to Delhi for "re-education" after their forced conversion. Later, the two "retrained" brothers were sent back to the south to suppress the rebellious Hoysaḷas. Upon their arrival in their homeland they conveniently reverted to their Hindu heritage, established their own kingdom, subdued the Hoysaḷas and expanded their holdings into a growing empire of size and strength — i.e. Vijayanagara.

Imperial Vijayanagara[25] with the Royal Centre was not devoid of a suitable amount of water as it was constructed adjacent to the Tuṅgabhadrā River (**Appendix C, Plate 80**). What makes this city unusual is the number of major tanks that appear to have been exclusively employed for bathing — specifically the Geometric Tank (**Plate 21, S, Plate 23**) sometimes referred to as the Puṣkaraṇī, the Great Tank (**Plate 21, T, Plate 22**), the Square Water Pavilion (**Plate 21, V, Plate 24**), the Octagonal Bath (**Plate 21, Y, Plate 25**), possibly the Floating Mahal (**Plate 21, G**) and the numerous tanks attached to the various residences (**Plate 21, W, Plate 26, Appendix C, Plate 79**). These tanks were supplied with water via aqueducts and channels from the Tuṅgabhadrā River.

Drinking water was supplied by step wells similar to the octagonal well found in Malpannagaḍi to the south (**Plate 21, Z, Plate 27**). That is not to say that the major tanks were not utilized for domestic use, but they most probably were not used to provide safe drinking water.

**A** Elephant Stables
**B** Rectangular Pavilion
**C** Parade Ground
**D** Vaulted Pavilion
**E** Padma Mahal
**F** *Prāsāda*
**G** Floating Mahal
**H** Square Watchtower
**I** Square Watchtower (ruins)
**J** Octagonal Watchtower
**K** Mādhava Temple
**L** Ellammā Temple
**M** Domed Watchtower
**N** Rāmacandra Temple
**O** Nine-domed Pavilion
**P** Octagonal Pavilion
**Q** 100-Columned Pavilion
**R** Mahānavamī Platform
**S** Geometric Tank (Puṣkaraṇī)
**T** Great Tank
**U** Octagonal Fountain
**V** Square Water Pavilion
**W** Residences
**X** Buildings
**Y** Octagonal Bath
**Z** Octagonal Well

[25] The plan is redrawn using Adobe PageMaker 6.0 from George Michell, *The Vijayanagara Courtly Style*, New Delhi: 1992, p. 88.

# Plate 21: Imperial Vijayanagara, Royal Centre

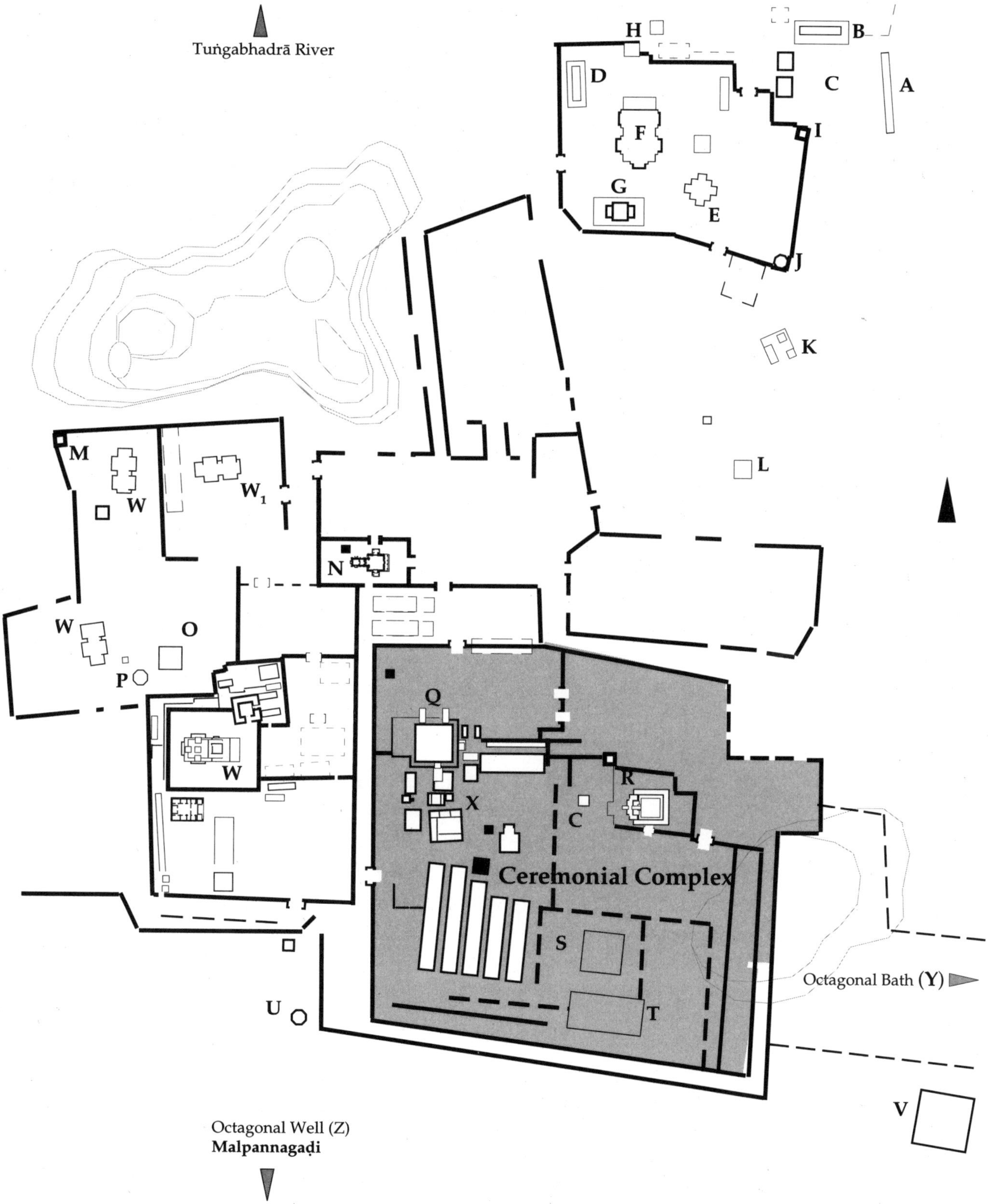

## Great Tank, Imperial Vijayanagara (a. CE 1400)

The Great Tank[26] of Imperial Vijayanagara is noteworthy for two important reasons. First, its size which is considerable, and second, it is not located near a temple but within the Ceremonial Complex of the Imperial city (**Plate 21, T**). It is found to the south of the Mahānavamī Platform and to the south of the famed Geometric Tank (**Plate 21, S, Plate 23**). The royal chapel, the Rāmacandra Temple, rests a distance to the north and within its own *prākāra*. For this reason, one may assume that if it were employed in ritual ablution prior to the performance of prayers, it would have been so used for the members of the Imperial Court exclusively who would have had to proceed a distance to the chapel.

The Great Tank, 222 ft long by 73 ft wide, consists of a large rectangular, stepped area with varying levels of water throughout the year (**Figure 27**). The Great Tank could easily have accommodated hundreds of bathers, although, since it was deep within the Ceremonial Complex, it was probably utilized only by members of the Imperial Court. Entrances are via four colonnaded gateways centred on each side of the tank. A broad terrace surrounds the tank proper. Exposed to the glaring sun, it is surmised by many that the terrace may well have been covered with canopies shading the Imperial bathers. The volume that this tank holds was easily supplied, via channels and aqueducts from the Tuṅgabhadrā River.

In addition, the overall size of the Great Tank is a potent icon for the power and wealth of the Imperial Court of Vijayanagara. Not only its size is iconic, but also because it is adjacent to the impressive Geometric Tank. In fact, Vijayanagara suffered from a plethora of water which added to its power.

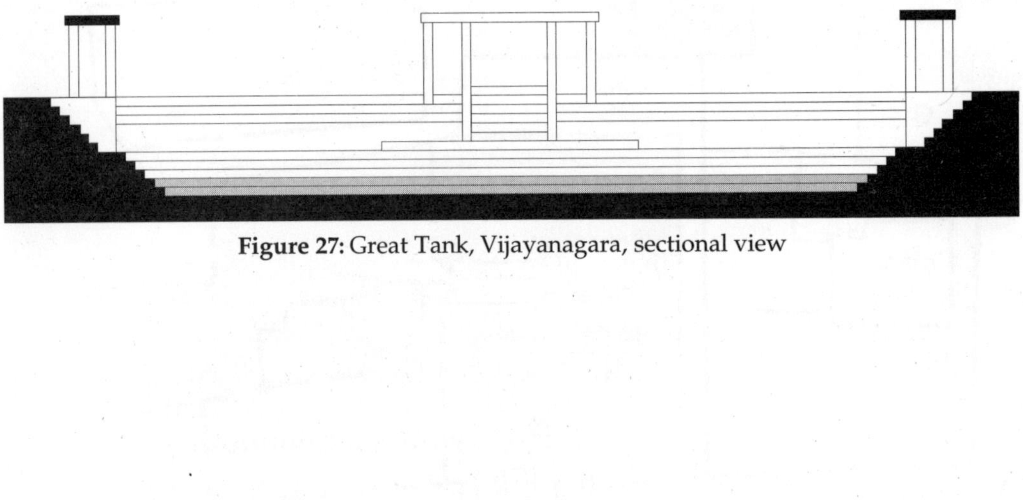

**Figure 27:** Great Tank, Vijayanagara, sectional view

[26] The plan is redrawn using Adobe PageMaker 6.0 from Internet.

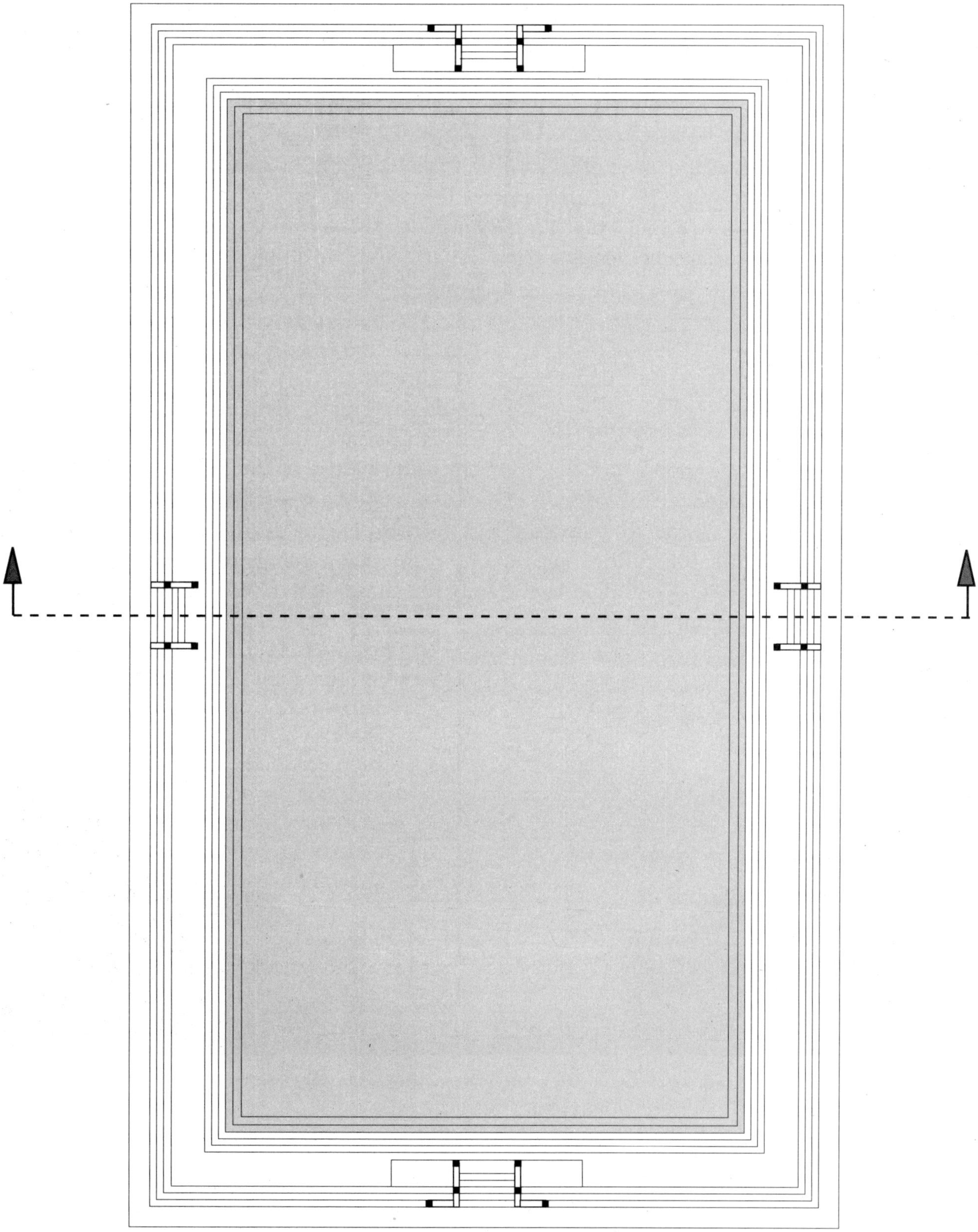

## Geometric Tank, Imperial Vijayanagara (a. CE 1400)

Adjacent to the Great Tank to the north and within the Ceremonial Complex (**Plate 21, S**) of Vijayanagara is found the impressive Geometric Tank,[27] sometimes referred to as the Puṣkaraṇī. It does not stand out from other tanks of India due to its size, but it is exemplary for its complicated and pleasing design. The tank is constructed of green diorite, is square, *en plan*, and descends in a series of four terraces surrounded by a repetitive stepped element (**Figure 28**) to the water.

The stepped elements of the Geometric Tank consists of a rather eccentric central section with redented corners, five steps high. These are found at the cardinal points and are flanked by simple rectilinear elements, also of five steps, repeating themselves to the corners where they merge. Each of these elements consists of six steps leading from the terrace above down to the terrace upon which each of the elements rests.

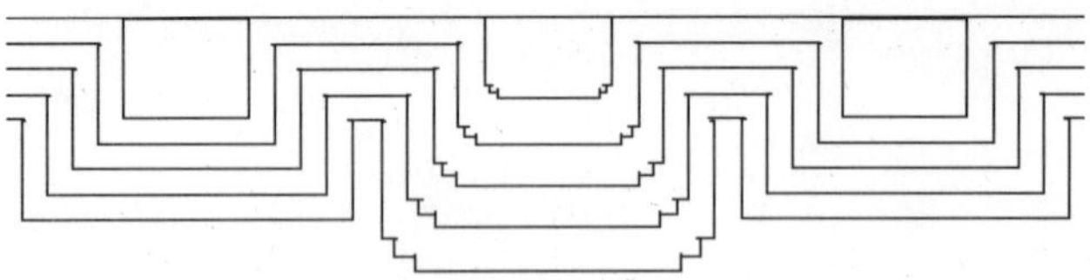

**Figure 28:** Step element found in the Geometric Tank

The angle of descent of the Geometric Tank is not as severe as found in other tanks or step wells (**Figure 29**). In part, this is due to the rise/tread ratio particularly of the central, rebated elements which is nearly 1:2. This adds to the inviting and pleasing nature of the tank's appearance. This commodious descent, coupled with the design of the Geometric Tank would lead one to suspect that it was constructed more as a locale for relaxation and the conspicuous display of power than merely a *bāolī* (simple bathing tank). It is further surmised by many that awnings may have covered much of the terraces when the city was occupied by the emperors and that the Geometric Tank was for the exclusive use of the court or imperial family.

With no outlet, the tank was constantly replenished via an aqueduct from the Tuṅgabhadrā River. It seems probable that the Geometric Tank, as well as the Great Tank, was exclusively used for bathing rather than a source of drinking water.

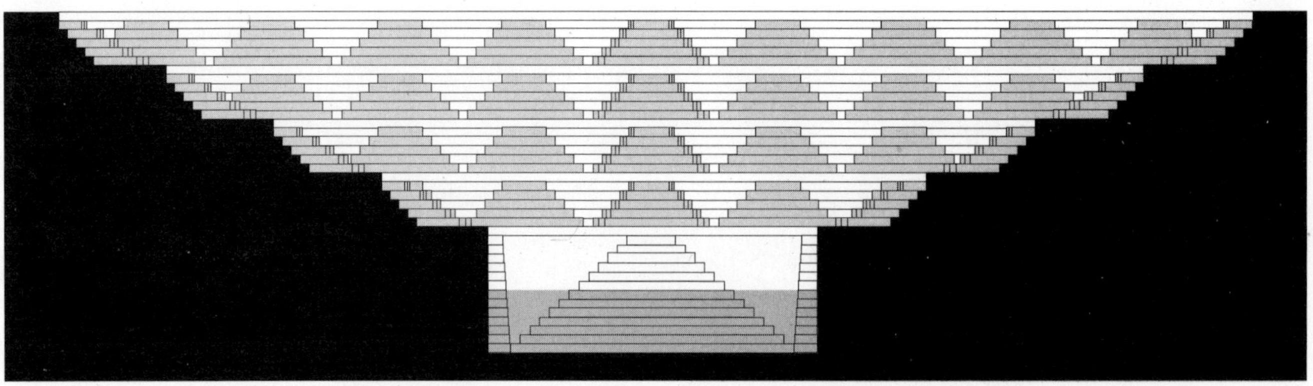

**Figure 29:** Geometric Tank, Vijayanagara, sectional view

27 The plan is redrawn using Adobe PageMaker 6.0 from the Internet.

# Plate 23: Geometric Tank plan

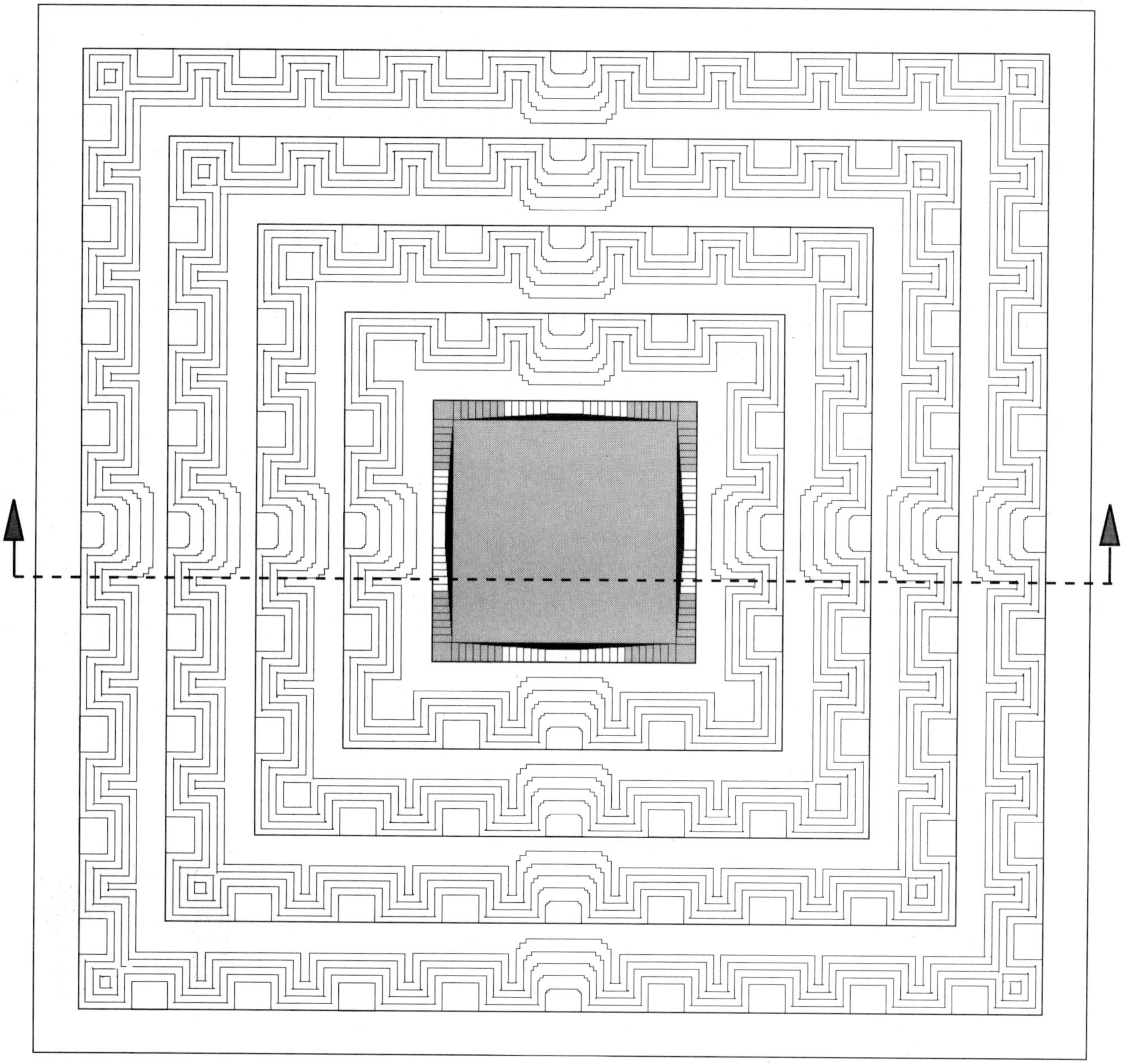

## Square Water Pavilion (Queen's Bath), Imperial Vijayanagara (a. CE 1400)

To the south-east, outside of the Royal Complex of Vijayanagara lies the Square Water Pavilion[28] sometimes referred to as the Queen's Bath. Unlike the other tanks and baths of Vijayanagara, it is completely enclosed — i.e. high walls with narrow windows and few means of ingress — and further surrounded by a shallow moat creating a potentially very private space. This may be the reason that it is referred to as the Queen's Bath. However, it is quite a distance from any of the Royal Complex areas that could be designated as the *zenānā* and most certainly would not have been a structure for the exclusive use of the women of the harem (**Plate 21, V**). The Square Water Pavilion being some distance from the Ceremonial Complex may have been the exclusive domain of the royal family and/or court, as some posit. Certainly, many members of the court may have desired some privacy in this enclosed bath. As an area for relaxation and delight away from prying eyes, this would have been a prime locale. None the less, the Square Water Pavilion is a structure of undeniable interest, uniqueness and beauty.

The Water Pavilion is square, *en plan*, with an unimposing entrance to the south. Surrounding the Square Water Pavilion is a wide, shallow moat which would have provided little security from any armed attack. However, it would have provided a visual barrier keeping any prying eyes at a distance and one in which they could be readily observed. Three, wide, arched openings punctuate the middle of the east, west and north façades. These would have provided an unimpeded flow of air and may well have been draped for additional privacy. A deep, vaulted corridor surrounds a square opening and the tank proper. On the north, south and west sides of this open space and overlooking the water are elaborate balconies (**Figure 30**). To the east are decorative spouts, four on each side of the main water channel that feed this bath. A narrow staircase just to the east of the entrance leads to the roof.

The exterior is simple, even austere and belies the sumptuous detail of the interior. The Water Pavilion seems less a bath than a place for private encounters. None the less, the whole underlines the wealth and power that exemplified the Vijayanagara Court.

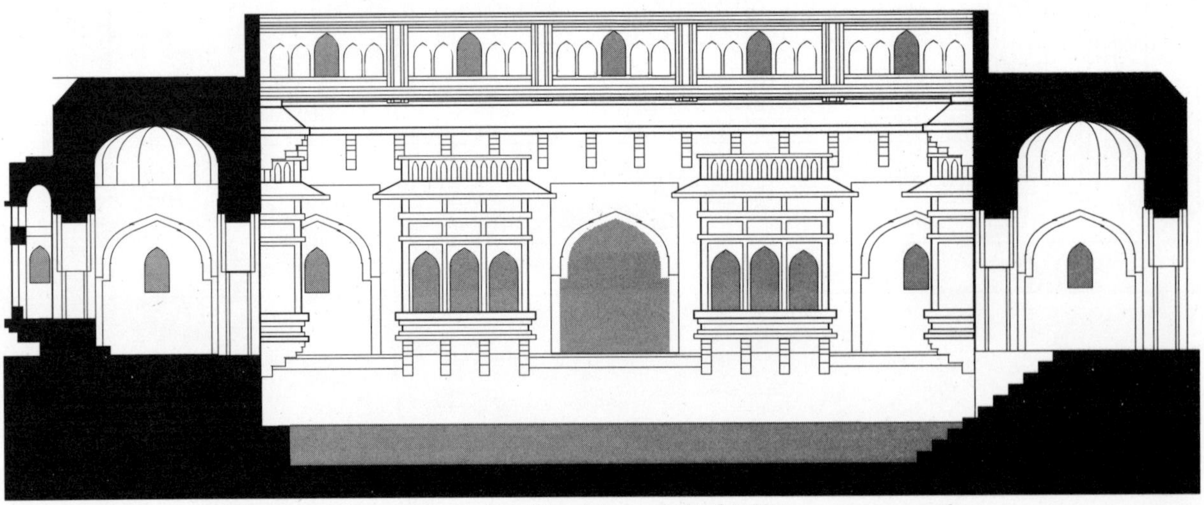

**Figure 30:** Square Water Pavilion (Queen's Bath), Vijayanagara, section

[28] The plan is redrawn using Adobe PageMaker 6.0 from George Michell, *The Vijayanagara Courtly Style*, New Delhi: 1992, p. 108.

## Plate 24: Square Water Pavilion (Queen's Bath) plan

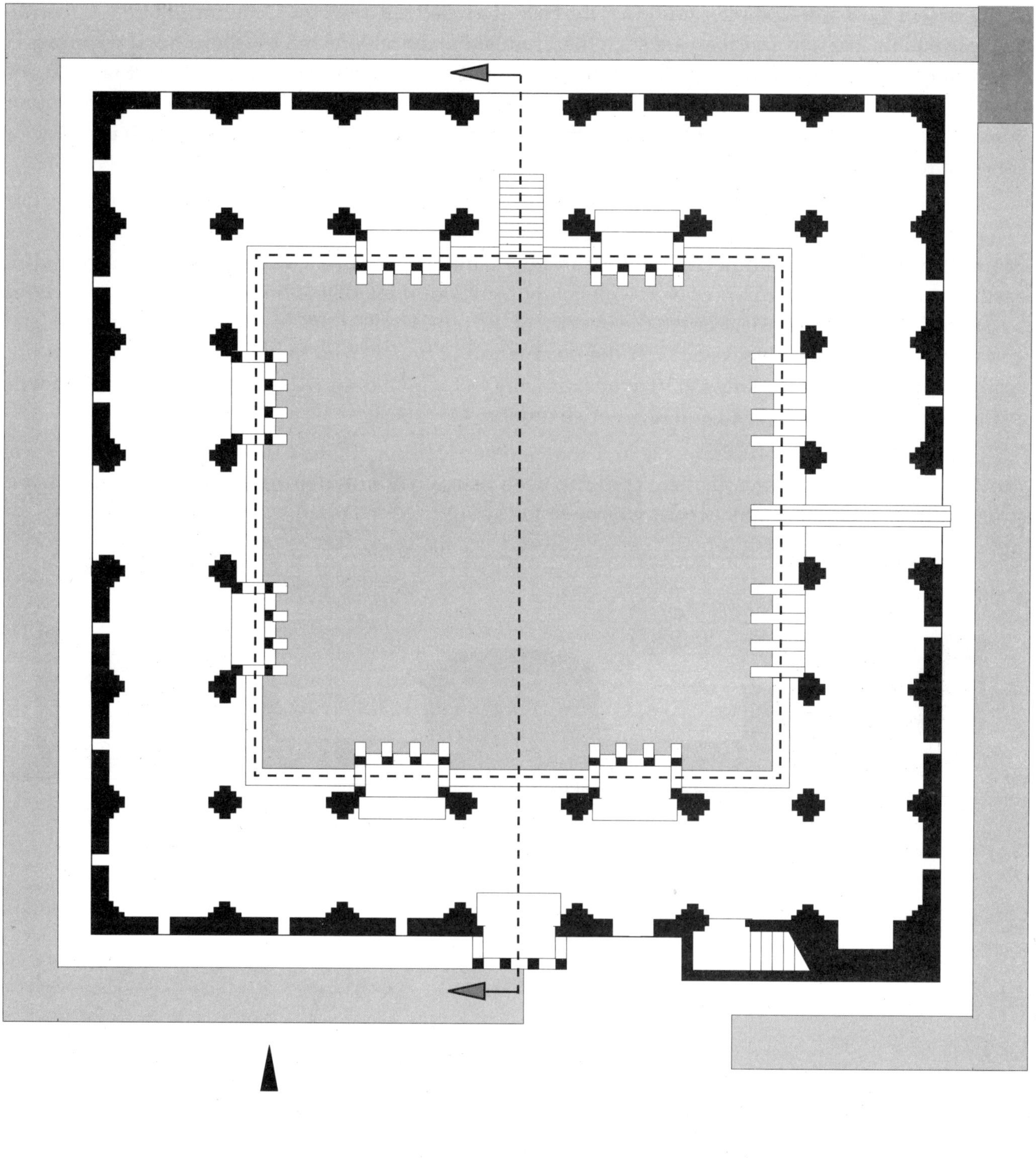

To the east of the Royal Complex, nestled at the base of a small hill, rests the Octagonal Bath[29] (**Appendix C, Plate 80**). Its location and plan are such that it must have been erected for the general populous of Vijayanagara rather than the exclusive use of the Royal Court. Unlike the Great Tank (**Plate 22**) and the Geometric Tank (**Plate 23**) which are found within the Royal Complex and the elaborate Square Water Pavilion (Queen's Bath) (**Plate 24**) within its own curtain wall, the Octagonal Bath is public on concept. It was truly a place for bathing.

The Octagonal Bath is of substantial size with an entrance to the south-east. The upper portion — i.e. a low wall, short supporting perimeter columns and the roofed ambulatory — is above the ground level. The surrounding ambulatory and the tank proper are excavated from the surrounding earth. Even though the tank proper is well below the ground level, it is not deep enough to be filled by ground water (**Figure 31**). Rather, the interior of the tank would have been waterproofed with clay and tightly fitting stone blocks to limit the percolation of water from the tank into the surrounding earth. As with most of the tanks in Vijayanagara, the Octagonal Bath would have been supplied with a constant water source via an aqueduct or channel.

Unlike the Great Tank (**Plate 22**) and the Geometric Tank (**Plate 23**), the Octagonal Bath was provided with a roofed area that covered the deep octagonal ambulatory and shielded the bathers from the bright sun. The tank proper was open to the sun.

**Figure 31:** Octagonal Bath, Vijayanagara, sectional view

29 The plan is redrawn using Adobe PageMaker 6.0 from George Michell, *The Vijayanagara Courtly Style*, New Delhi: 1992, p. 120.

## Plate 25: Octagonal Bath plan

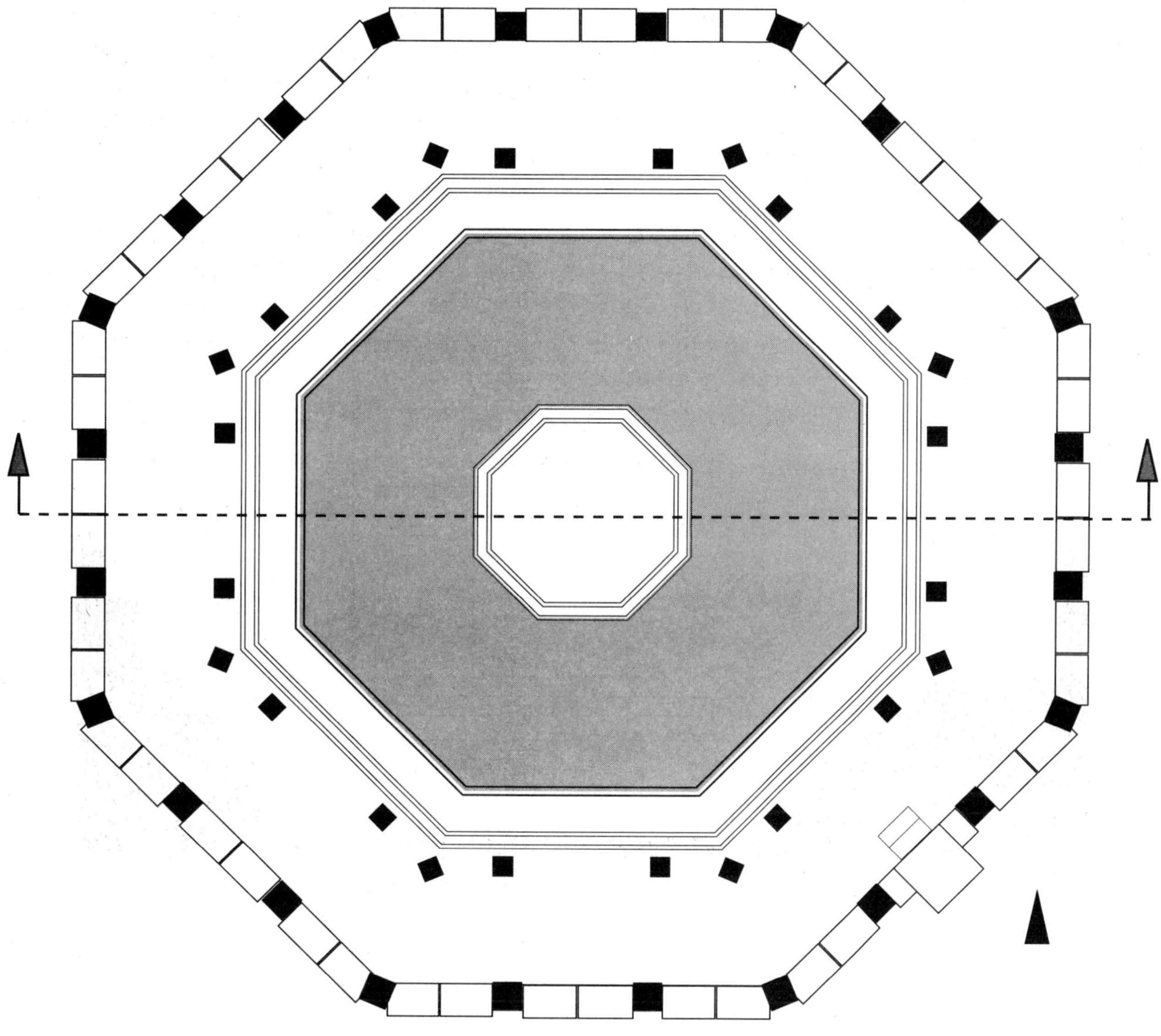

# Residence Tank (North-west of Rāmacandra Temple [$W_1$]) Imperial Vijayanagara

To the north-west of the Ceremonial Complex of Imperial Vijayanagara lies an area that is composed of what appear to be a number of residences (**Plate 21, W**). Since these structures were within the Royal Centre, it may be assumed that they were residences or palaces for members of the royal family or high-ranking nobility. Nearly everyone of these structures was supplied with a small Residence Tank.[30] These residential tanks would have been for the exclusive use of the members of the household to which they were attached.

Small in comparison to the major tanks they were visually sheltered (**Appendix C, Plate 79**), frequently at the back of the particular residence. The tanks were provided with a constant supply of water via the various aqueducts and channels that crisscrossed the Royal Centre. These private, residential tanks indicate the plenteous supply of water that graced this city and are but one more iconic indication of the wealth and power of Vijayanagara.

*En plan*, the Residence Tank is composed of two flights of opposing stairs leading to a common landing (**Figure 32**). A broader flight of steps, at right angle to the two, leads down to the bottom level of the stepped tank.

**Figure 32:** Residence Tank, Vijayanagara, sectional view

[30] The plan is redrawn using Adobe PageMaker 6.0 from the Internet.

**Plate 26:** Residence Tank (North-west of Rāmacandra Temple [$W_1$]) plan

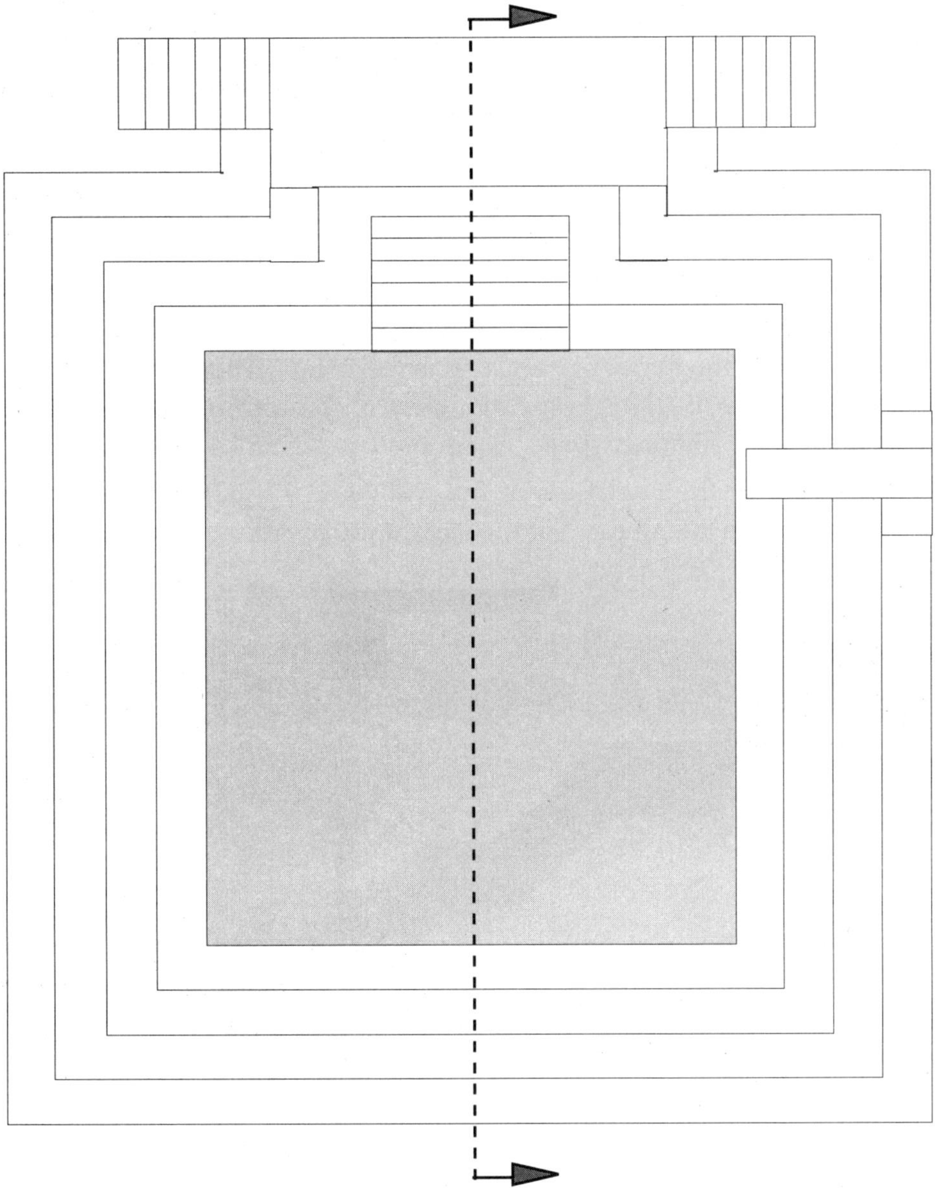

# Octagonal Well, Malpannagaḍi (Vijayanagara), Karnataka (a. CE 1400)

A short distance to the south of the Imperial City of Vijayanagara is the small suburb of Malpannagaḍi (**Appendix C, Plate** 80). Here is found a lovely step-well known as the Octagonal Well.[31] This well (*bāolī*) is deep enough to be effected by the ground water in the area and supplied this community with much needed drinking water.

The Octagonal Well consists of a trench with a series of stairways and five landings leading down to the water level. Part of the lower space of the *bāolī* was octagonal. However, much of the lower portion of this space was square or cubic in dimension (**Figure 33**). At the fourth landing are two sets of narrow stairs leading up to opposite doorways. These doors opened on to a narrow passageway and thence to an octagonal ambulatory punctuated with arched openings. These archways looked out into the octagonal space with the sky above and the water below. The ambulatory was undoubtedly employed as a place of cool relaxation for the members of the community. There is some evidence that at least part of the octagonal, inner space, was roofed.

The octagonal shape plus the heaviness of the walls and supporting arch members provided sufficient strength to offset the lateral force of the surrounding ground (**Appendix C, Plate 76**).

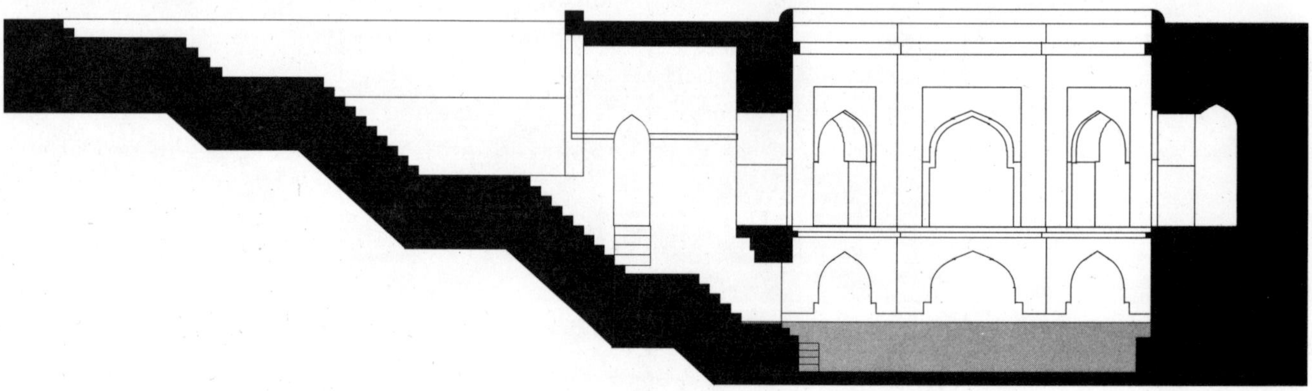

**Figure 33:** Octagonal Well, Malpannaguḍi, sectional view

[31] The plan is redrawn using Adobe PageMaker 6.0 from George Michell, *The Vijayanagara Courtly Style,* New Delhi: 1992, p. 126

## Plate 27: Octagonal Well plan

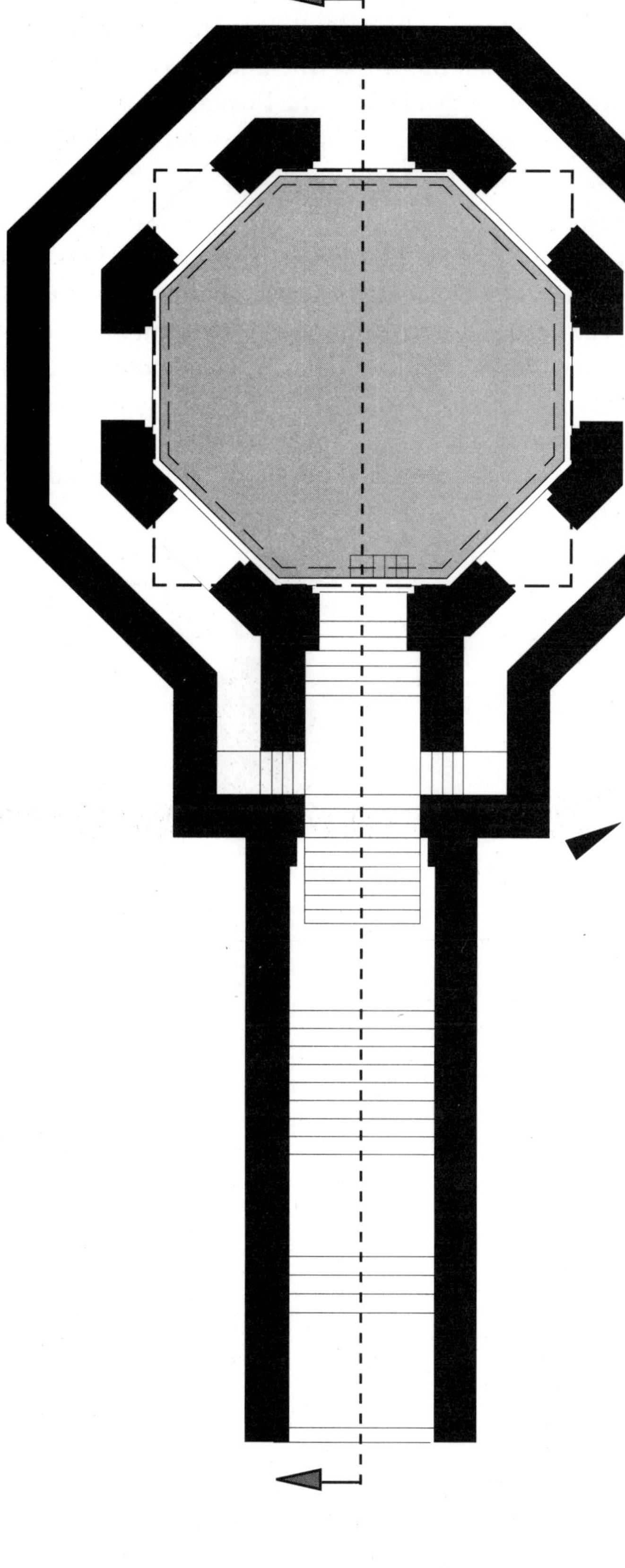

## Rāmatīrtha Puṣkariṇī, Santhebennūr, Karnataka (a. CE 1400)

Connected with the brilliant Vijayanagara Empire, the town of Santhebennūr possesses one of the most impressive tanks. It is said to have been built by Senga Hanumanthappa Nāyak, the governor of Santhebennūr. This was at the time that the town was known as Raṅgapura and a vassal state to Imperial Vijayanagara. The lower levels of the Rāmatīrtha Puṣkariṇī[32] must surely reach the aquifer as it has never been without water even during a prolonged drought. For that reason it is often referred to as being an *akṣaya pātra* — i.e. a vessel which never dries up.

Square, *en plan,* the Rāmatīrtha Puṣkariṇī drops quickly to the water level. A broad flight of steps leads down to the water level where a narrow platform surrounds the tank, barely wide enough for two persons. Unlike other tanks, access to the water is safely gained only via this flight of steps — i.e., unless one wishes to risk the very perilous, narrow and steep descent around the sides (**Figure 34**). A pleasing entrance treatment is provided with stone benches for relaxation. Centred in the Rāmatīrtha Puṣkariṇī is an elaborate, multistoried *nirazhi maṇḍapa* known locally as the *vasanta maṇḍapa. En plan* and elevation, this pavilion indicates a debt to the Imperial Vijayanagara overlord. Centred in the east and west of the massive tank are two-columned pavilions. Here, one could rest under the shade provided by the roof.

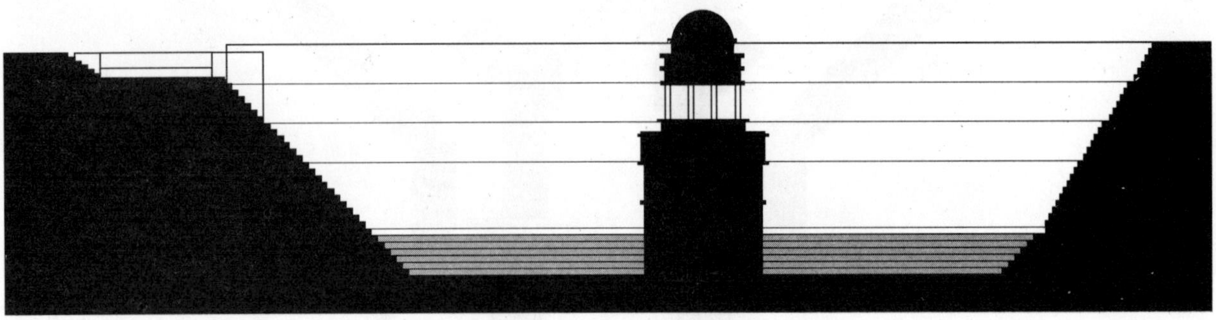

**Figure 34:** Rāmatīrtha Puṣkariṇī, sectional view

[32] The plan is redrawn using Adobe PageMaker 6.0 from the Internet.

## **Plate 28:** Rāmatīrtha Puṣkariṇī plan

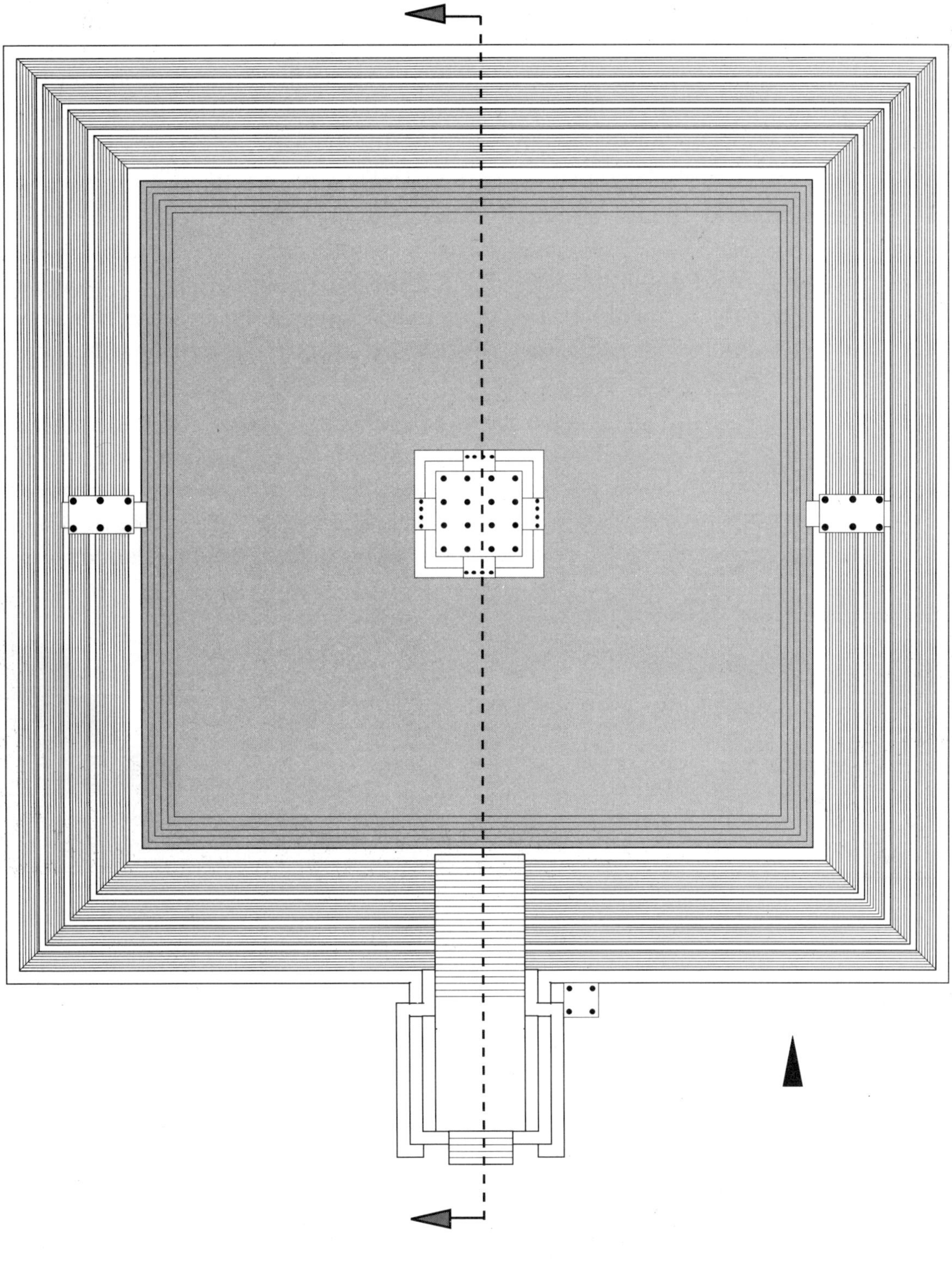

# Jāmī Masjid Haud, Campāner, Gujarat (a. 1485)

Seventy odd miles south and a bit east of Ahmedabad lies Campāner. In the fifteenth century the powerful Sultān Nasīr-ud-Dīn Mahmūd Shāh I Begarah (r. CE 1459–1511) captured the citadel of Campāner from the Hindu ruler Jaysingh Patai Rawal (**Appendix D**) and soon constructed the impressive Masjid-ī Jāmī (**Appendix C, Plate 90**). It becomes the cap stone to the life of Sultān Mahmūd Begarah and sets an example for subsequent members of this dynasty.

Associated with the grand Jāmī Masjid is the Jāmī Masjid Haud.[33] It is the sacred pool where ritual ablution is performed prior to the five required Islamic prayers in the mosque. Normally the *haud* (ablution pools) of Indian *masjid* are rather small, virtually insignificant in size and design. More often than not they are encountered within the *sahn* (open courtyard) which fronts the sanctuary of the *masjid*. The Jāmī Masjid Haud is found outside the sacred confines of the mosque. This may suggest an unusual philosophic application in that, for many, the precincts of the *masjid* itself are considered sacred. To enter therein, one must be ritually clean.[34]

None the less, the Jāmī Masjid Haud is unique, not only because it is attached to an Islamic place of worship, but also due to its unusually sophisticated design. Based upon an octagon, the sacred Islamic shape, and consisting of five levels, the Jāmī Masjid Haud includes elements that are found in Solaṅkī tanks (**Plates 7-9**) — the double-stepped elements on each terrace (**Figure 35**). The Jāmī Masjid Haud is a superlative example of the adaptation of Hindu design to an Islamic requirement.

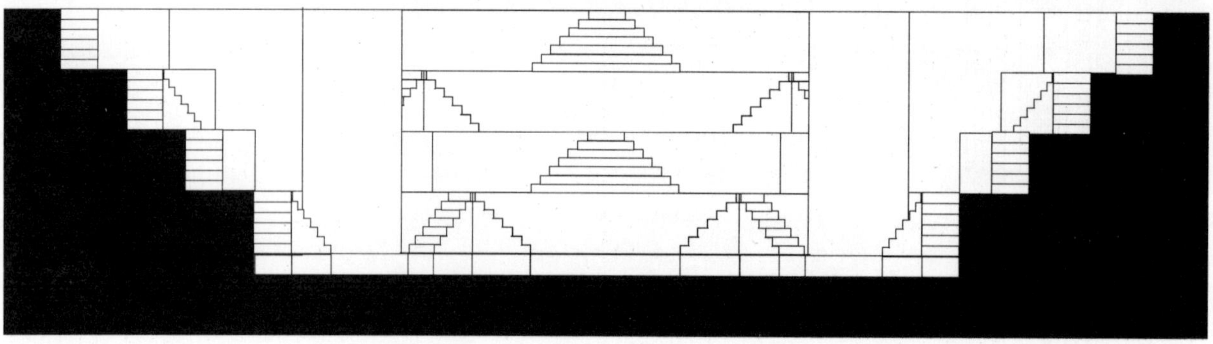

**Figure 35:** Jāmī Masjid Haud, Campāner, sectional view

[33] The plan is redrawn using Adobe PageMaker 6.0 from the Internet.

[34] For example in Malaysia, the Masjid Negara (National Mosque) in Kuala Lumpur is open to members of all faiths. This author has twice attended Friday prayers with Muslim friends with no restrictions. On the other hand, in the Masjid Wilayah (Federal Territory Mosque) non-Muslims are restricted to only the outer courts, as also is the case with the Masjid Negri of Selangor, Shāh Alam.

# Plate 29: Jāmī Masjid Haud, Campāner, plan

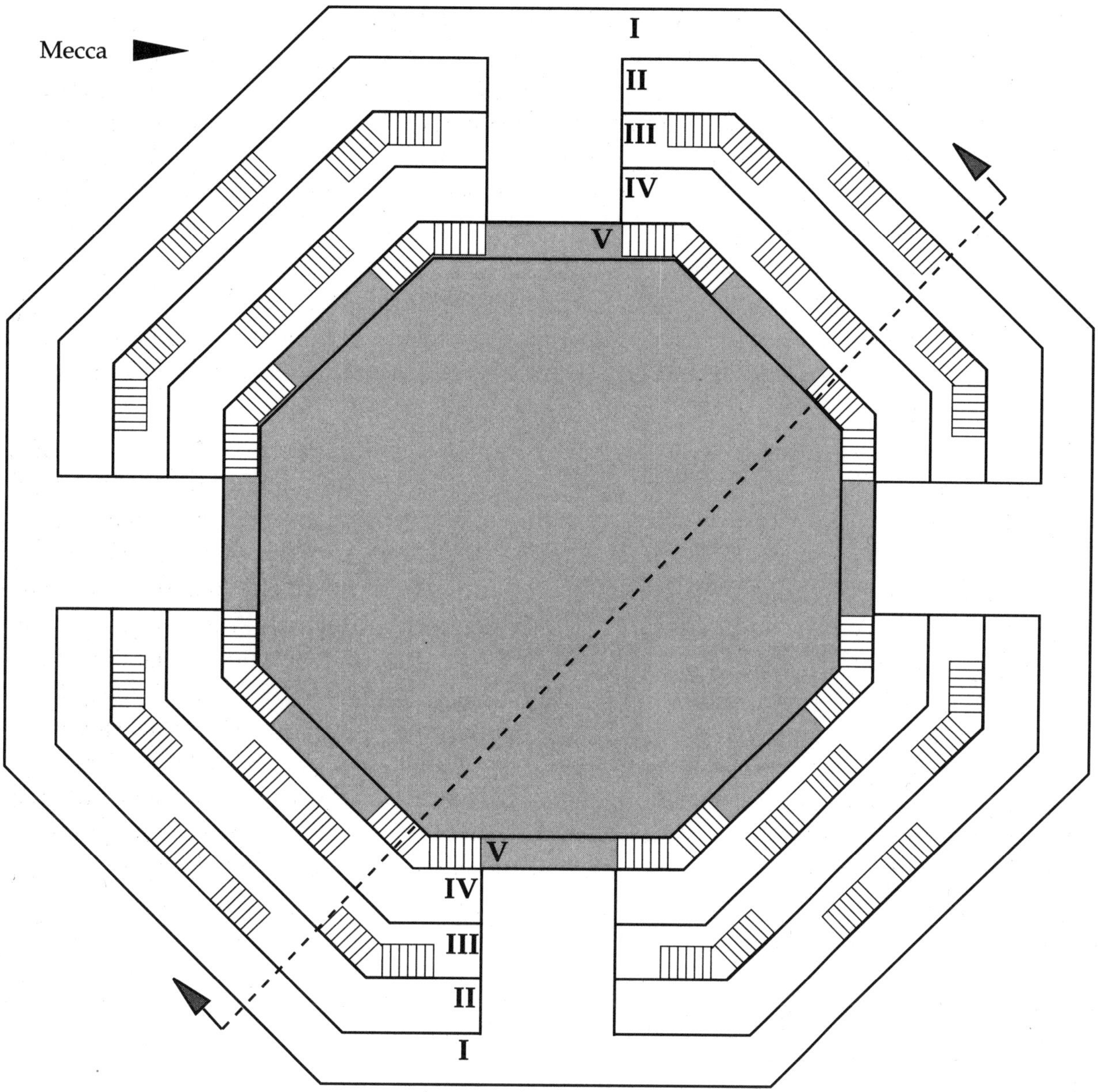


**I** First Terrace
**II** Second Terrace
**III** Third Terrace
**IV** Fourth Terrace
**V** Fifth Terrace

# Virūpākṣa Temple Complex, Hampī, Karnataka (b. CE 1510)

A few kilometers north and west of Imperial Vijayanagara and its Royal Complex is the Sacred Centre (**Appendix C, Plate 78**). Here a number of temple complexes are to be found. The largest and in some ways the most important is the Virūpākṣa (Pampāpatī) Temple Complex.[35] As with many temple complexes in southern India this rather imposing ensemble was constructed over several centuries. Certainly the Virūpākṣa (Pampāpatī) Temple was a pilgrimage site before the establishment of the Vijayanagara Empire.[36] However, it is known to have been extensively renovated in 1510 for the coronation of Kr̥ṣṇadevarāya (r. CE 1509-30) (**Appendix D**), the greatest of the Tuluva monarchs.

The main area of the temple consists of two *prākāra* aligned generally east to west. To the north of the main temple proper and its courtyard, beyond the girdle wall are found myriad small temples and shrines. And, to the south of this conglomeration of shrines is the large Manmatha Tank. The tank is of an eccentric design, *en plan*, but occupies a large enclosure contiguous to the girdle wall of the Virūpākṣa (Pampāpatī) Temple.

Several of the tanks associated with the other complexes within the Sacred Centre are of some distance from the temple proper to which they are attached (**Plates 32-35**). This is doubtless due to the particular design features of the major temples of Vijayanagara — i.e., there is a long, broad Processional Avenue leading up to the main entrance or *gopura* (**Appendix C, Plate 82**). In several cases these avenues are bordered by a colonnade, creating an impressive focus on the temple. It is a sacred Processional Avenue and analogous to the Imperial Processional Way that is encountered in a number of palaces throughout the world and especially in Asia,[37] and not unlike the Avenue of Sphinges that leads to the main entrance of the Temple of Amun at Karnak, Egypt. Over the years this avenue has hosted myriad stalls and shops and has assumed the name of a *bazaar*. However, it is questionable that this avenue was initially designed for commercial purposes. Rather, it lends itself to a sacred Processional Avenue. In addition this Processional Avenue as well as those of Viṭṭhalapura (**Plate 32**) and Kr̥ṣṇapura (**Plate 34**) are all three precisely oriented east-southeast — i.e. 13° south of east. This similarity between these three temples' Processional Avenue is a conundrum.[38]

[35] The plan is redrawn using Adobe PageMaker 6.0 from K.M. Suresh, *Temples of Hampī and Its Environs*, Delhi: 2003, p. 147 (this plan of the Virūpākṣa Temple Complex is shown on p. 76); Vijayanagara Research Project: Virūpaksha; www.vijayanagara.org.

[36] There is evidence that as early as the late seventh century, this site was known as Pampā-tīrtha.

[37] The Imperial Processional Way is discussed by Henry Stierlin, *Architecture of the World: China*, pp. 98-99.

[38] See J. McK. Malville, "Cosmic Landscape and Urban Layout", *New Light on Hampī: Recent Research at Vijayanagara*, 2001: pp. 112-25.

# **Plate 30:** Virūpākṣa Temple Complex

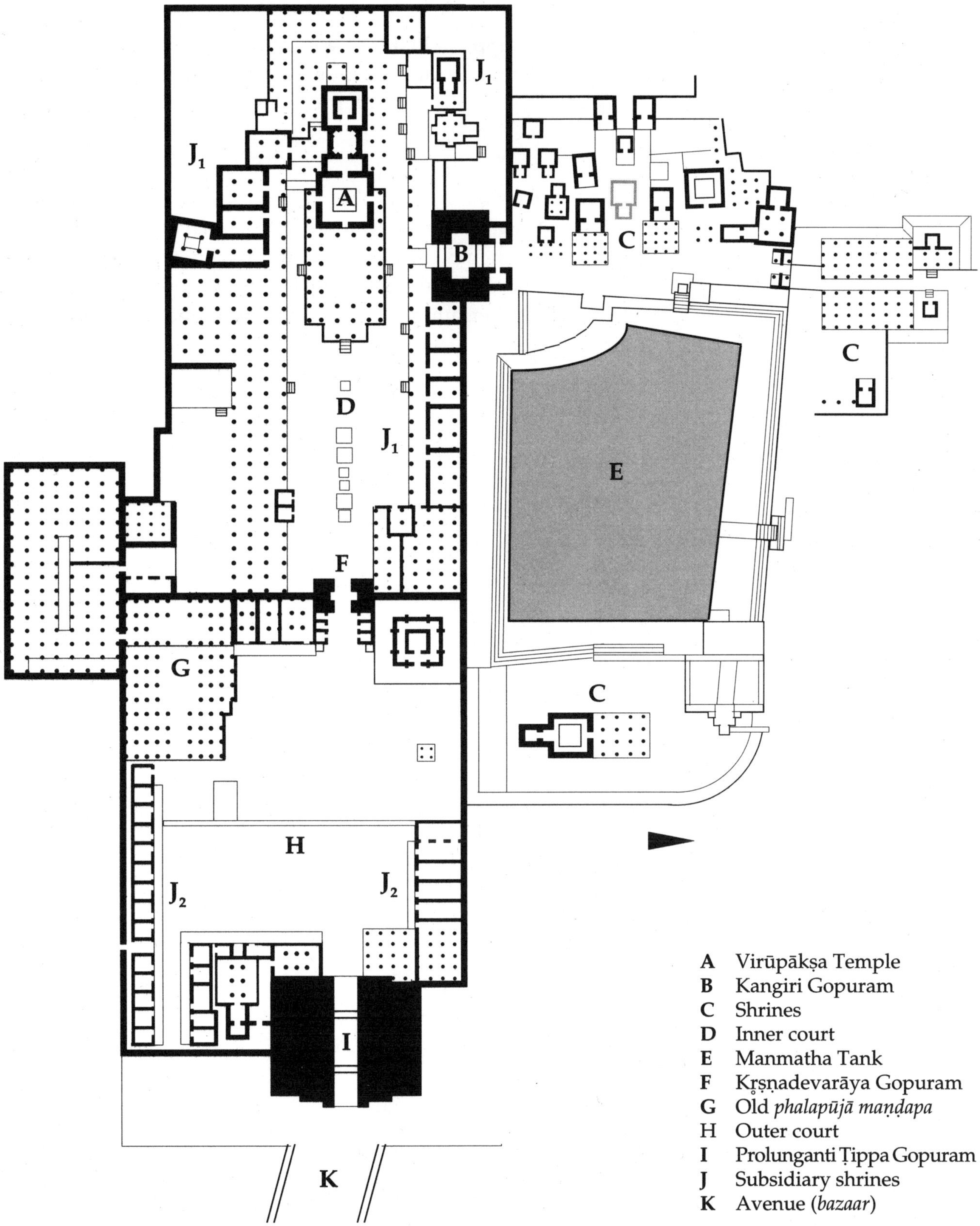


**A** Virūpākṣa Temple
**B** Kangiri Gopuram
**C** Shrines
**D** Inner court
**E** Manmatha Tank
**F** Kr̥ṣṇadevarāya Gopuram
**G** Old *phalapūjā maṇḍapa*
**H** Outer court
**I** Prolunganti Ṭippa Gopuram
**J** Subsidiary shrines
**K** Avenue (*bazaar*)

The Manmatha Tank[39] to the north of the temple precincts is eccentric, *en plan*. Unlike other temple cities — Śrī Raṅganāthaswāmi Temple at Tiruchirappalli (**Plates 45-46**) or the Aruṇācala (Annamalaiar) Temple of Tiruvannamalai (**Plate 43**) — in which the tanks are found within one of the rectangular-walled *prākāra*s that surround the complex, the Manmatha Tank is bounded by its own curtain wall to the north of the main complex. Its size is, in comparison, larger than the other tanks of the Sacred Centre of Vijayanagara.

The tank consists of broad terraces and steps of varying number leading down to the water's edge. Channels and acquaducts had once supplied the tank with water, but these have long since fallen into disuse. There are entrances to the west, north and east. The tank is basically rectilinear with an interesting variation — the south-west corner of the Manmatha Tank consists of a series of steps which are arc-like, *en forme*. This curious design, this aberration may well be due to the underlying rock formation as much of the Sacred Centre is set upon and within the capricious geological formations of the area.

[39] The plan is redrawn using Adobe PageMaker 6.0 from K.M. Suresh, *Temples of Hampī and Its Environs*, Delhi: 2003, p. 147 (this plan of the Manmatha Tank Complex is shown on p. 79).

# Plate 31: Manmatha Tank plan

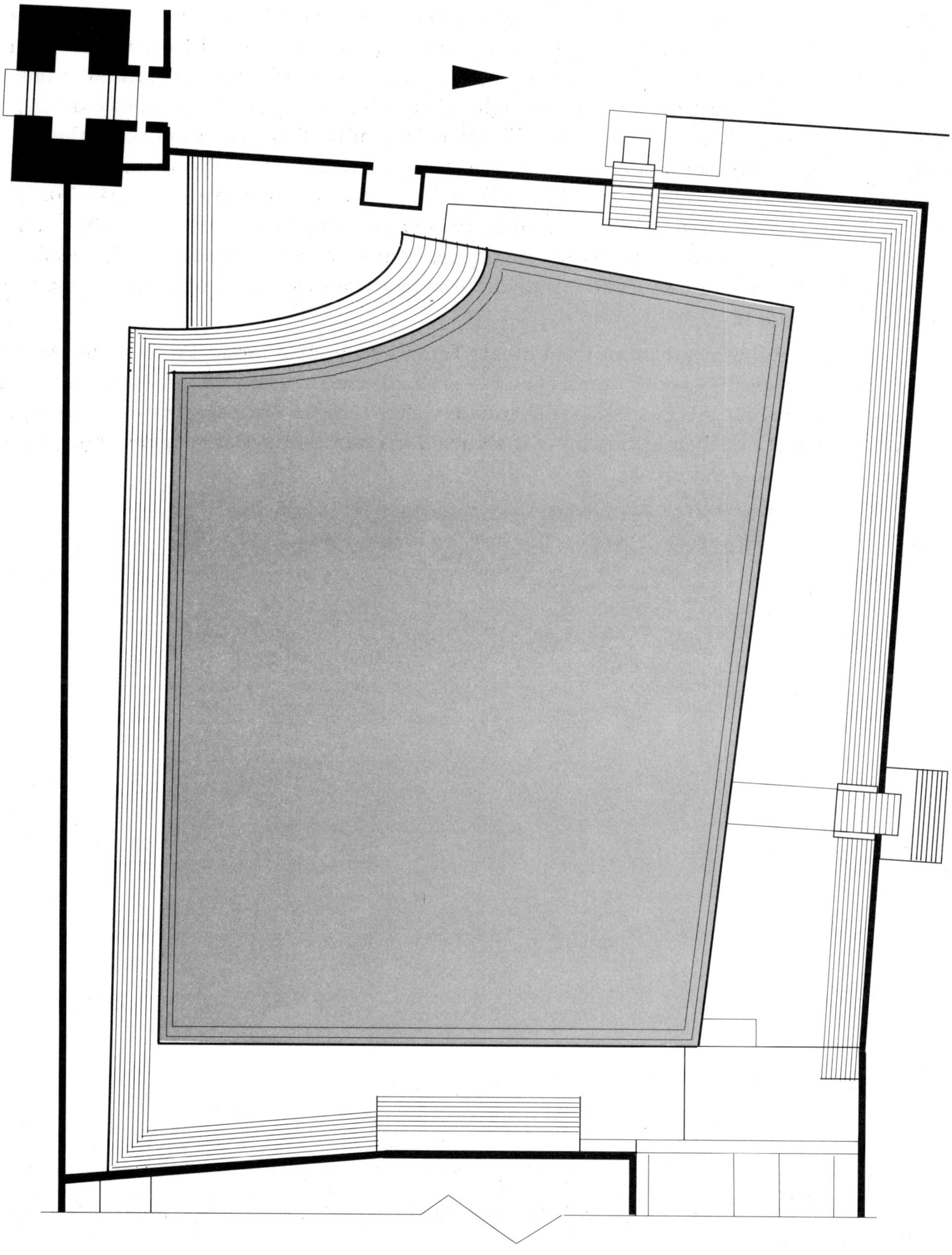

# Viṭṭhalapura, Hampī, Karnataka (a. CE 1506)

Viṭṭhalapura[40] with the Viṭṭhala Temple Complex (**Appendix C, Plate 83**), also known as the Viṭhobā Temple, is one of the architectural masterpieces of the Vijayanagara Sacred Centre. The Viṭṭhala Temple Complex is smaller than the Virūpākṣa (Pampāpatī) Temple Complex (**Plate 30**), none the less, it is architecturally spectacular in its own right. There is evidence that it was started during the reign of Vīra Narasiṁha (Naresh Tuluva) (r. CE 1505-09) (**Appendix D**), the first monarch of the Tuluva Dynasty. Part of the complex is known to have been constructed late in the Vijayanagara ascendency under the auspices of one Avubhalaraju during the reign of the Sadāśivarāya (r. CE 1542-76). Other shrines were raised over a period of years around the temple complex proper by wealthy courtiers to the point that the whole ensemble became a conglomeration of shrines, small temples, walkways and terraces. Within the girdle wall of the Viṭṭhala Temple, there was a sense of order and the whole stands in regal splendour.

From the east, entrance *gopura* of the Viṭṭhala Temple extends a broad Processional Avenue, eastward and a little south — i.e. 13° south of east — often referred to as the *bazaar*. Again, the precise orientation or alignment of this Processional Avenue as well as those of Virūpākṣapura and Kṛṣṇapura, causes one to suspect that their iconography is involved in cosmic–imperial–religious meaning, not merely a commercial addendum.

[40] The plan is redrawn using Adobe PageMaker 6.0 from Henri Stierlin, *Hindu India: From Khajuraho to the Temple City of Madurai*, Cologne: 2002, p. 196; John M. Fritz et al., *New Light on Hampi*, Mumbai: 2001, p. 28; Vijayanagara Research Project: Virūpākṣa: www.vijayanagara.org; K.M. Suresh, *Temples of Hampi and Its Environs*, Delhi: 2003, p. 166 (this plan of the Virūpākṣa Temple complex is shown on p. 81).

## Plate 32: Viṭṭhalapura with Tank and Processional Avenue

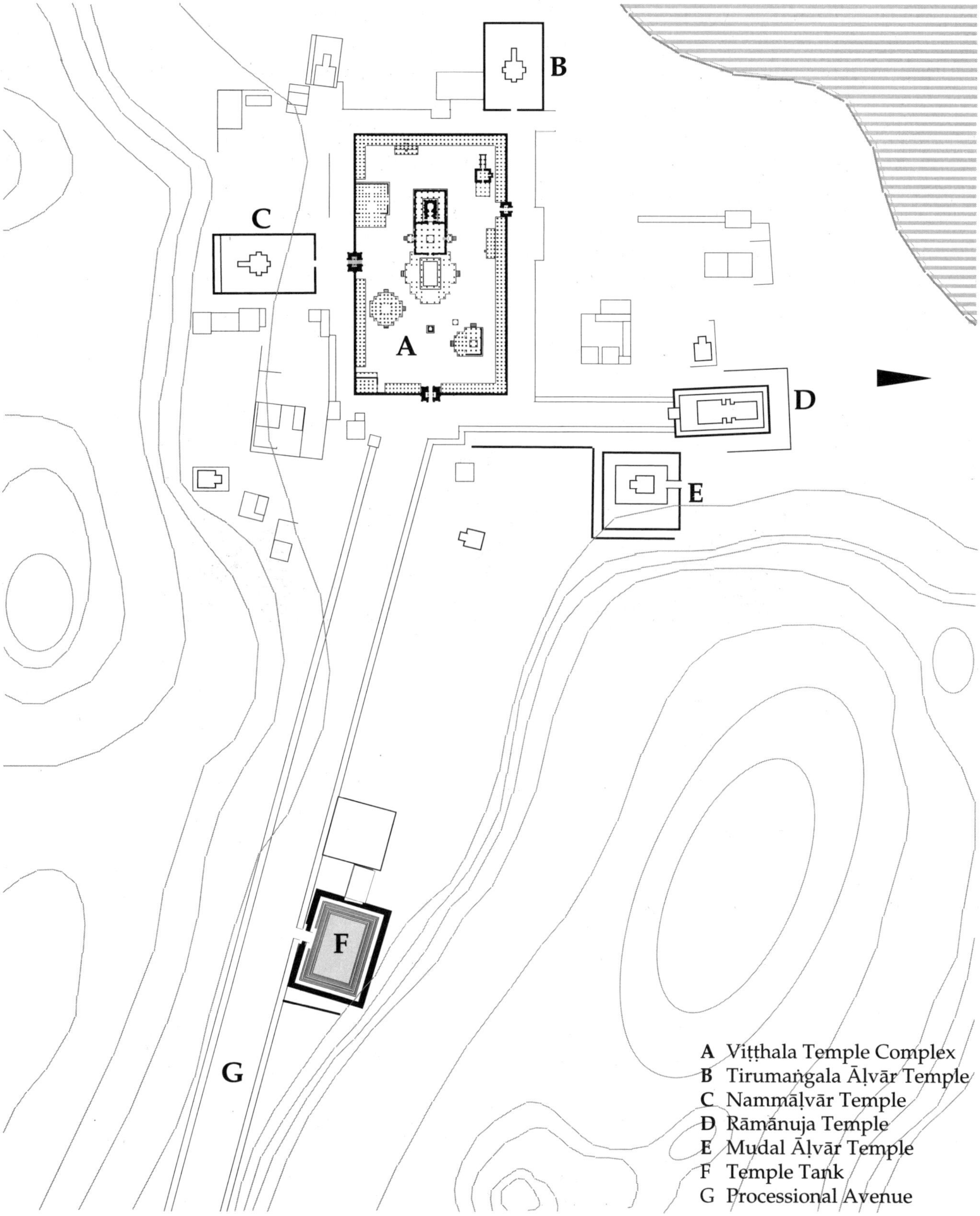

# Viṭṭhalapura Tank

To the east, along the Processional Avenue from the Viṭṭhala Temple Complex and to the north of the colonnaded avenue, is found the Viṭṭhala Temple Tank.[41] It is not as large, in proportion to the Viṭṭhala Temple Complex as the Manmatha Tank is to the Virūpākṣa Temple Complex (**Plate 30**), but it is, none the less, worthy of consideration.

The Viṭṭhala Temple Tank is rectangular and surrounded by a low wall. Inside this wall is a covered colonnade surrounding the tank proper. It is an area where the devotees can rest out of the direct rays of the sun, particularly during the hot months (**Figure 36**). Towards the west is a terrace that leads to a small, simple shrine. The inclusion of a shrine in conjunction with the tank is not a usual feature. As with other tanks at Vijayanagara, specifically the Kṛṣṇa Temple Tank, a small shrine rested in the middle of the tank — i.e. a *nirazhi maṇḍapa*. It was here that images of the presiding deity would have been brought during high holy days.

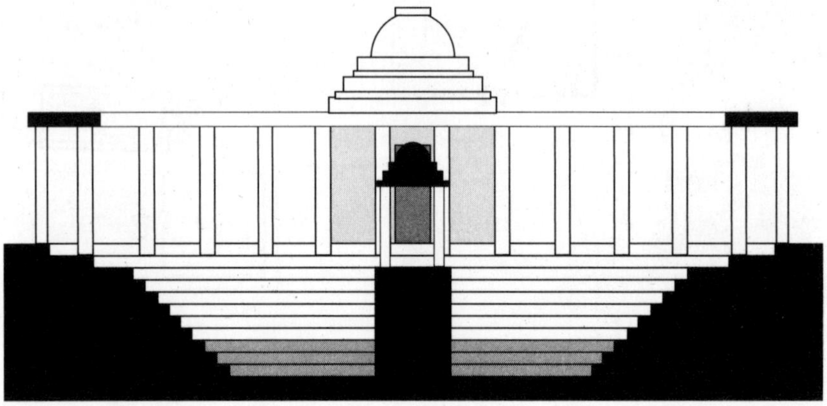

**Figure 36:** Viṭṭhala Temple Tank, Hampī, sectional view

[41] The plan is redrawn using Adobe PageMaker 6.0 from Henri Stierlin, *Hindu India: From Khajuraho to the Temple City of Madurai*, Cologne: 2002, p. 196; John M. Fritz et al., *New Light on Hampī*, Mumbai: 2001, p. 28 and the Internet.

# Plate 33: Viṭṭhalapura Tank plan

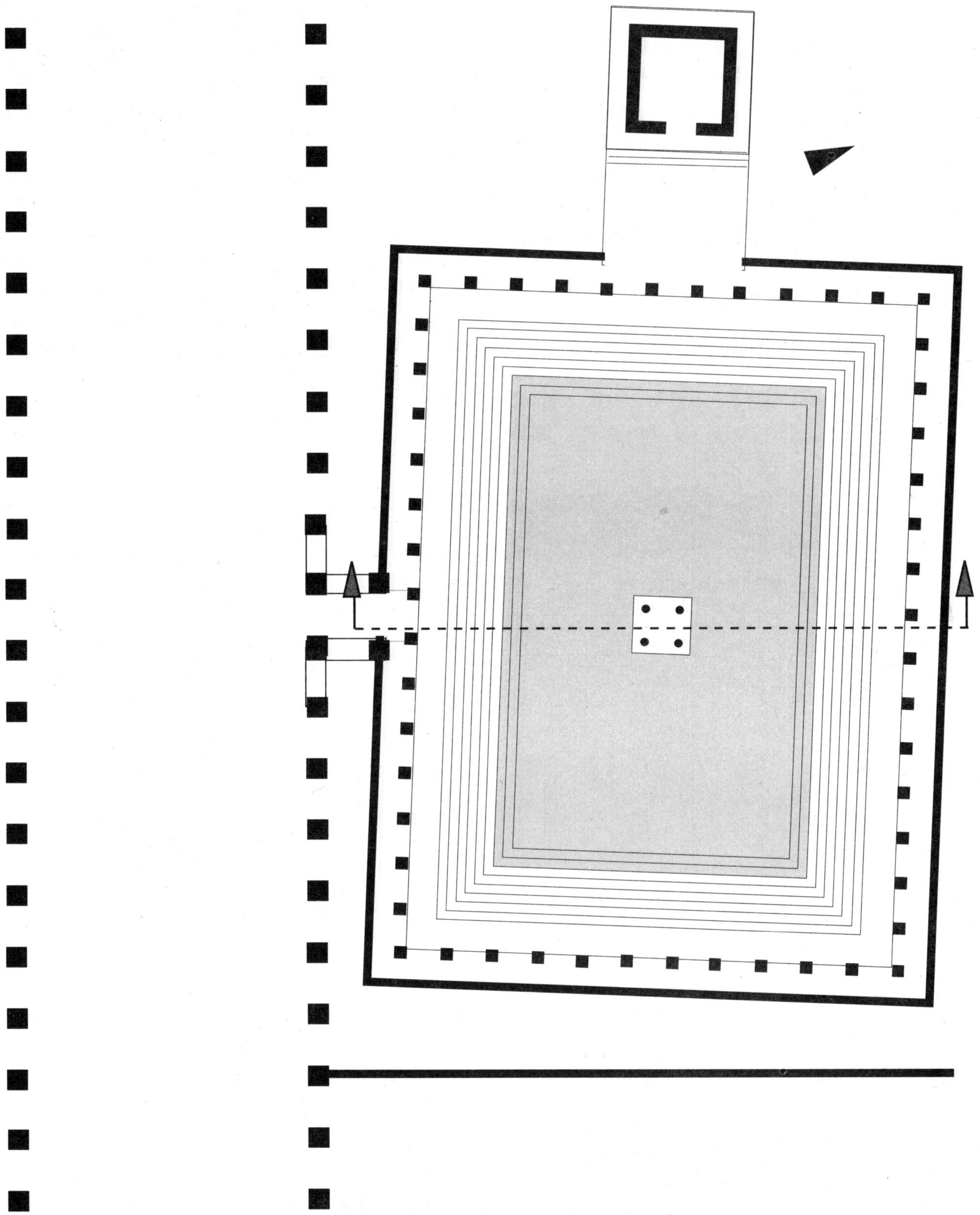

# Kṛṣṇapura, Hampī, Karnataka (c. CE 1513)

A short distance to the south of the Virūpākṣa (Pampāpatī) Temple Complex rests the Kṛṣṇa Temple Complex[42] (**Appendix C, Plate 84**). The Kṛṣṇa Temple Complex was constructed by Kṛṣṇadevarāya (r. CE 1509-30) (**Appendix D**) after his victory over Orissa during the reign of Prataparudra (r. CE 1497–1540). The Lord Kṛṣṇa was the patron deity of Kṛṣṇadevarāya and the dedication of the temple to this deity was quite natural.

Although it is one of the smaller temple complexes associated with Vijayanagara, it possesses several notable features. Firs the Processional Avenue that leads to the entrance *gopura* is exceptionally wide and bounded by a triple colonnade (**Appendix C, Plate 84**). As noted above in conjunction with the Virūpākṣa (Pampāpatī) Temple Complex, this avenue must be seen to serve a ceremonial purpose. Second, the tank associated with this temple, like the Acyutarāya Temple Complex and the Viṭṭhala Temple Complex (**Plate 32**), is a distance away from the main gate on the approach, the Processional Avenue. Like many of the temples in the Sacred Centre, the girdle walls are eccentric in shape and angle with few parallel walls, more so with the Kṛṣṇa Temple Complex. Within the inner court the buildings are generally aligned with the *vimāna*.

[42] The plan is redrawn using Adobe PageMaker 6.0 from K.M. Suresh, *Temples of Hampi and Its Environs*, Delhi: 2003, p. 120.

## Plate 34: Kr̥ṣṇapura with Kr̥ṣṇa Temple Tank and Processional Avenue

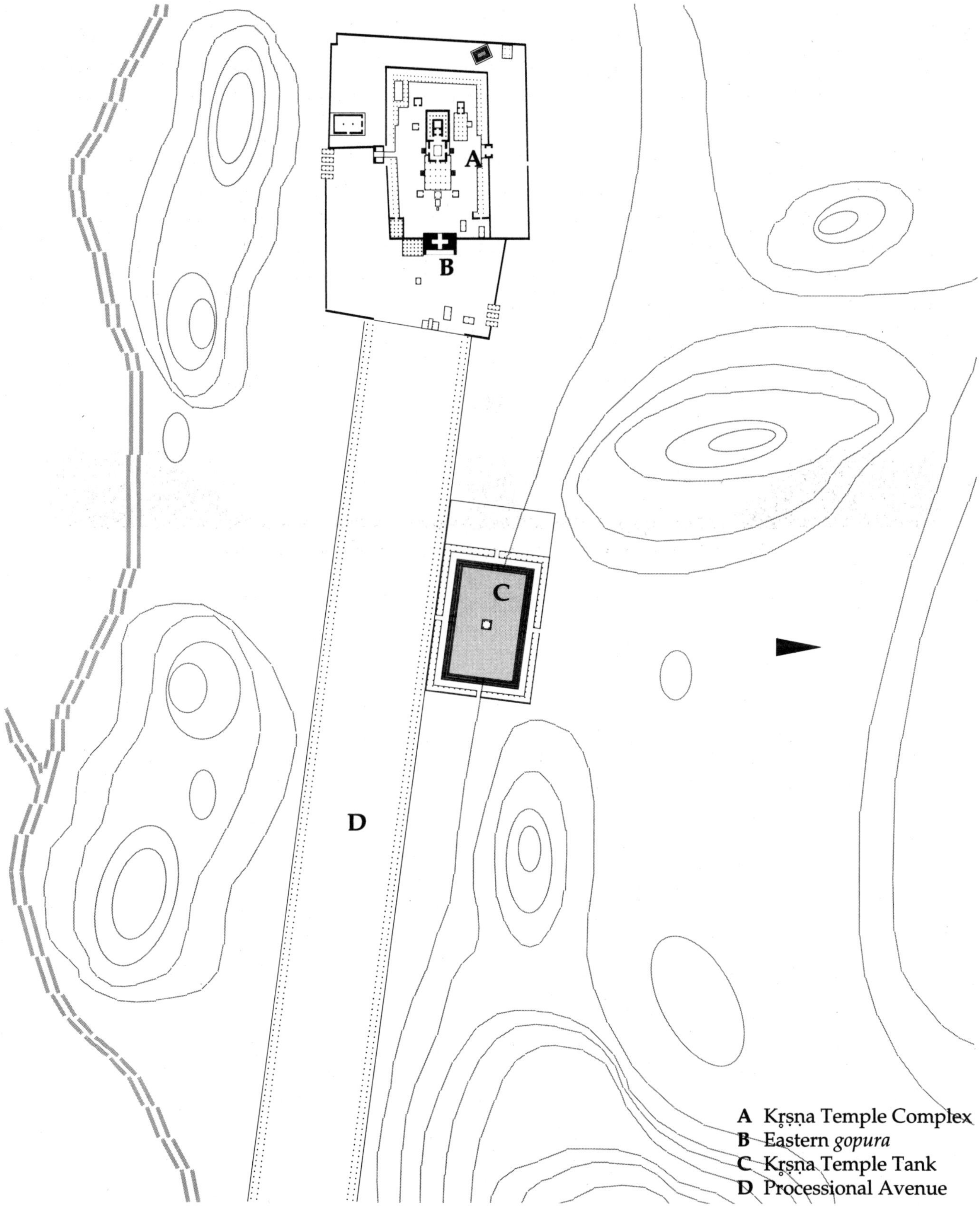

# Kṛṣṇa Temple Tank, Hampī (Vijayanagara), Karnataka

As noted above, to the south of the Virūpākṣa Temple Complex is the Kṛṣṇa Temple Complex. Although diminutive in comparison, it, none the less, is a complex of some beauty. Part of the ensemble is the Kṛṣṇa Temple Tank.[43] In comparison to the other tanks associated with temple complexes in the Sacred Centre, the Kṛṣṇa Temple Tank is not small! It was certainly large enough to accommodate hundreds of bathers at the same time.

The rectangular tank is surrounded by a commodious terrace. In turn, the terrace is bounded by a raised, double, covered colonnade with entrances in the middle of each side, at the cardinal points of the compass with the principle entrance in the west. There, staircases with broad rests lead down to the water's edge. The main entrance to the tank is via an impressive triple colonnaded gate at the west end of the tank, towards the temple complex, proper. In the centre of the tank is a small, square, artificial island with a tiered-roof shrine-like *śālā*, a *nirazhi maṇḍapa* (**Figure 37**). Here the image of the deity would be brought on high holy days for special *pūja*.

**Figure 37:** Kṛṣṇa Temple Tank, Hampī (Vijayanagara), sectional view

43 The plan is redrawn using Adobe PageMaker 6.0 from the Internet.

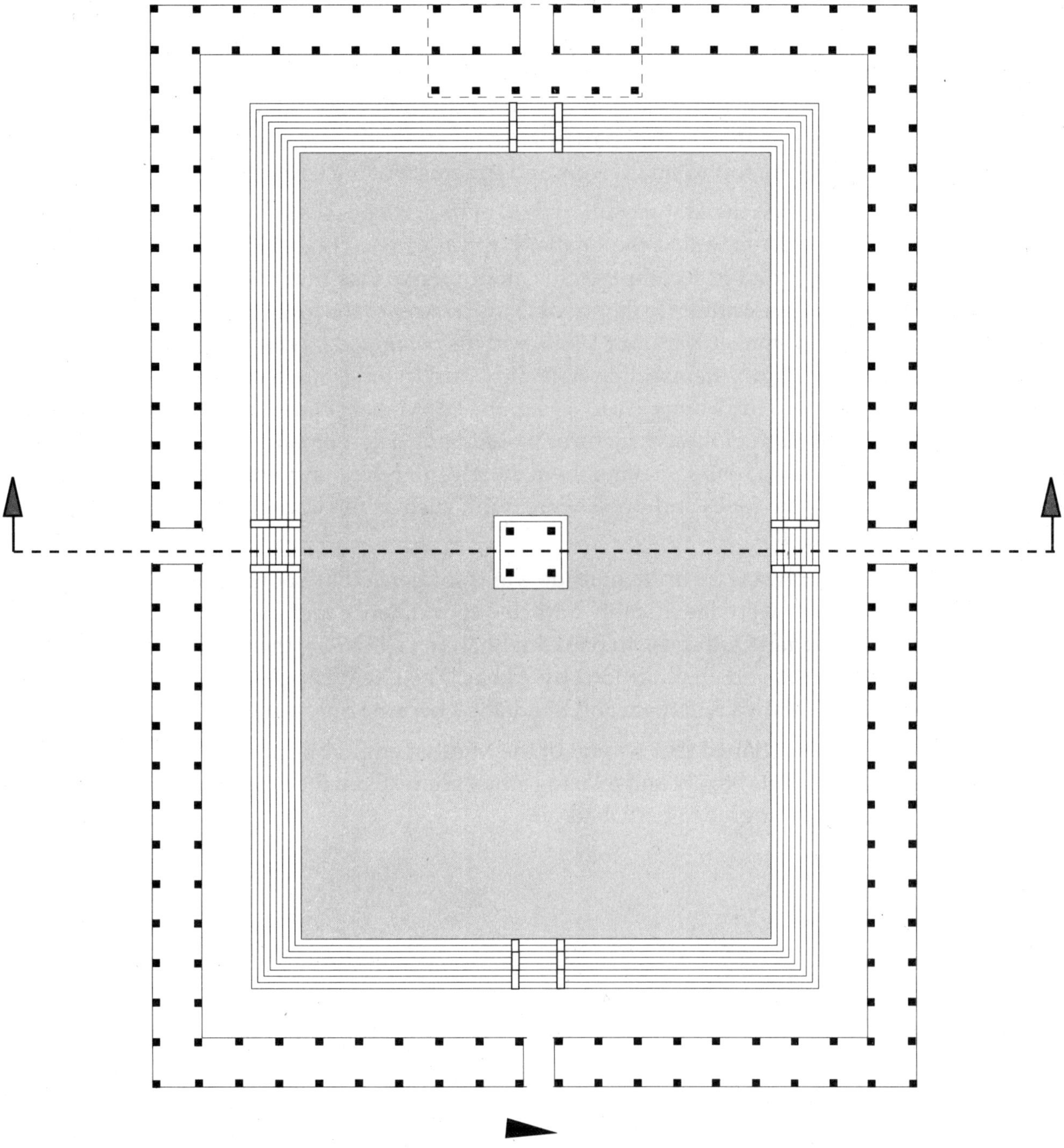

# Shādiābād, Māṇḍu, Madhya Pradesh (a. CE 1410)

The Mālwā Sultanate was established after the first third of the fourteenth century in the Deccan. Soon Shādiābād[44] of Māṇḍu was enlarged by the Khalji Dynasty of the Mālwā Sultanate (**Appendix D**). Located on the top of a precipitous plateau, numerous defensive and residential structures were raised that underlined the importance and wealth of this powerful Sultanate. As with many locales in this part of India, there was a dearth of water, particularly on a raised, defensive location as Shādiābād. Frequently, in the valleys were found small rivers and streams, but not so on the upper reaches of land.

The wealth and power that the Mālwā Sultāns had at their disposal allowed them to construct the vast Mūñja Tālāo (**Plate 36, L**) as well as the smaller Kāphūra Tālāo (**Plate 36, M**). The time and effort that would have been expended to accomplish this monumental task must have been staggering, but the fact that it was accomplished attests to the strength and power of the Mālwā Sultāns. Both the Mūñja Tālāo and the adjacent, but smaller Kāphūra Tālāo, with its octagonal pavilion, were for the exclusive use and enjoyment of the luxury-demanding Sultānic Court. Numerous palaces and pavilions were constructed on or adjacent to the Mūñja Tālāo — e.g. the Lāl Mahal (**Plate 36, I**) and the Jahāẓ Mahal (**Plate 36, J - Plates 37-38**). Both of these structures would have been cooled by the breezes of the *tālāo*, especially in the early evening hours, making them delightful, refreshing retreats for the members of the court. Small, very private *bāolī*s and pools were built, such as the Campā Bāolī (**Plate 36, N**) and the Rewā Kuṇḍa (**Plate 36, F**).

The early Sultāns of Mālwā were brilliant leaders and soldiers. They soon expanded their domain. However, the latter Sultāns were inextricably addicted to indolence and excess. In 1531 Mālwā was defeated by the Gujarat Sultān Qutb-ud-Dīn Bahādur Shāh (r. CE 1526-35) then by Sher Khan (Sūr Sher Shāh) (r. CE 1540-45) in 1542, and again in 1562 by Akbar, The Great (**Appendix D**). After the defeat by the Great Mughal, the Mālwā Sultanate and Shādiābād became only a memory.

So commodious was Shādiābād that several of the Mughal emperors, after they had subdued the region, availed themselves of its beauty and relaxing atmosphere. Even the Emperor Jahāṅgīr expended large amounts of monies to upgrade Shādiābād.

[44] The plan is redrawn using Adobe PageMaker 6.0 from Christopher Tadgell, *The History of Architecture in India*, London: 1990, p. 190 and George Michell, *The Royal Palaces of India*, London: 1998, p. 83.

# Plate 36: Shādiābād

- **A** Delhi Darwāẓā
- **B** Hāthī Pol
- **C** Outer Court
- **D** Mosque of Dilāwar Khan
- **E** Palace of Bāẓ Bahādur
- **F** Rewā Kuṇḍa
- **G** Palace of Gada Shāh
- **H** Hindola Mahal
- **I** Lāl Mahal
- **J** Jahāẓ Mahal
- **K** Inner Court
- **L** Mūñja Tālāo
- **M** Kāphūra Tālāo
- **N** Campā Bāolī

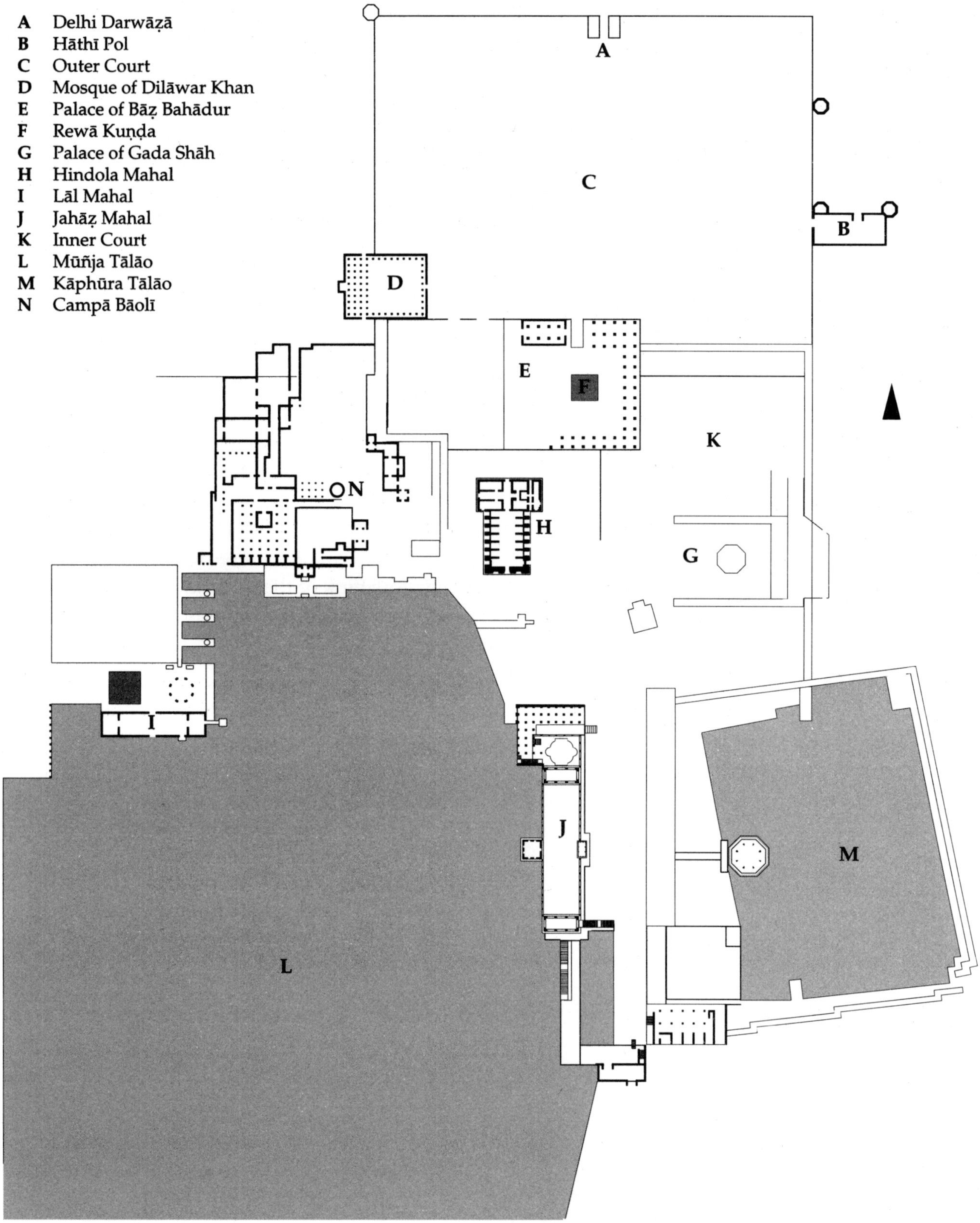

# Jahāẓ Mahal, Shādiābād (*c.* CE 1460)

Shādiābād is noteworthy because of its numerous and varied structures. Not the least being the Jahāẓ Mahal.[45] The "Ship Palace", a translation of Jahāẓ Mahal, is unique in many aspects. Being constructed by the Sultān Ghiyās-ud-Dīn (Ghiyāth) Shāh (r. CE 1469–1500) (**Appendix D**) on the large reservoir, the Mūñja Tālāo, and close to the smaller Kāphūra Tālāo, this pleasure pavilion appears to float upon the surface of the Mūñja Tālāo in glistening splendour. We have long heard of the joys the Mughals found in their perfected gardens, shimmering waterways and bubbling fountains. Yet, here, tucked away, south of imperial Delhi, a glittering example of architecture worthy of any court or capital is encountered. The Jahāẓ Mahal was raised to feed the insatiable desire for luxury that the indolent Sultāns of this line seemed to require.

The Jahāẓ Mahal is strongly horizontal as it sweeps from north to south with low steady rhythms of openings and decorative borders. The pleasure palace consists of a series of open air chambers on the first level reserved for private bathing and relaxation. These spaces give way to either the Mūñja Tālāo to the west or a narrow garden and the Kāphūra Tālāo to the east. An elaborate tank is found on the first level as well as on the terrace level. Sheltered from the sun within these chambers, the members of the royal family were treated to the breezes which blew across the reservoir during the hot dry months as well as being shaded from the unremitting sun of Māṇḍu. Such luxury when others toiled ceaselessly in the sun feed the utter indolence of these Sultāns.

The southern part of the terrace level was punctuated on either end with rectangular pavilions, a domed, square pavilion jutting over the reservoir and opposite a small, square-domed kiosk. On this level a small tank, baroque *en plan*, mirrored the larger shallow tank of the ground level. Here, raised above the hurly-burly of courtly necessity and where the gentle, cooling breezes of the evening could waft, the Sultān would be able to pursue all manners of pleasure.

The Jahāẓ Mahal, with its two eccentrically designed tanks and the *tālāo*s to bracket it, reflects the physical necessity of water, its pleasing quality, and its iconic manifestation of power and wealth. Parenthetically, as brilliant as the Khaljī Mālwās were, their predilection for luxury and indolence was the cause for their ultimate downfall.[46]

---

45 The plan is redrawn using Adobe PageMaker 6.0 from Christopher Tadgell, *The History of Architecture in India*, London: 1990, p. 190.

46 Part of this description is taken from the author's *Royal Palaces, Residences and Pavilions of India: An Iconographic Consideration*, New Delhi: 2006, p. 52.

# Plate 37: Jahāẓ Mahal, Shādiābād, Terrace Level plan

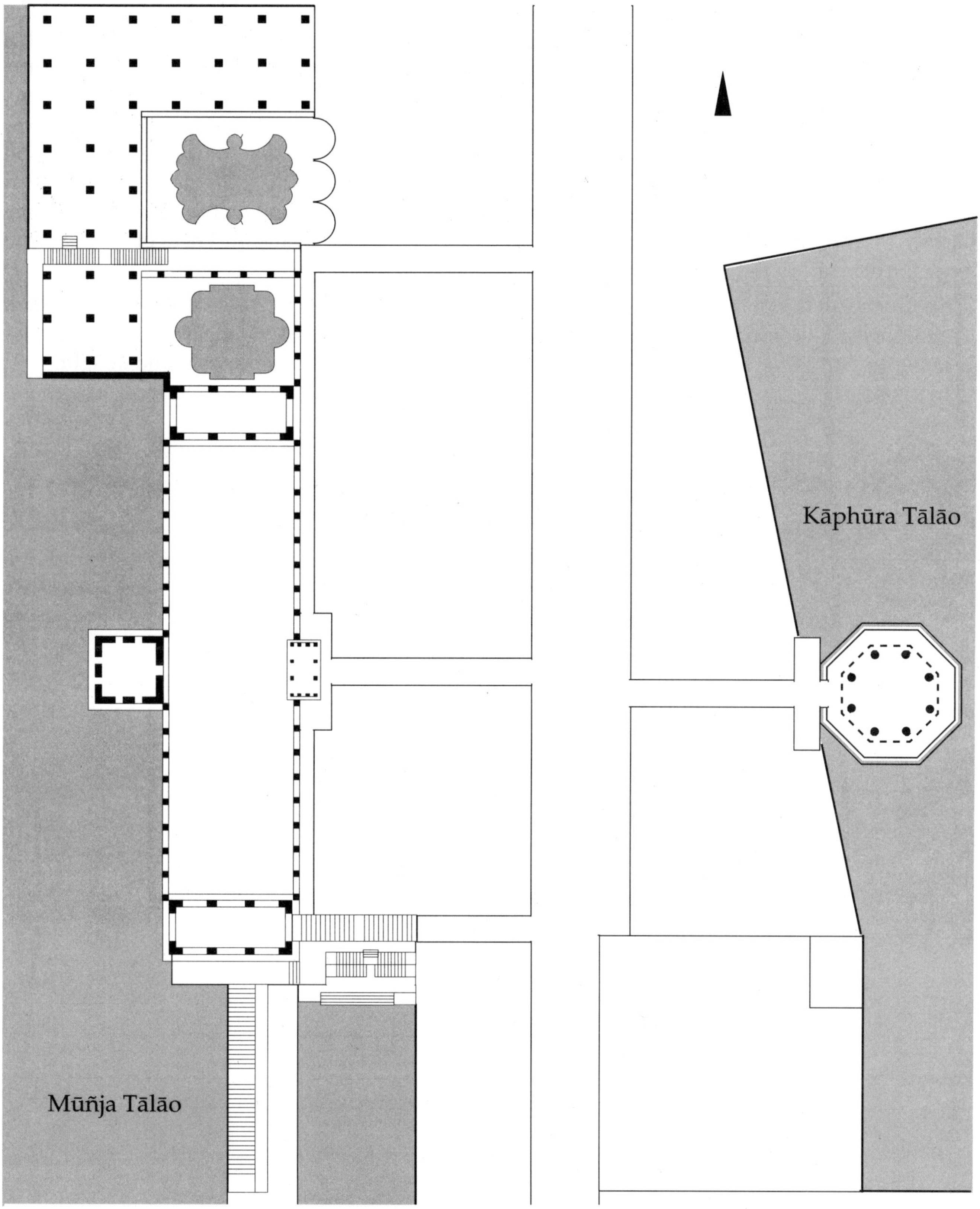

# Jahāẓ Mahal Pools

Shādiābād possessed a number of remarkable buildings. It also enjoyed a number of pools for the delight of the Sultānic court. The two pools of the Jahāẓ Mahal[47] at Māṇḍu (**Plate 37**) were for the exclusive bathing and delight of the Sultāns and their immediate court. In this amazing structure, the pools were separated by one storey. One, the simpler of the two, was situated on the upper terrace (**Plate 38,** bottom) and the other, the more ornate pool, was enclosed on three sides and on the lower level (**Plate 38,** top). The lower pool, florid, *en plan,* could be visually isolated from all, except those on the upper level, by the introduction of a curtain across the east part of this space. It would have been a perfect locale for the ladies of the court to bathe and be observed by any on the upper level if they wished. Surrounding the lower pool were rooms and spaces cooled by the thickness of the masonry, shaded from the sun for additional privacy and comfort.

A number of the tanks and pools within this brief study were associated with the accomplishment of certain rituals as well as cleansing the body. The two pools of the Jahāẓ Mahal were not conceived as being employed for such utilitarian purposes. They were primarily for pleasure and delight.

[47] The plan is redrawn using Adobe PageMaker 6.0 from Christopher Tadgell, *The History of Architecture in India*, London: 1990, p. 190 and the Internet.

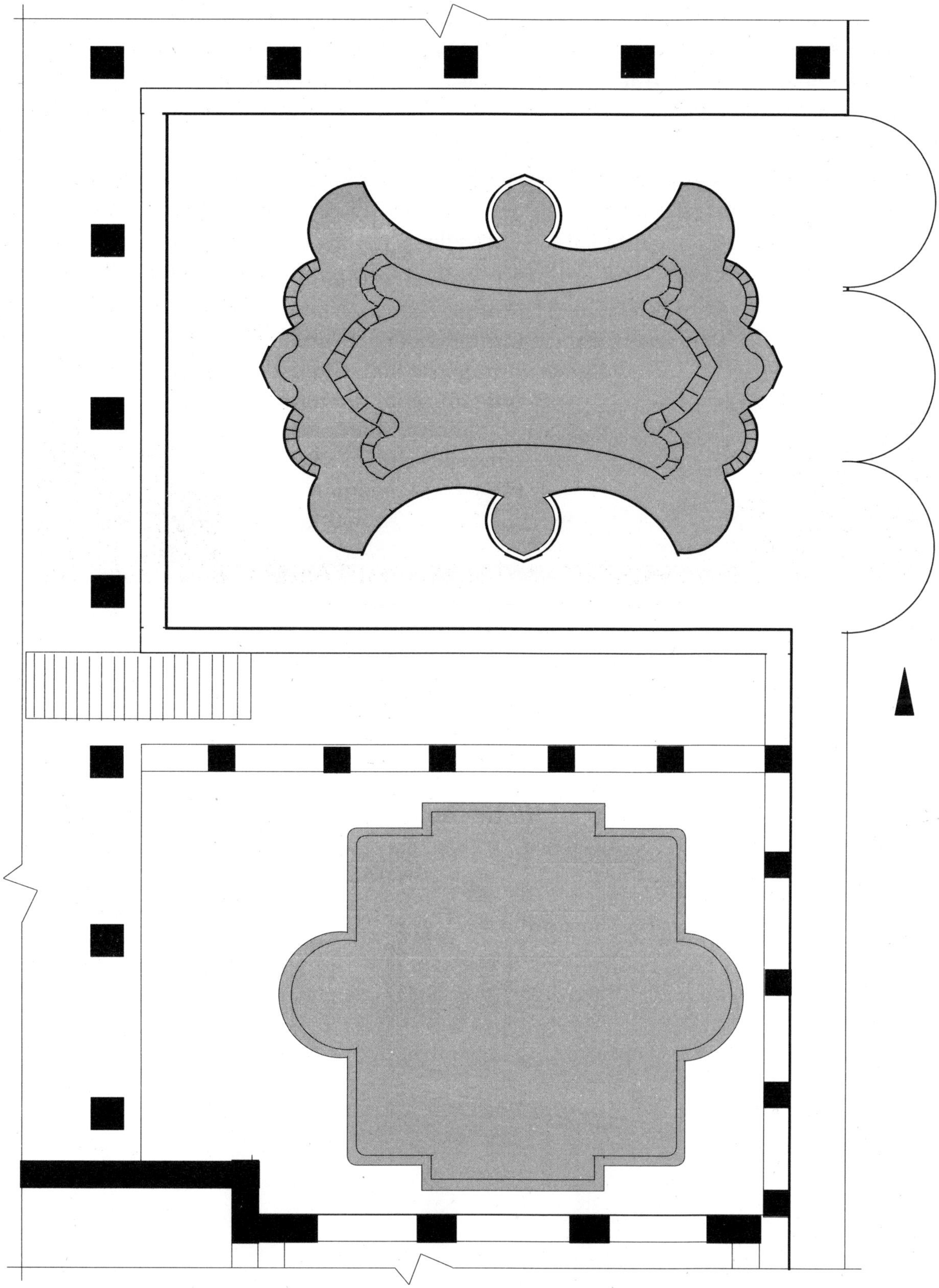

## Rewā Kuṇḍa, Shādiābād (a. CE 1555)

There are many stories of great loves in India. One of them revolves around the love of the ill-stared Sultān Miyāṅ Bayezid Bāẓ Bahādur (r. CE 1555-62) (**Appendix D**) for Rānī Rūpamatī. Unfortunately the Sultān Baẓ Bahādur ascended the throne at the worse possible time. But, that did not stop him from constructing a suitable palace in which he built a tank for the Rānī Rūpamatī's comfort, the Rewā Kuṇḍa[48] (**Plate 36, F**).

By many standards this was not a lavish pool, but its location on one of the terraces of his palace commanded a incomparable view through numerous arches (**Plate 39** [bottom part]). The Rewā Kuṇḍa is square, *en plan*, with three steps, doubled at right angles in the corners leading down to the broad, first level of the *kuṇḍa*. Then, centred on each sides are double stairways which descend to the next level of the pool. Here, the Sultān Bāẓ Bahādur and his beloved Rānī Rūpamatī may have escaped the heat of the day. As conjectured about the Great Tank and Vijayanagara (**Plate 22**), the Rewā Kuṇḍa may have been covered by awnings, further sheltering those who would have sought the pool's cooling effects.

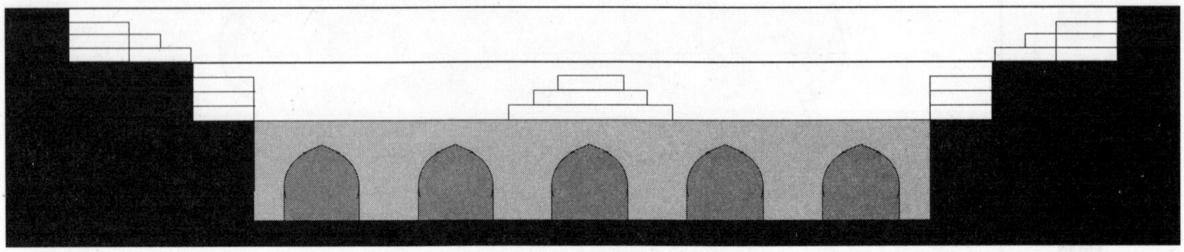

**Figure 38:** Rewā Kuṇḍa, sectional view

[48] The plan is redrawn using Adobe PageMaker 6.0 from the Internet.

**Plate 39:** Rewā Kuṇḍa plan

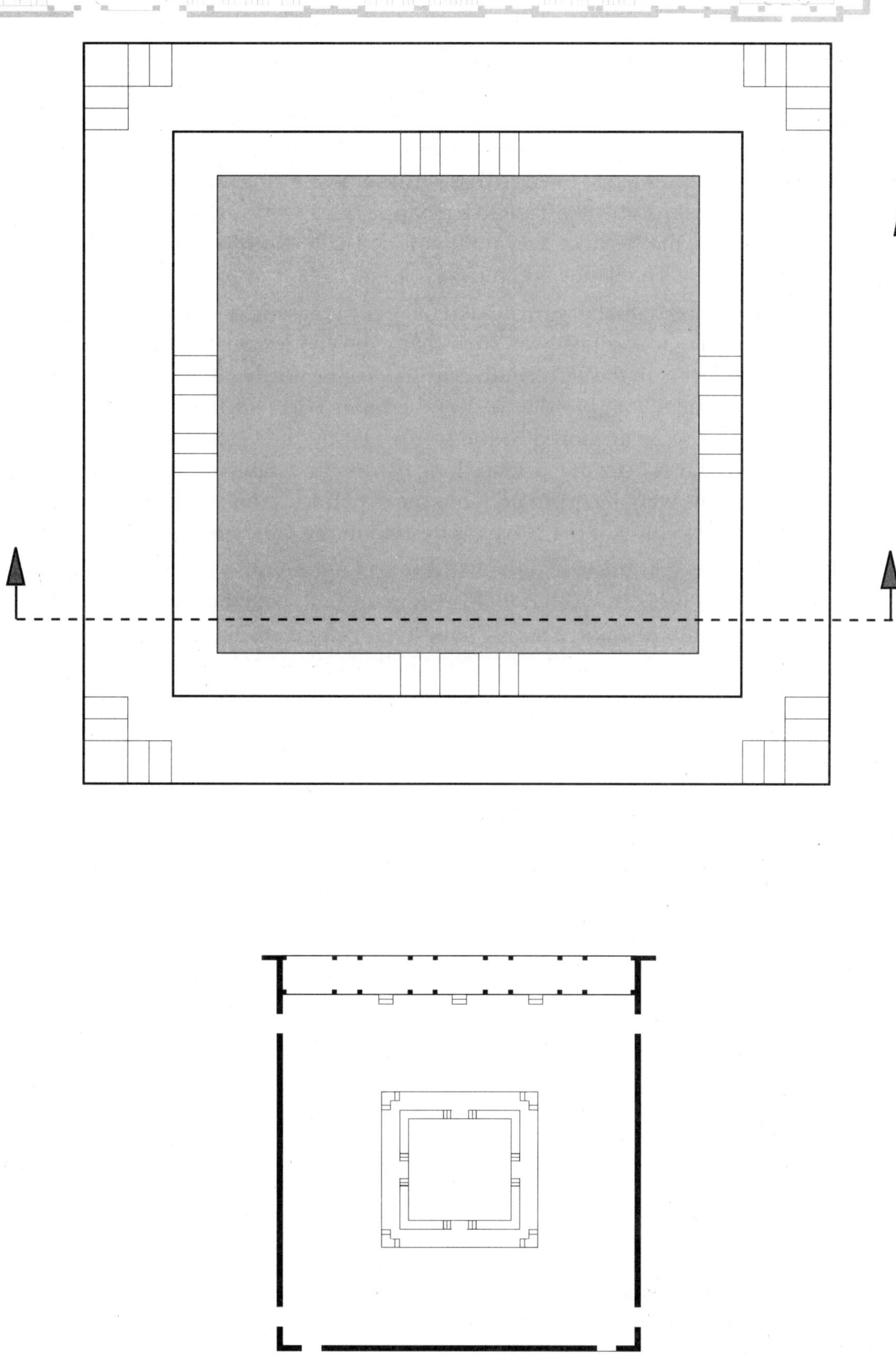

# Inner Fort, Gingee, Tamil Nadu

The responsibilities and daily duties of the Imperial Court at Vijayanagara expanded exponentially as the Empire grew. Soon the administration of the various polities which had been defeated and swept under the spreading wings of Vijayanagara grew in number. Many of them were at quite a distance from the Imperial capital. The daily administration of these regions became difficult and unwieldy. Governors, Nāyakas were appointed to oversee these areas as providing an effective way of governing them. The major governors were: Nāyaka Tubaki Kṛṣṇappā (r. CE 1490–1521) (**Appendix D**) of Gingee who was appointed in 1490; the Nāyaka Viśvanātha (r. CE 1529-64) of Madurai was named in 1529; and Sevappā (r. CE 1549-72) of Thañjāvur became Nāyaka in 1549.

Gingee, to the east of Tirvannamalai, surrounded by rocky crags, developed into a defensive centre of no mean import. Its strategic location was such that even the French undertook renovation and strengthening of its battlements in the eighteenth century. To the north and east of the citadel rose the Kṛṣṇagiri (**Appendix C, Plate 87**) and to the west of the Inner Fort[49] sprung up the soaring Rājagiri. These two precipitous hills were anchored to the south-east by the Candragiri. Battlements on the western mount, the Rājagiri had been in existence long before the establishment of the Nāyakas' rule. These extensive fortifications were incorporated into the expanded plans of the governors. The Inner Fort, main citadel and the dwelling of the Nāyakas nestled in the valley between these three hills.

Unlike its Imperial overlord, Vijayanagara, Gingee was not favoured with a fast running river. However, it did possess a number of reservoirs that fed its many tanks and there were a number of natural fissures that opened to amounts of water. Bathing tanks were dotted around the vast reaches of Gingee. The major tank (*kuḷam*) within the Inner Fort is the Anaikuḷam and the Kamalakanniammaṅkuḷam serviced the towering Rājagiri. Both provided year-round bathing for the occupants of Gingee. The Kalyāṇa Mahal (**Plate 40, A**) thought to be the "Queen's residence", the *zenānā* possessed its own diminutive tank for the women of the court.

49 The plan is redrawn using Adobe PageMaker 6.0 from Jean Deloche, *Senji (Gingee): A Fortified City in the Tamil Country*, Pondicherry: 2000, pp. 77 and 88.

# **Plate 40:** Inner Fort, Gingee

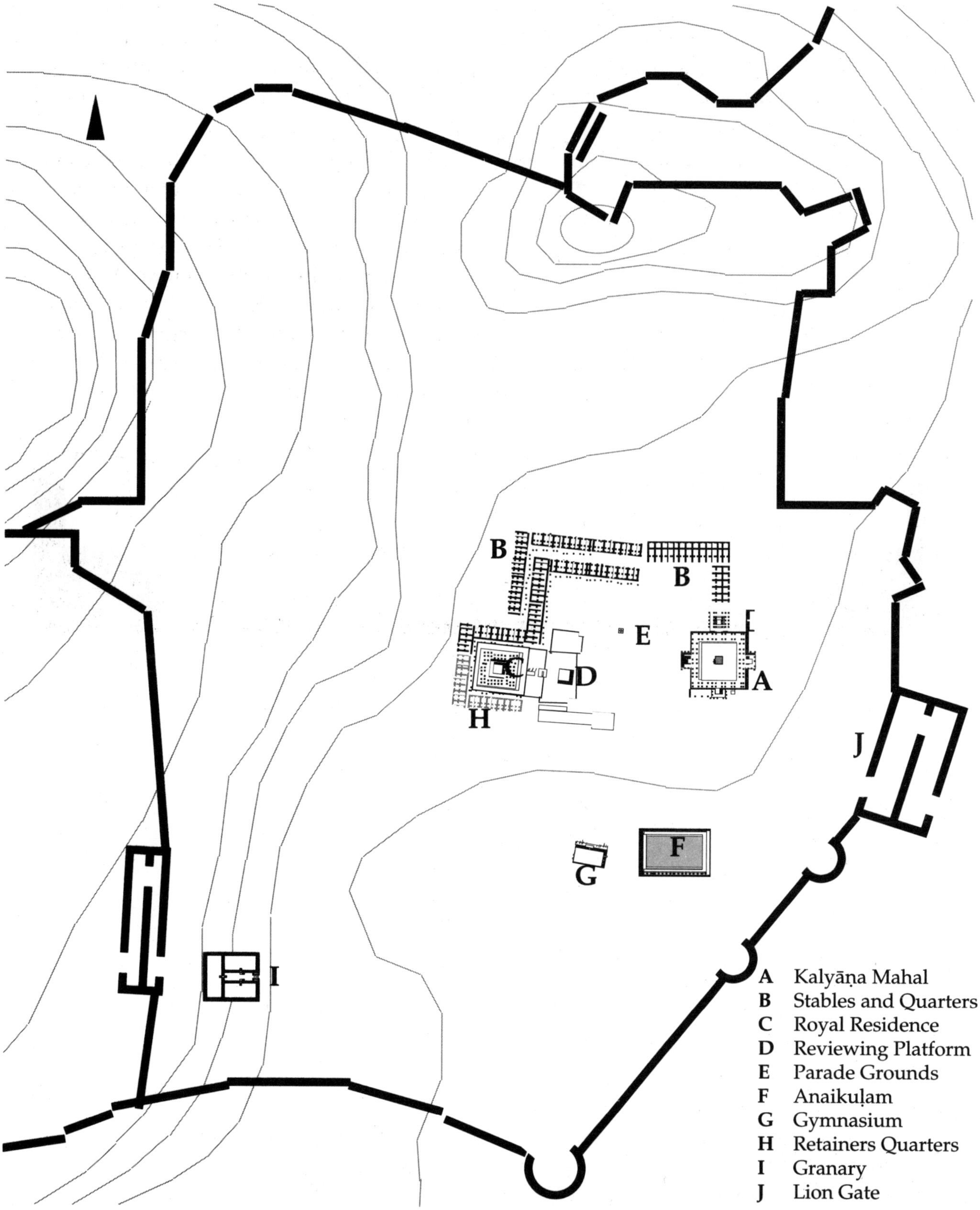

# Anaikuḷam (Bathing Tank), Inner Fort, Gingee, Tamil Nadu

The large, commodious Anaikuḷam[50] (trans: "elephant tank")[51] was the focal tank of the Inner Fort. It rested to the south of the residential complex of this citadel (**Plate 40, F**) and was complemented by the Chinnataṅgai *eri*, the Chinnataṅgai pond to the north. However, Anaikuḷam was exclusively utilized for ablution for the residents of the Inner Fort, whereas, the Chinnataṅgai *eri* was a source of water during the dry months.

Rectangular, *en forme*, the whole is sunk into the earth, creating a retaining wall nearly 2 m high around the external perimeter of the tank (**Figure 39**). In this area of Gingee, the water table is relatively close to the surface and this, in part, allows the Anaikuḷam to contain water for the whole year. The tank is entered via a short flight of steps in western part of the northern wall. These steps lead down to a wide terrace which surrounds the upper step of the tank. From this northern entrance and at the edge of the terrace, a set of stairs, integrated into the descending steps of the tank, leads down to the bottom of the tank. The surrounding terrace exhibits a covered colonnade which extends around the entire perimeter of the tank proper and shorter columns rest upon the retaining wall. This covered colonnade provides much needed relief from the blazing sun particularly during the hot, dry months of the year.

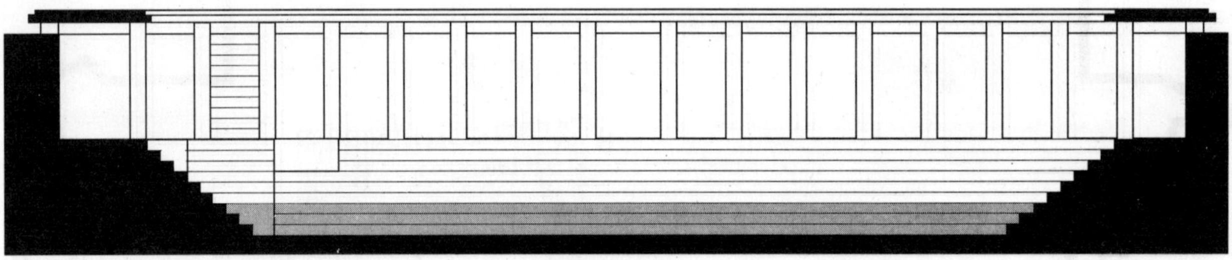

**Figure 39:** Anaikuḷam Tank, Gingee, sectional view

[50] The plan is redrawn using Adobe PageMaker 6.0 from Jean Deloche, *Senji (Gingee): A Fortified City in the Tamil Country*, Pondicherry: 2000, p. 183.

[51] The term "elephant tank" refers to the size and not that it was used to bathe elephants.

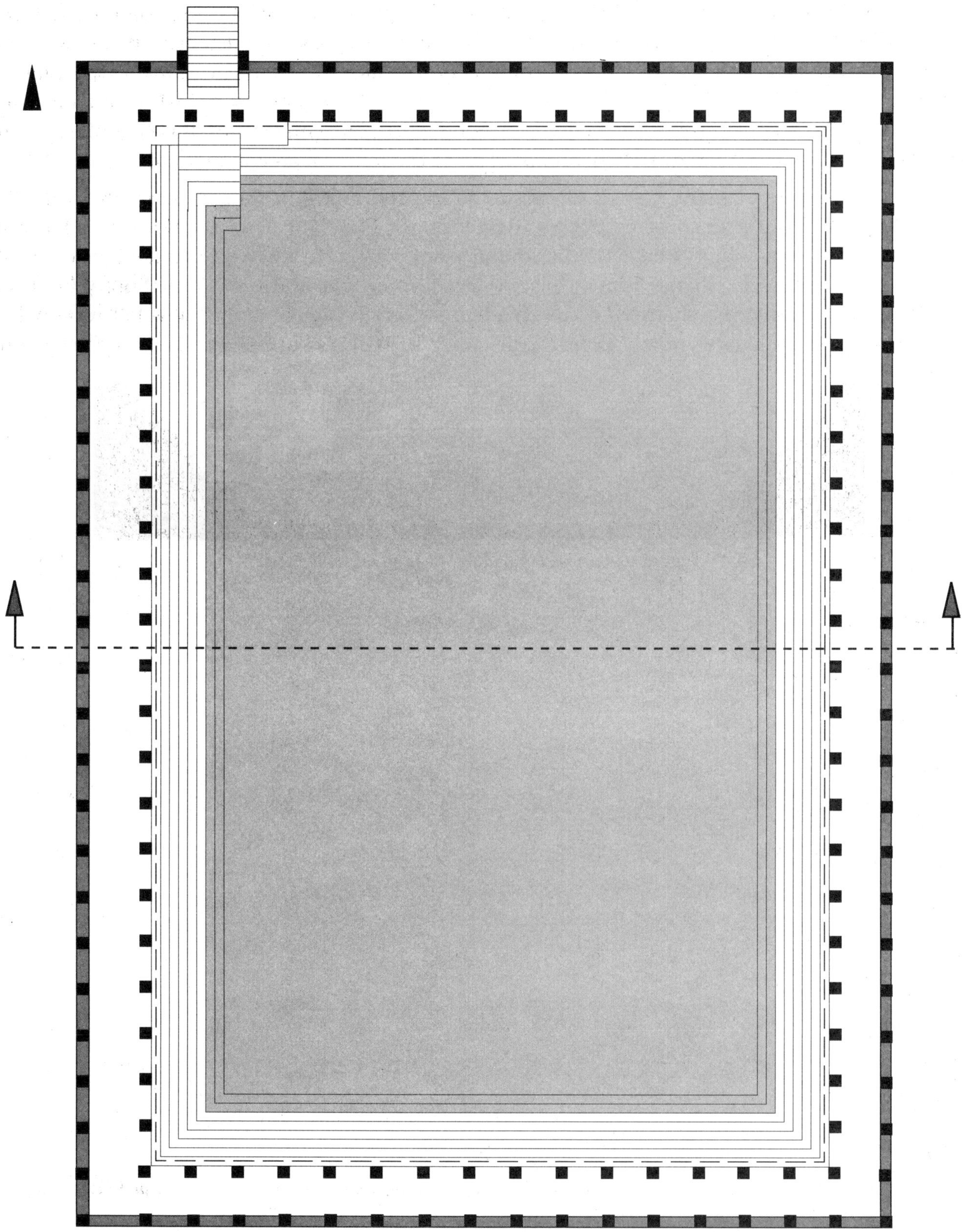

# Kamalakanniammaṅkuḷam (Bathing Tank), Rājagiri, Gingee

Due west of the Inner Fort rises the towering hill, the Rājagiri. It is one of the three sentinel-hills that surround and protect Gingee. The Kr̥ṣṇagiri and the Candragiri possessed defensive battlements and various other structures to provide for the comfort of the defenders. However, the fortifications of Rājagiri were far more developed and extensive than the other two mounts. Its history as a defensive position began in the early thirteenth century. Among the accoutrements atop this crag, the Rājagiri, were a temple, audience hall, warehouse, barracks and the Kamalakanniammaṅkuḷam.[52]

The Kamalakanniammaṅkuḷam is found just to the south of Rājagiri's most secure bastion or keep on a relatively flat expanse of ground (**Appendix D, Plate 88**). The tank (kuḷam) is basically rectangular in shape with a "notch" in the southeastern corner. To the west is a double colonnade which is covered, offering shelter from the blazing sun to those who utilize this bath (**Figure 40**). There is some indication that this sheltered area extended around the whole tank. The water level within the Kamalakanniammaṅkuḷam was at the capricious nature of the weather and the monsoonal rains.

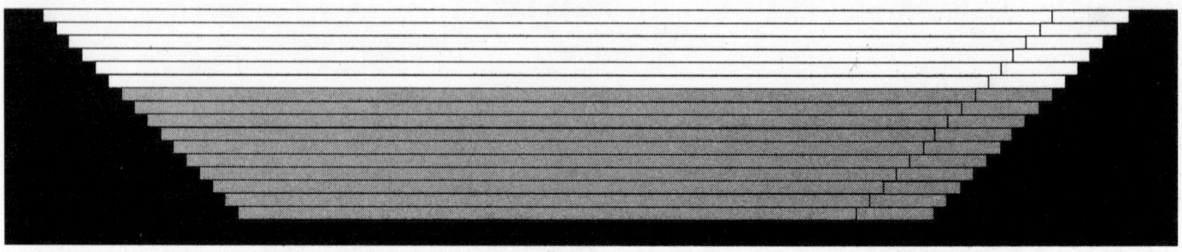

**Figure 40:** Kamalakanniammaṅkuḷam, sectional view

[52] The plan is redrawn using Adobe PageMaker 6.0 from Jean Deloche, *Senji (Gingee): A Fortified City in the Tamil Country*, Pondicherry: 2000, p. 178.

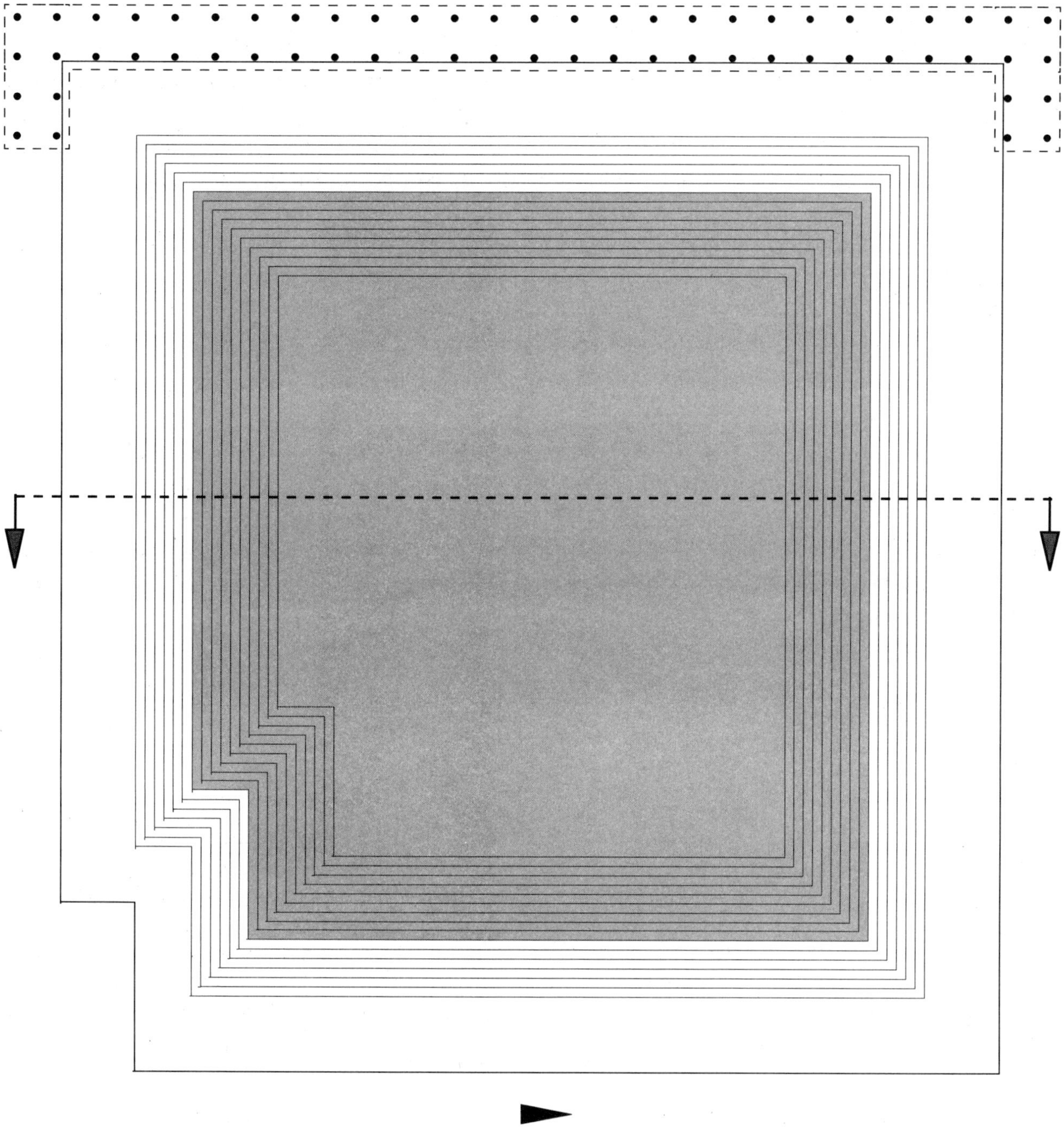

# Aruṇācala (Annamalaiar) Temple, Tiruvannamalai, Tamil Nadu

Tamil Nadu is the locus for a number of large, extensive temple cities, sacred cities. Expansive, gigantic, impressive, awesome are some of the adjectives that are used to describe the Śrī Raṅganāthaswāmi Temple (**Plates 45-46**) of Śrīraṅgam, Jambukeśvara Temple at Tiruchirapalli, Mīnākṣī Sundareśvara Temple (**Plates 56-57**) of Madurai, the Śrī Ekāmbareśvara Temple (**Plate 48**) at Kāñcīpuram and the Aruṇācala (Annamalaiar) Temple[53] at Tiruvannamalai. These are the most well-known sacred cities in the rich cultural agar of the south. Each in its own way is spectacular, and underlines the manifest devout nature and the wealth of the south.

Unlike most of the other temple cities, the Aruṇācala (Annamalaiar) Temple presents an ordered, symmetrical layout, *en plan*. The three outer walls (*prākāra*) are punctuated with nine towering *gopura*s — i.e. four each on the outer two walls and one on the southern wall of the third. Each of these three high walls is strictly orthogonal/rectilinear and nests one within the other. Deep within these protecting walls, the Lord Śiva, the unchangeable Aruṇa or sun, personified as Annamalaiyar, is represented in the form of a *liṅga*.

Two tanks are to be found in the Aruṇācala (Annamalaiar) Temple. The largest and the most important one, the Śivagaṅgā Tank (**Plate 43, K**), is to be found in the south-eastern quadrant of the outer *prākāra*, just to the south as one passes through the Vallāla or Rāja Gopuram (**Plate 43, A**), and across the entrance avenue from the Thousand-Pillared Hall. The second smaller tank, the Brahma Tīrtha (**Plate 43, J**), is immediately to the south of the entrance *gopura* (Ilayanar Koil Gopuram) (**Plate 43, E**) of the second *prākāra*.

**A** Vallāla or Rāja Gopuram
**B** Ammaniammal Gopuram
**C** Thirumanjuna Gopuram
**D** Pey (Mel) Gopuram
**E** Ilayanar Koil Gopuram
**F** Second *prākāra* south *gopuram*
**G** Second *prākāra* west *gopuram*
**H** Second *prākāra* north *gopuram*
**I** Third *prākāra* (Kili *gopuram*)
**J** Brahma Tīrtha
**K** Śivagaṅgā Tank
**L** Thousand-Pillared Hall
**M** Kambathu Ilayanar
**N** Śivagaṅgai Vinayagar
**O** Bairavar Temple
**P** Katchi Maṇḍapa
**Q** Sanctum

[53] The plan is redrawn using Adobe PageMaker 6.0 from the Internet.

# Plate 43: Aruṇācala (Annamalaiar) Temple

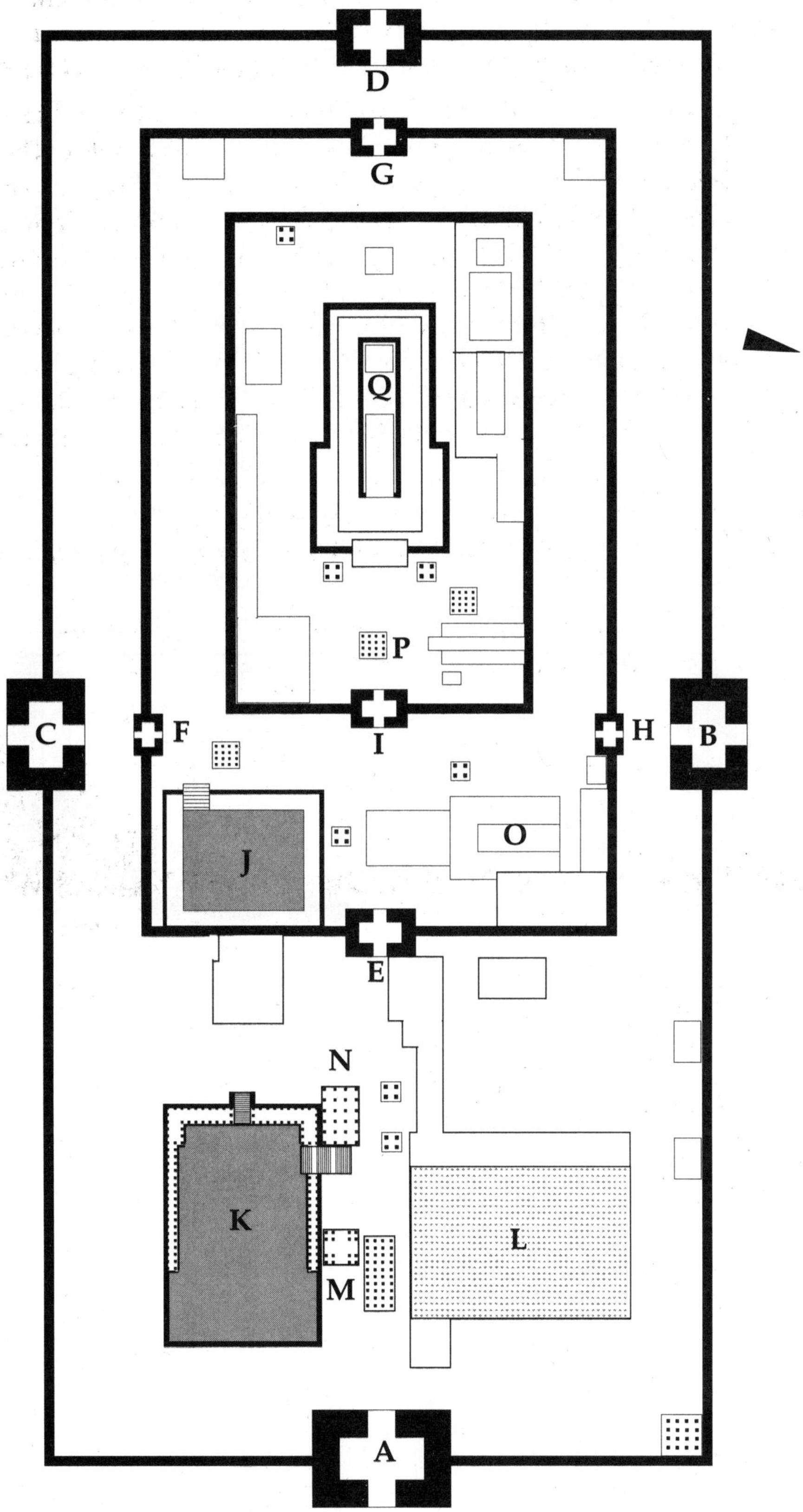

# Śivagaṅgā Tank

The substantial size of the Śivagaṅgā Tank[54] befits a temple as large, important and developed as the Aruṇācala (Annamalaiar) Temple. As noted above Śivagaṅgā Tank is one of two tanks within the sacred complex; the other being the smaller Brahma Tank of the second *prākāra* (**Plate 43**).

The Śivagaṅgā Tank was excavated within this outer *prākāra* and found just to the south of the towering Vallāla or Rāja Gopuram, the main entrance-way to the complex. There are three access-stairways of varying widths which lead from ground level down to the water's edge — i.e., two along the northern border and one centred in the western side. Part way down from the ground level is a covered, colonnaded terrace. This terrace runs along part of the north and south and all of the west sides of the tank (**Figure 41**). This is covered colonnade is a feature that is encountered in a number of temple tanks and provides relief for the devotees from the blazing sun. However, unlike most of the other temple tanks, these terraces do not step down to the water's edge and below, rather, they drop to the bottom of the tank. The three entrance stairs are the only access pilgrims have to the water, unless, during the rainy season, the water level has risen to flood the terrace level.

The Aruṇācala (Annamalaiar) Temple is an important Śaiva shrine and the complex receives many devotees on a regular basis. The alternative to bathing in the Śivagaṅgā Tank is to enter the second *prākāra* and utilize the smaller Brahma Tank. This tank is so designed that pilgrims have easy access to the water on all four sides.

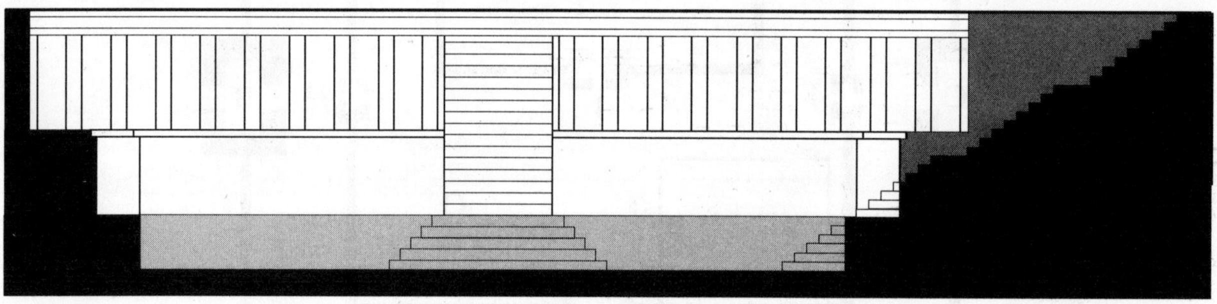

**Figure 41:** Śivagaṅgā Tank, Aruṇācala (Annamalaiar) Temple, section indicating major northern staircase

54 The plan is redrawn using Adobe PageMaker 6.0, from the Internet

# Plate 44: Śivagaṅgā Tank plan

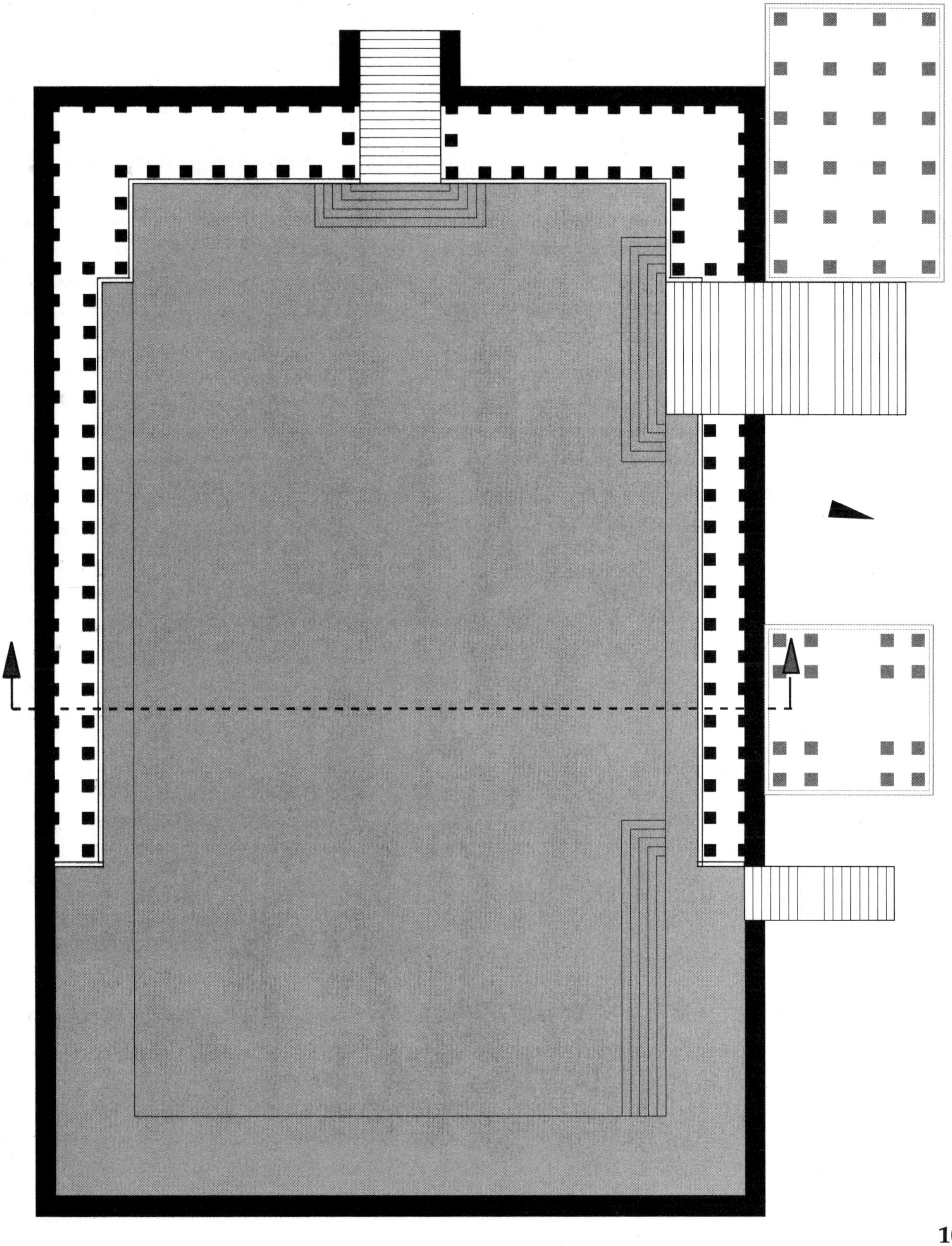

# Śrī Raṅganāthaswāmi Temple, Tiruchirappalli, Tamil Nadu (from eleventh century CE)

The Śrī Raṅganāthaswāmi Temple[55] of Śrīraṅgam island between the Kāverī and Kollidam rivers, near Tiruchirappalli, is one of the most notable temple ensembles in India. It is the largest of all complexes and holds scores of shrines — e.g. shrines of Viśvaksena, Rāma, Kṛṣṇa, Nachiyar, Chakrataḷvār, Garuḍa, Hanumān and Āṇḍāla — and *maṇḍapa*s covering 156 acres. It is the foremost shrine among all the 108 Vaiṣṇava *divyadeśa*s.

The shrine of this Vaiṣṇava temple is nested within seven high impressive girdle walls — seven *prākāras*. Its construction spanned centuries and was contributed to by the Cera, Coḷa, Pāṇḍya, Hoysaḷa, Vijayanagara, and Nāyaka dynasties (**Appendix D**). Within the three outer *prākāra*s are surrounding avenues, *pradakṣiṇāpatha*s named Adaivavalaindan, Cittra and Uttara from the outermost to the third *prākāra*. These were employed in the *pradakṣiṇā* (circumambulating) that was required for worship (*pūja*). Entrances, some with towering *gopura,* surround the *vimāna* — i.e. there are seven entry ways to the south, six to the north, five to the east and three to the west.

The Śrī Raṅganāthaswāmi Temple contained a number of tanks for ablution and *abhiṣeka*. One, the Theppakuḷam was a very short distance to the north-west of the western Adaivavalaindan Gopura. Here, the devotees could perform their ritual ablution before entering the complex.

**A** Shrine
**B** First Wall
**C** Second Wall
**D** Third Wall
**E** Fourth Wall
**F** Fifth Wall
**G** Sixth Wall
**H** Seventh Wall
**I** Streets
**J** Adaivavalaindan Gopura
**K** Cittra Gopura
**L** Uttara Gopura

[55] The plan is redrawn using Adobe PageMaker 6.0 from photocopies V.N. Hari Rao, *The Srirangam Temple: Art and Architecture,* 1967.

**Plate 45:** Śrī Raṅganāthaswāmi Temple overall plan

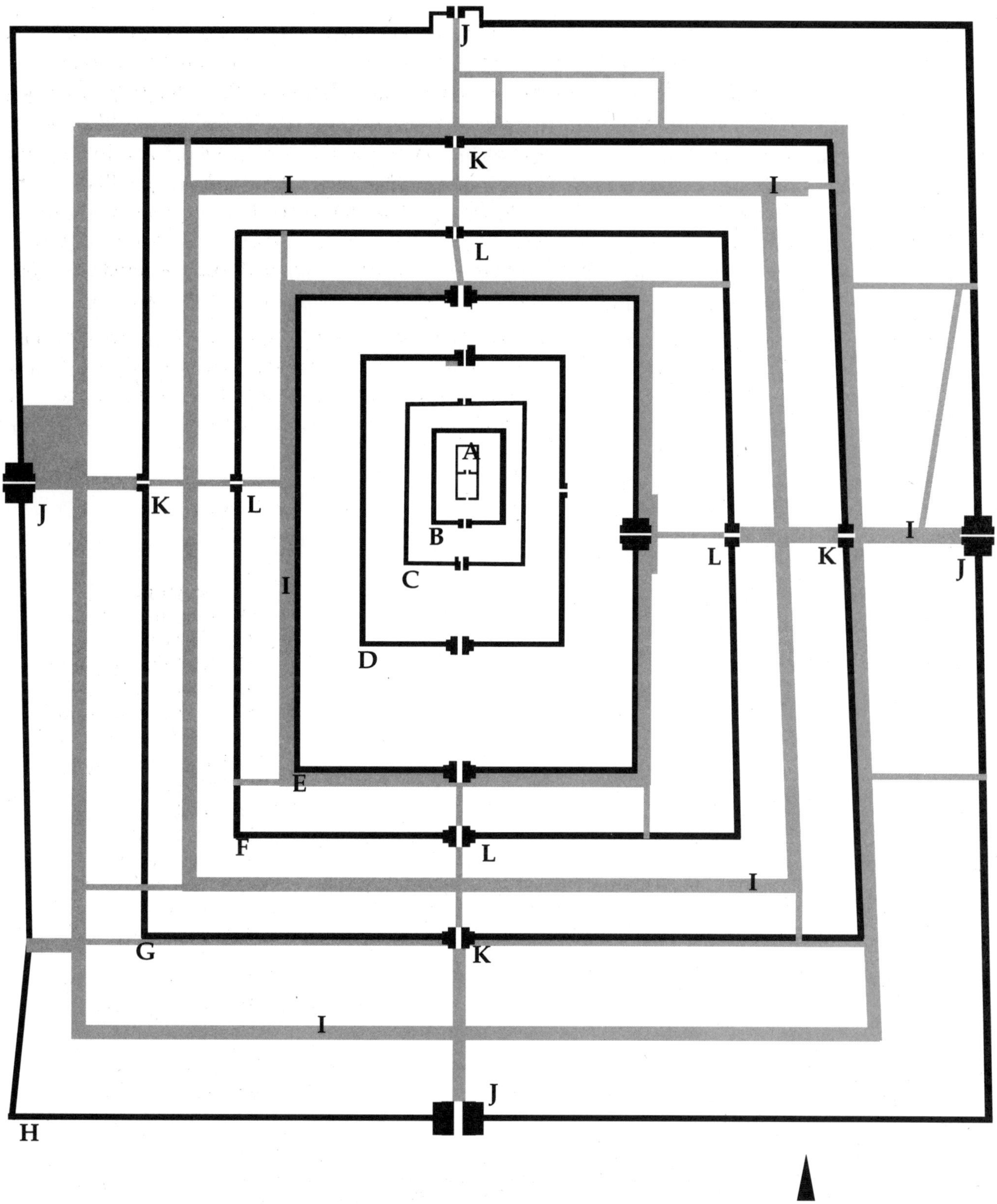

# Śrī Raṅganāthaswāmi Temple

It is within the fourth and subsequent girdle walls where one encounters the myriad shrines, *maṇḍapa*s and auxiliary structures that are part of the Śrī Raṅganāthaswāmi Temple.[56] Foremost in sheer size, the Thousand-Pillared Maṇḍapa (**Plate 46, N**), the smaller Śeṣagiri Maṇḍapa (**Plate 46, P**), the Kambar Maṇḍapa (**Plate 46, Q**), the Vasanta Maṇḍapa (**Plate 46, R**), and the Raṅgavilāsa Maṇḍapa (**Plate 46, T**) are found within the fourth *prākāra*. These set the stage for what is to be encountered within the next three *prākāra*. The fifth *prākāra* holds two important monuments — the Garuḍa Maṇḍapa (**Plate 46, S**) and the Candrapuṣkariṇī (aka. Candra Tīrtha) (**Plate 46, O**). It is also within this *prākāra* that the small Sūryapuṣkariṇī is found (**Plate 46, U**). Within the sixth *prākāra*, one is confronted with a bewildering number of shrines, columns and *maṇḍapa*s, running one into the other. The plethora of sacred elements found in the sixth *prākāra* of the Śrī Raṅganāthaswāmi Temple are duplicated in the seventh.

The sanctum, the *vimāna* of Śrī Raṅganāthaswāmi Temple, is a substantial structure, rectangular, *en plan*, and divided in half. It is in the northern half that the shrine (**Plate 46, A**) to the reclining Viṣṇu is found. The most sacred shrine is *kumbha*-shaped, an unusual form for a Vaiṣṇava shrine. Within the Lord Viṣṇu reclines on the coils of a serpent which is partially submerged in water flowing from a sacred spring.

**A** Sanctum
**B** Innermost entrance
**C** Inner entrance
**D** Kārttikai Gopuram
**E** Nanningan Gopuram
**F** Parāmapada vaial
**G** Aindu vaial
**H** Thayar Sannidhi Gopuram
**I** Vellai Gopuram
**J** First wall
**K** Second wall
**L** Third wall
**M** Fourth wall
**N** Thousand-Pillared Maṇḍapa
**O** Candrapuṣkariṇī
**P** Śeṣagiri Maṇḍapa
**Q** Kambar Maṇḍapa
**R** Vasanta Maṇḍapa
**S** Garuḍa Maṇḍapa
**T** Raṅgavilāsa Maṇḍapa
**U** Sūryapuṣkariṇī

[56] The plan is redrawn using Adobe PageMaker 6.0 from photocopies V.N. Hari Rao, *The Srirangam Temple: Art and Architecture*, 1967 and the mislabelled plan from Henri Stierlin, *Hindu India: From Khajuraho to the Temple City of Madurai*, 2002, p. 210.

## **Plate 46:** Śrī Raṅganāthaswāmi Temple plan

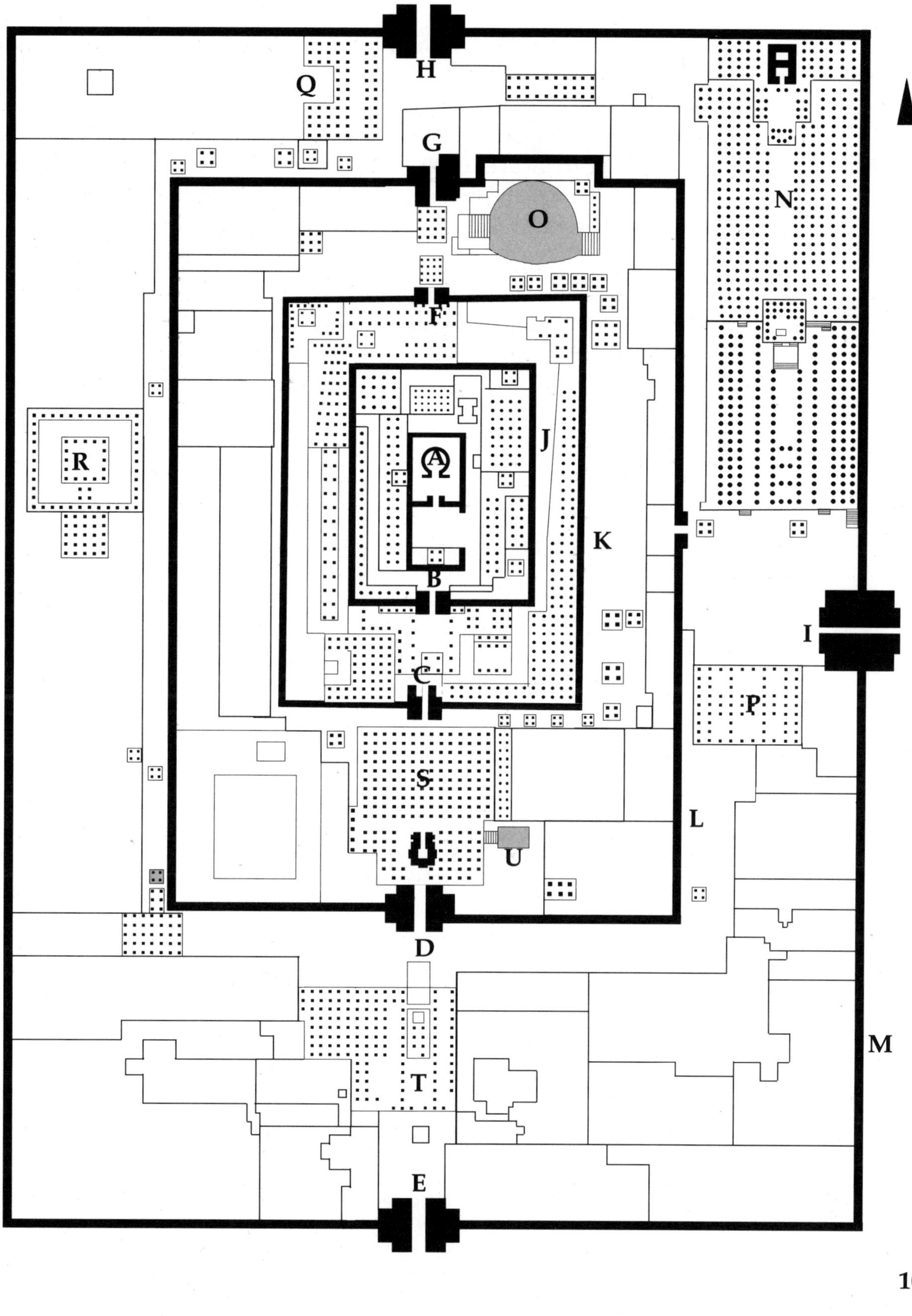

Śrī Ranganāthaswāmi Temple, as many larger temple complexes, possesses a number of tanks. Three are noteworthy — the Theppakuḷam just outside the western Adaivavalaindan Gopura, the Sūryapuṣkariṇī in the sixth *prākāra* and the Candrapuṣkariṇī[57] in the same *prākāra*. The Candrapuṣkariṇī (aka. Candra Tīrtha) (**Plate 46, O**) is a temple tank of uncommon design. Not rectilinear, as are most temple tanks, the eccentric plan of the Candrapuṣkariṇī can be characterized as conforming to an eccentric circular shape. Found in the north-eastern quadrant of the fifth *prākāra*, it is bounded on the north by an extention of the fifth *prākāra*'s wall to accommodate the tank. The Kodardanrāma Shrine is immediately to the east and a complex of small structures leading to the Ainduvaial borders the west.

Access to the Candrapuṣkariṇī is from stairs leading down to the water's edge from the east and the west of the tank. Due to the location of the Śrī Raṅganāthaswāmi Temple on the island Śrīraṅgam, at the confluence of the Kāverī and Kollidam rivers, the water table is relatively stable since the temple is in close proximity to both the Kāverī and Kollidam rivers. Even with the aquifer close to the surface, the tank is known to be especially deep. This may be, a some state, because the tank was fed by a natural spring.

[57] The plan is redrawn using Adobe PageMaker 6.0 from photocopies of V.N. Hari Rao, *The Srirangam Temple: Art and Architecture*, 1967 the mislabelled plan from Henri Stierlin, *Hindu India: From Khajuraho to the Temple City of Madurai*, 2002, p. 210 and the Internet.

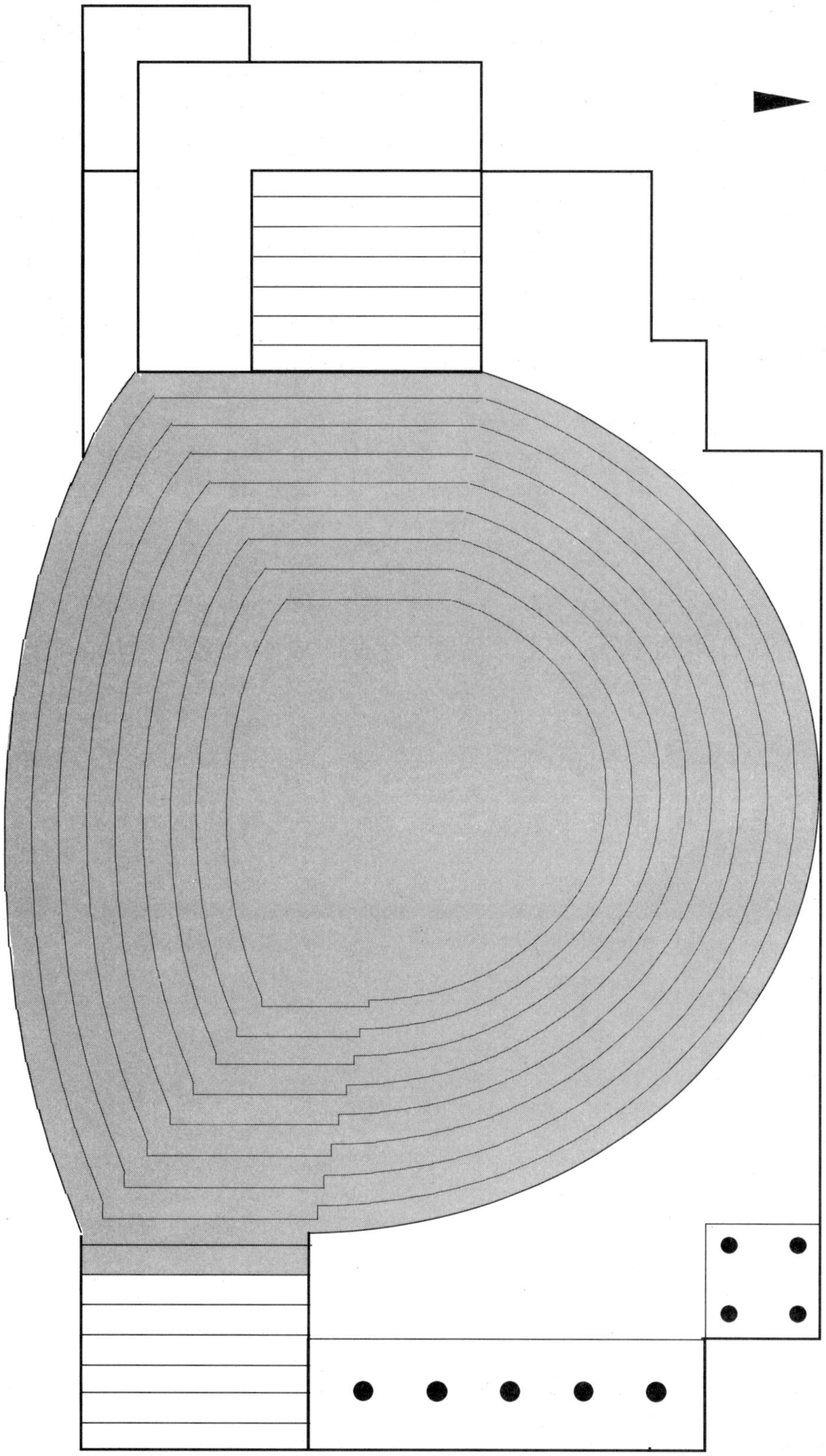

# Ekāmbareśvara Temple Tank, Kāñcīpuram, Tamil Nadu (*c.* CE 1500)

Kāñcīpuram is one of the most important pilgrimage cities in India. It possesses major sacred temples of the three major Hindu sects — the Vaiṣṇava Devarājaswāmi Temple or Varadarāja Temple, the Śakti Kāmākṣī-Amman Temple and the Śaiva Ekāmbareśvara Temple. Each of these significant sites possesses a complex of shrines, *maṇḍapa*s and tanks.

To the north of the entrance *gopura* of the Śaiva Ekāmbareśvara Temple is found the Ekāmbareśvara Temple Tank.[58] It is a tank of significant size and complex design. Two *śālā*s, six columns wide, announce and provide a dramatic entrance to the confines of the tank proper. The steps of the tank descend in a 1:1.5 ratio, making them more ergonomically desirable. They drop ten steps to reach a broader terrace-like area surrounding the whole and then continue downward. To the west of the tank are two structures, shrines, one plain and one complex in design which open, not onto the tank, but to the west. Across from the two large *śālā*s is a covered arcade, four columns deep. This rises from the broad terrace-like step of the first series of steps (**Figure 42**). During excessively heavy monsoons, the water has been known to reach this level. While to the east is a set of steps that lead up to a higher terrace with access to the north to the other areas of the complex on that level. In the centre of the tank is a small, four pillared *nirazhi maṇḍapa*, the *axis mundi*. Broad steps surround and lead down to the water's edge.

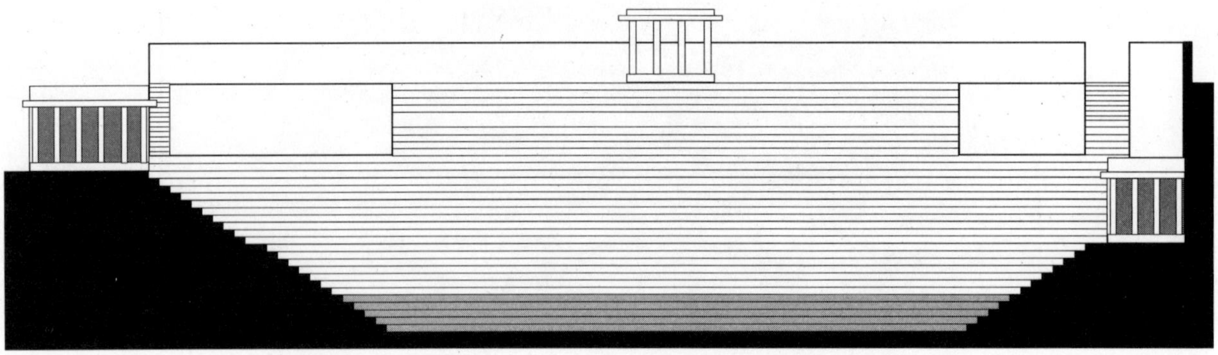

**Figure 42:** Ekāmbareśvara Temple Tank, sectional view

58 The plan is redrawn using Adobe PageMaker 6.0 from the Internet.

## **Plate 48:** Ekāmbareśvara Temple Tank plan

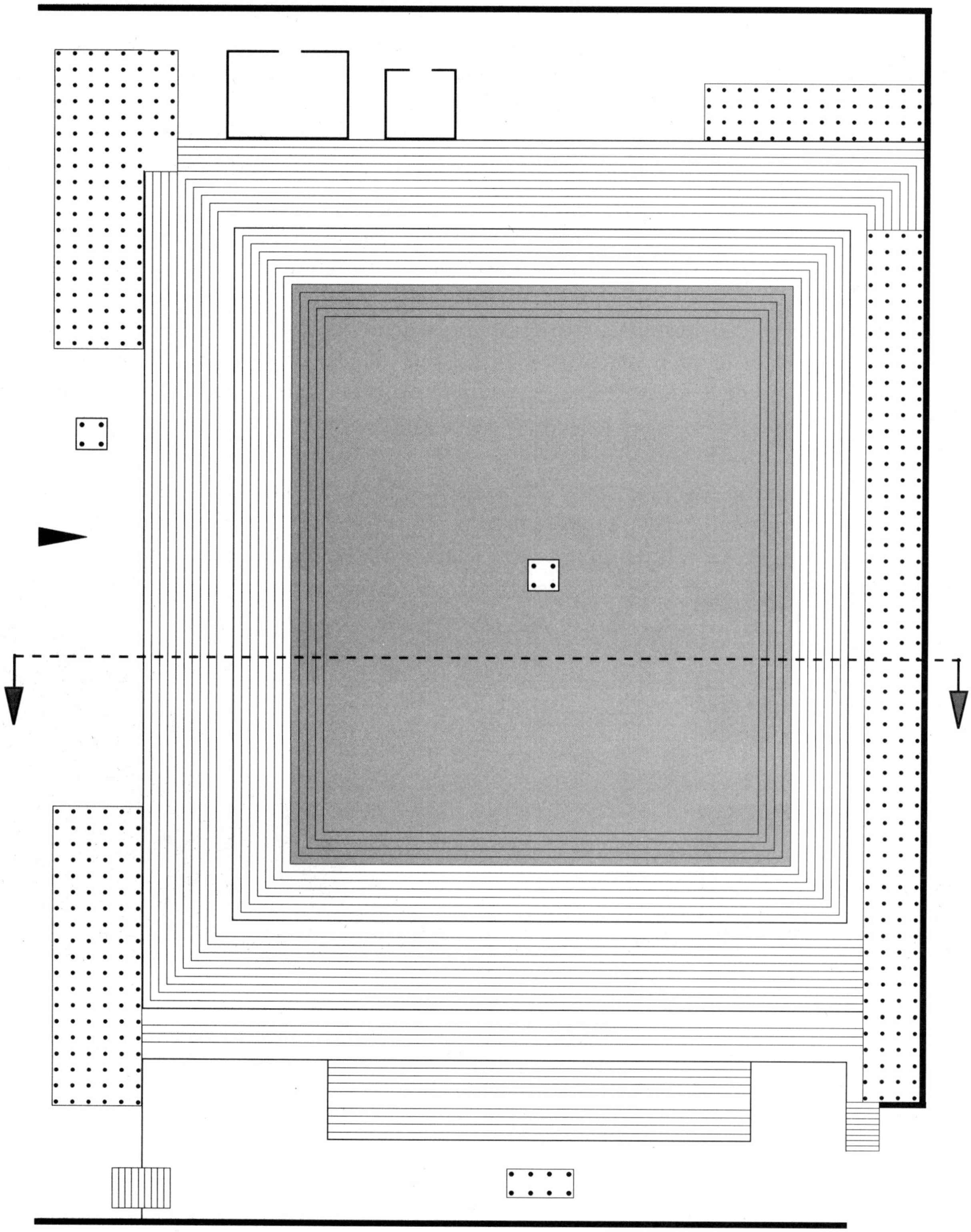

# Rajoṅ-kī Bāolī, Delhi (b. CE 1526)

The Rajoṅ-kī Bāolī[59] was constructed in Delhi during the short-lived reign of the Lodī Dynasty (CE 1451–1526) (**Appendix D**). As with a majority of the dynasties of the Delhi Sultānate, the Lodis were not of Mother India, but Afghanis. They were in essence Muslim foreigners ruling a Hindu land. As with many of the foreign rulers, they undertook an extensive building program. *Bāolīs* were a small, but important, part of their legacy. They might have built lavish palaces and monumental tombs for themselves, but the *bāolīs* were for the common people. In this way, they attempted to make their domination more palatable to the Hindu populous.

The Rajoṅ-kī Bāolī is found near Adam Khan's Tomb and, *en plan*, it is typical of many *bāolīs*. A short flight of steps descend down to a terrace and a further flight descends and gives access to the first storey. Here a covered arcade with pointed, Saracen-style arches, gives shade and relief from the hot Delhi sun. The next flight of stairs, narrower than the first two, descends to the second storey. Here a narrow walkway leads to three arched openings and the cool shade within. The stairs continue to descend without a landing to the lowest storey and thence to the water level below (**Figure 43**). From the entrance terrace to the bottom, the steps descend without a landing. From the top, this gives a dizzying perspective, broken only by the terraces of the ever-narrowing storeys.

An octagonal shaft was dug to allow water to be drawn without descending the *bāolī*'s steps. It was dug a few feet from the rectangular trench, as was normal .

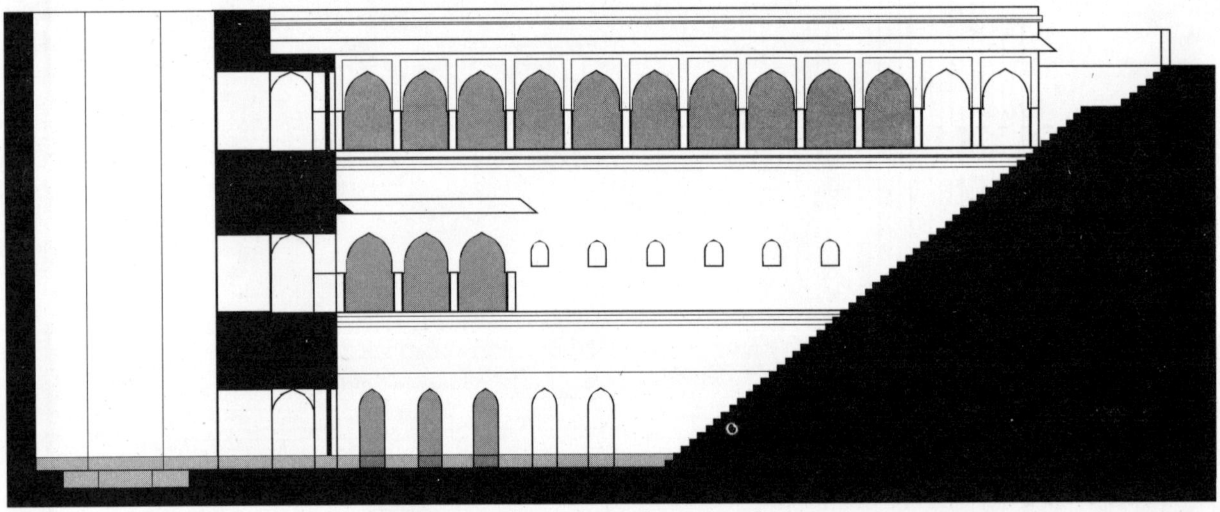

**Figure 43:** Rajoṅ-kī Bāolī, sectional view

[59] The plan is redrawn using Adobe PageMaker 6.0 from the Internet.

## Plate 49: Rajoṅ-kī Bāolī plan

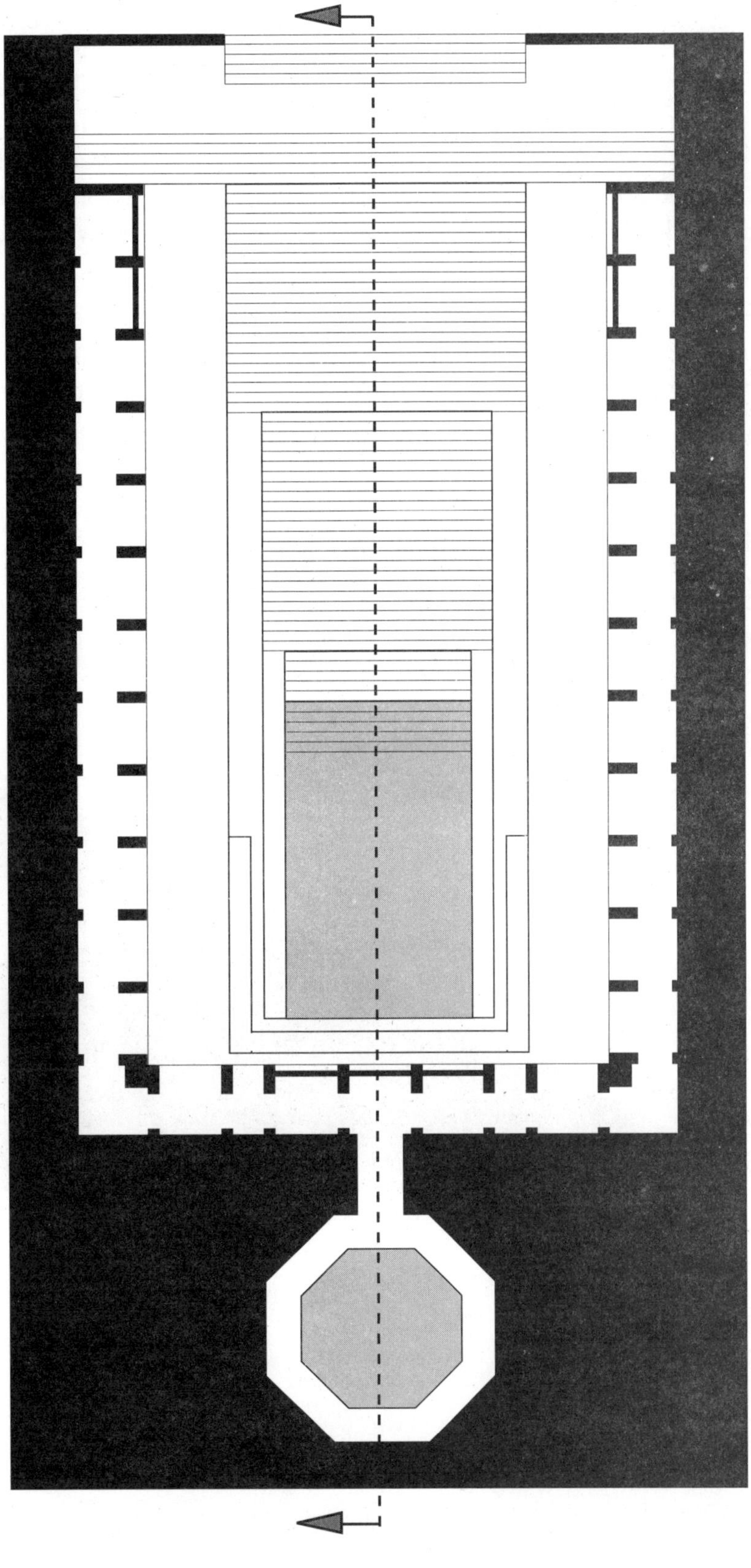

# East Well (Bāolī), Rohtās, Pakistan (a. CE 1540)

Rohtās, in present-day Pakistan, was an important and strategic town during the reigns of the early Mughals of the Timurid Dynasty and the Sūris (**Appendix D**). They lavished great amounts of money on its fortifications and infrastructure. Like most of this region of the subcontinent, rain came only during the monsoon, even though there were numerous streams from glacial runoff. None the less, there were long, dry months for which proper access to water was required for this fortified city built high on a rocky plateau to survive and prosper. The East Well[60] or Bāolī constructed in the south-eastern part of the city and, along with two other *bāolīs*, supplied the city with needed drinking water.

Its plan was typical. A circular shaft was sunk deep into the soil and to the west a slanted trench was excavated. A series of flights of stairs with landings was laid in ever-decreasing widths to parallel the narrowing of the five storeys. The earth into which the trench was dug was of such questionable stability that regularly-placed cross-braces were employed to maintain the trench's viability (**Figure 44**).

Unlike earlier *bāolī*, particularly those found in Gujarat, the East Well or Bāolī of Rohtās was elementally utilitarian. Here, no arcaded levels providing shade for the populous are to be found. This seems to reflect the strictly military, initially defensive quality for which the citadel of Rohtās was originally destined. The East Well or Bāolī of Rohtās was a source of water in a semi-arid land, pure and simple, *en plan*!

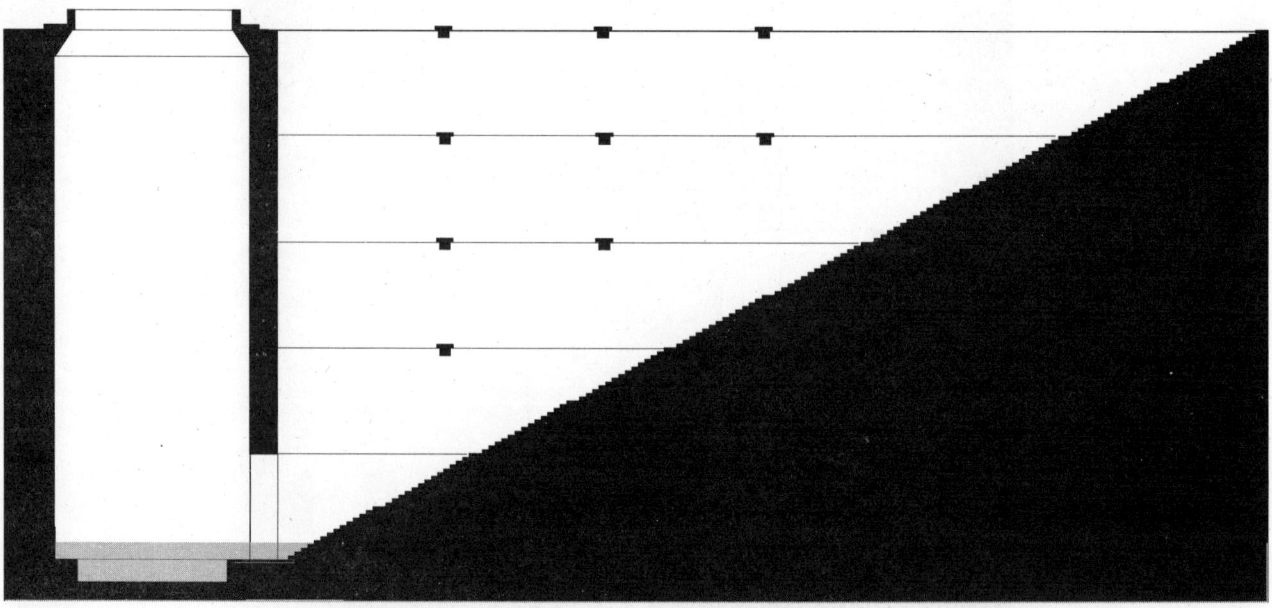

**Figure 44:** East Well (Bāolī), sectional view

60 The plan is redrawn using Adobe PageMaker 6.0 from the Internet.

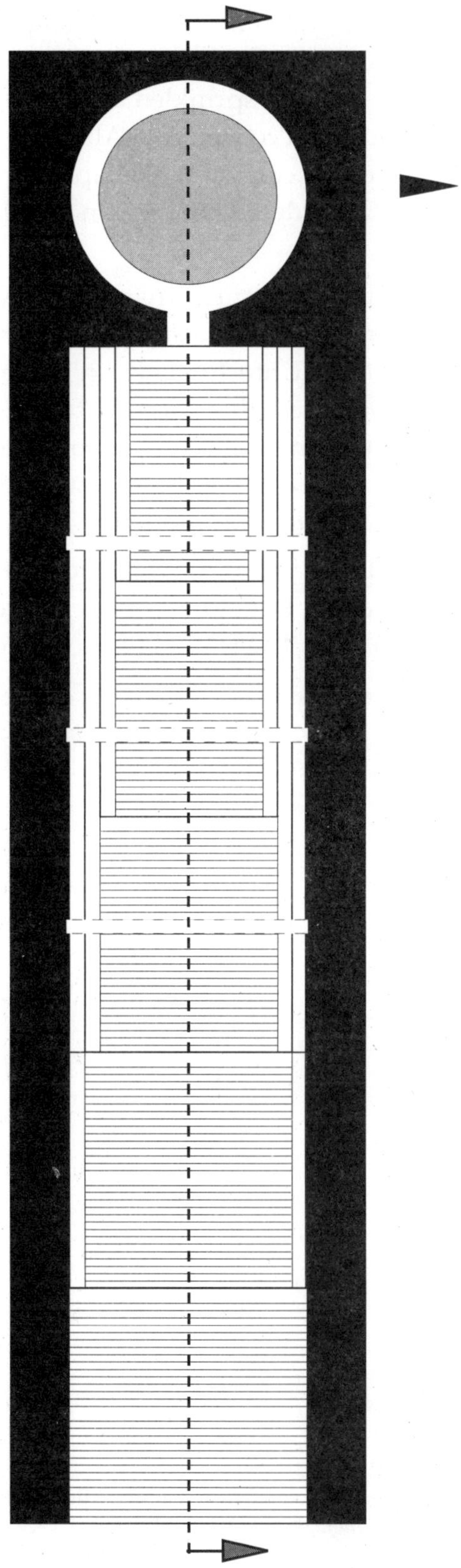

# Fatehpur Sīkrī, Uttar Pradesh (*c.* CE 1569-85)

From its inception to its abandonment, Akbar the Great's Fatehpur Sīkrī[61] was, and still is, a monument of imperial power and innovative design. Constructed on a low plateau, it was not to be another of the typical *gaṛh mahal* (citadel) that are seen sprinkled about the vast subcontinent. Rather, it was an imperial residence-cum-city of thoughtful conception. All of the buildings in the imperial compound are oriented to the cardinal points of the compass. No haphazard, meandering alleys and streets are found here. Fatehpur Sīkrī reveals an ordered conception which flowed from an obviously ordered mind.

Along its northern border meandered a small river which was to be the source of much needed water for the city. Dams were constructed, *tālāo*s were built to provide sources of water and *bāolī*s were dug to augment the river and the *tālāo*. Unfortunately, in the latter case — i.e. the *tālāo* — the dam gave way and was not replaced. Doubtless this was due to the fact that the suburbs of the city had ballooned to such an extent that the existing water supply could not sustain the burgeoning population in this semi-arid locale. As quickly as the city was built, it was abandoned and the spectacular dream of Akbar, the Great remains forsaken, but not forgotten.

Several *bāolī*s were constructed as a source of water for the Imperial city. One large, octagonal well was dug to the north of the famed Buland Darwāẓā (**Plate 51, X**) and another adjacent to the Hāthī Pol or Elephant Gate (**Plate 51, W**). The Hāthī Pol Bāolī is relatively unique as it does not follow traditional *bāolī* design (**Plate 53**).

| | |
|---|---|
| **A** | Court of Public Audience |
| **B** | Dīwān-ī-'Ām, Public Audience Hall |
| **C**[1] | Public Entrance |
| **C**[2] | Imperial Family Entrance |
| **D** | Garden of the Turkish Sultān |
| **E** | Ekastambha Prāsāda (aka. Dīwān-ī-Khās) |
| **F** | Āṅkh-ī-Michonī (aka. Āṅkh-ī-Khās, Daulat Khānā) or Dīwān-ī-Khās |
| **G** | Pacīsī Court |
| **H** | House of the Turkish Sultān |
| **I** | Madrasā |
| **J** | Khwābgāh, Emperor's Quarters |
| **K** | Anūpa Tālāo (Cār Caman) |
| **L** | Daftar Khānā (Admin. and Archives) |
| **M** | Hospital and Garden |
| **N** | Pañca Mahal |
| **O** | Sunahrā Makān (Miriam's House) |
| **P** | Miriam's Garden |
| **Q** | Jodhā Bāī Palace |
| **R** | Hawā Mahal |
| **S** | Nagīnā Masjid |
| **T** | Bīrbal Bhavan (House of Rājā Bīrbal) |
| **U** | Stables for Camels and Horses |
| **V** | Hammām |
| **W** | Hāthī Pol or Elephant Gate Bāolī |
| **X** | Buland Darwāẓā Bāolī |
| **Y** | Tālāo |

[61] The plan is redrawn using Adobe PageMaker 6.0 from George Michell, *The Royal Palaces of India*, London: 1998, p. 114; Satish Grover, *Islamic Architecture in India*, New Delhi: 1996, pp. 135-36 and Satish Grover, *Masterpieces of Traditional Indian Architecture*, New Delhi: 2004, pp. 96-97.

## Plate 51: Fatehpur Sīkrī, plan of royal compound

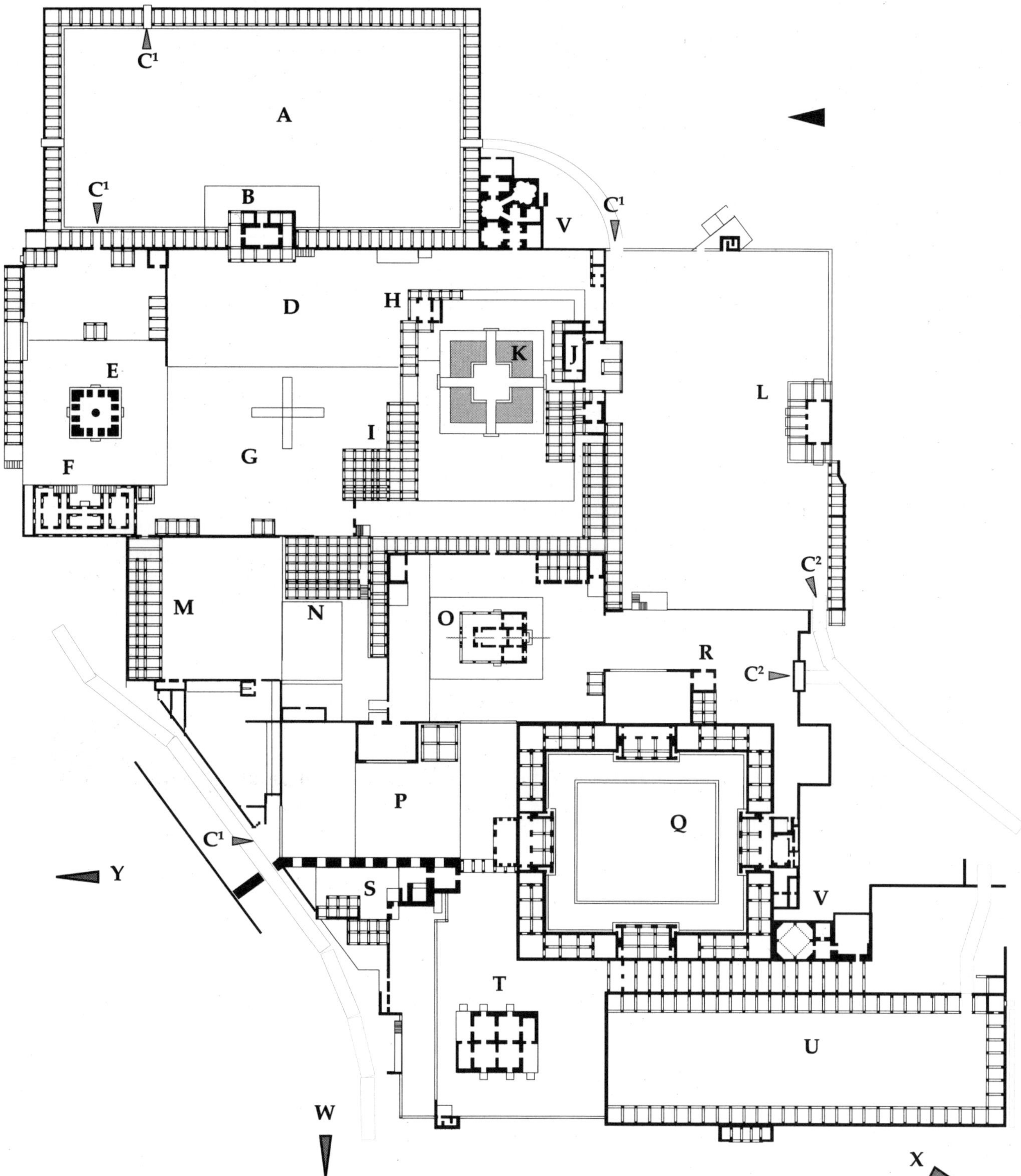

# Anūpa Tālāo (Cār Caman), Fatehpur Sīkrī, Uttar Pradesh (a. CE 1569)

The Anūpā Tālāo[62] (Cār Caman) of Fatehpur Sīkrī may be viewed as one of the ultimate, iconic expressions of power and wealth. Akbar the Great, being not pleased with the congestion of the Lal Qila of Agra, as well as befitting his supreme puissance and unlimited affluence, constructed his imperial city, Fatehpur Sīkrī. It was to become one of the ultimate examples of one man's power, wealth and aesthetic sensibility.

Built to the west and a little north of Agra, on a plateau, the site seemed appropriate as there was a small river that bordered the eastern area of the city. Soon Fatehpur Sīkrī sprung up and the imperial court moved there. In addition, thousands of others — i.e. government workers, servants, retainers, shopowners and those who felt it was beneficial to be near the imperial presence — migrated to that locale. Despite a large *tālāo* being constructed to the east of the city and many smaller reservoirs, the drain of thousands of people upon the water resources was simply too much! Within a short time Fatehpur Sīkrī was abandoned.

One of the lesser known components of Fatehpur Sīkrī, but one which holds immense iconic significance is the Anūpa Tālāo or the Cār Caman. It is within its own enclave, bounded by the House of the Turkish Sultān, the Madrasā and the Khwābgāh (**Plate 51, H, I-J**). In comparison to the whole, it appears almost insignificant. Neither tank nor *bāolī*, here is found a pool, a square pool with a raised, railed island in the Centre (**Figure 45**). This island is connected to the outer edges of the pool by four bridges at the cardinal points of the compass. Open to the sky, it may well have possessed a temporary canopy to shade the emperor, for this retreat was his alone. He was able to relax on an island surrounded by water. While others may have sought this life sustaining liquid for their existence, Akbar the Great utilized it for his pleasure. What more imperious icon could there be in a land where water was at a premium, more precious than gold?

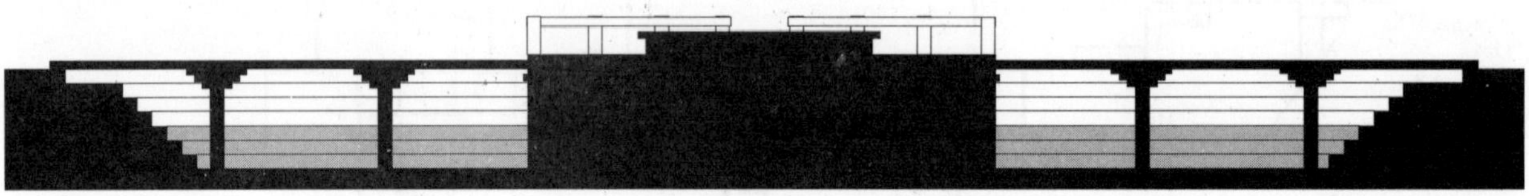

**Figure 45:** Anūpa Tālāo (Cār Caman), Fatehpur Sīkrī, sectional view

62 The plan is redrawn using Adobe PageMaker 6.0 from George Michell, *The Royal Palaces of India*, London: 1998, p. 114, Satish Grover, *Islamic Architecture in India*, New Delhi: 1996, pp. 135-36 and Satish Grover, *Masterpieces of Traditional Indian Architecture*, New Delhi: 2004, pp. 96-97.

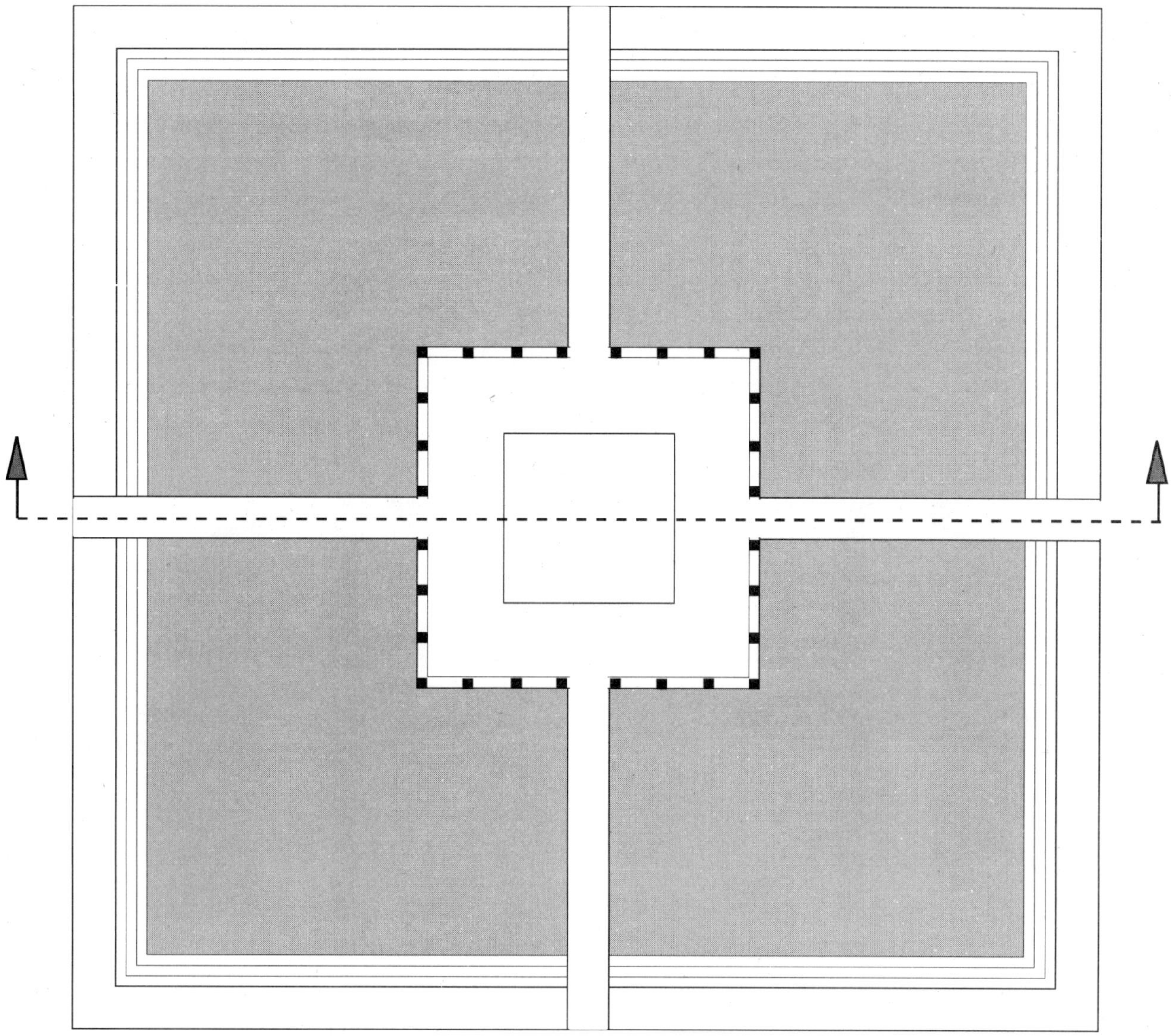

# Elephant Gate Bāolī, Fatehpur Sīkrī, Uttar Pradesh (a. CE 1569)

South-east of Fatehpur Sīkrī's imperial residential complex lies the Elephant Gate Bāolī.[63] It is a stepped well, but unique in design, differing from the more common shaft-trench form. Akbar the Great experimented successfully with a number of architectural forms and it is evident that this *bāolī* was one of them. The plan of the Elephant Gate Bāolī is essentially taken from the design of the Rudabāī-nī Wāv (**Plate 19**) and the Dādā Harīr Wāv (**Plate 20**) in that the double, spiral stairs were moved and incorporated on either side of the shaft. This plan eliminated the trench which was obviously deemed superfluous here.

In this *bāolī,* water could be drawn from three locations — i.e., directly from the shaft via the roof by which two sets of stairs were provided for access, by the circular staircases, and again from the shaft on the ground level within the shaded confines of the octagon-like block (**Figure 46**). The shaded levels of the Rudabāī-nī Wāv and the Dādā Harīr Wāv were incorporated in the enclosed, covered octagon-like block. Not only did this design accomplish all the aspects of the Gujarat *wāvs*, it was considerably less expensive to construct since the trench was eliminated.

**Figure 46:** Elephant Gate Bāolī, Fatehpur Sīkrī, sectional view

[63] Redrawn using PageMaker 6.0 from the Internet. There are a number of photographs of the Elephant Gate Bāolī found on the Internet, but there were no sources found that provided a plan. Even sources devoted to Fatehpur Sīkrī mention the *bāolī*, may even display a picture of the exterior, but no plan! The plan indicated in **Plate 51** is essentially conjectural — i.e. the disposition of the columns within the interior. The outer shape, the thickness of the walls, the size and shape of the two octagonal stair wells and the octagonal shaft are carefully reproduced from the pictures on the Internet and are relatively to scale. The sectional view is also relatively to scale.

## Plate 53: Elephant Gate Bāolī plan[64]

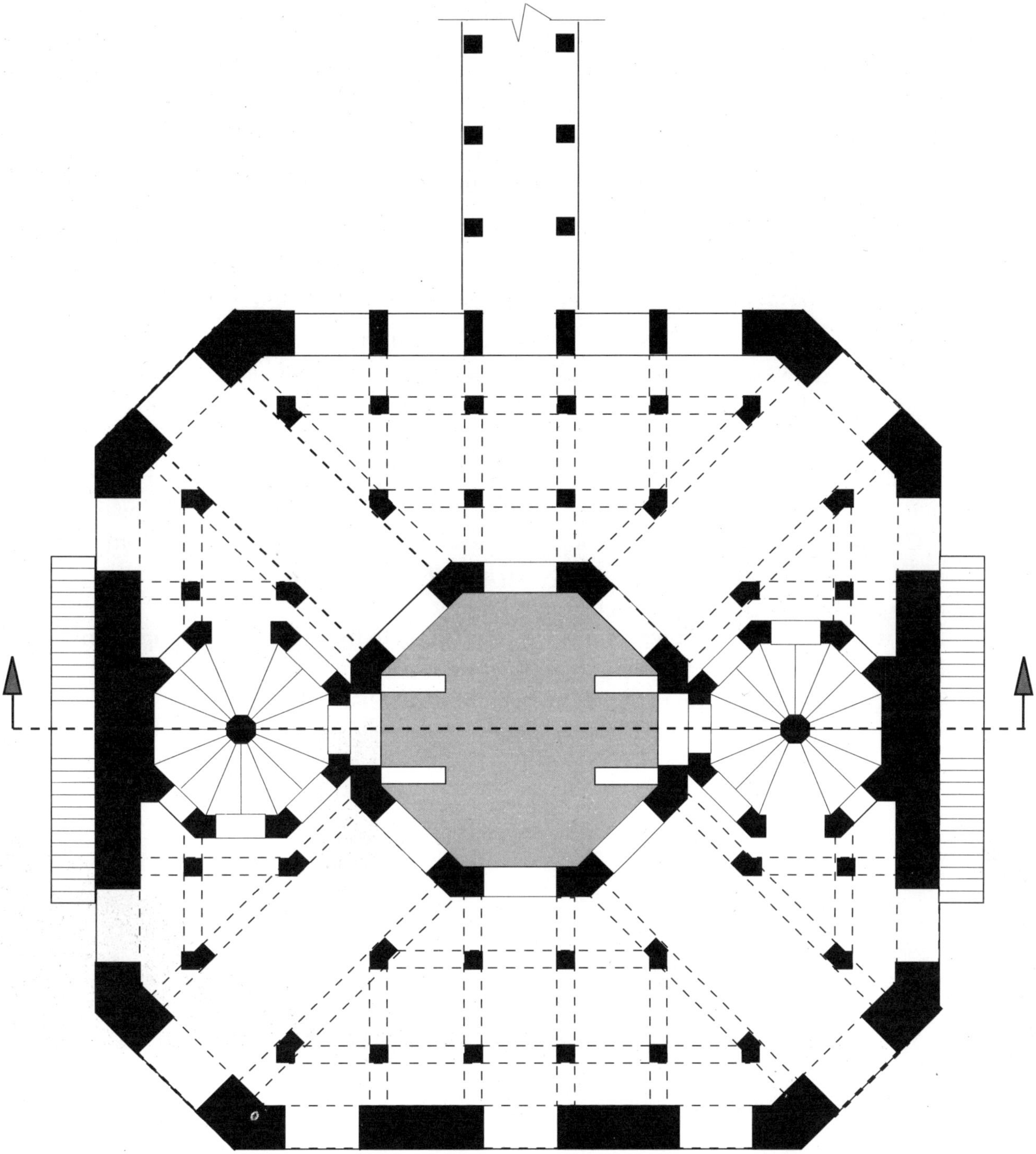

[64] The interior, especially the placement of the supporting columns, is conjectural. However, the central shaft, the two stairwells, the entrances and the exterior stairs are accurate.

# Cānda Bauṛī, Bījāpur, Karnataka (b. 1580)

The Cānda Bauṛī[65] was constructed during the reign of the Sultān Ali 'Adil Shāh I (r. CE 1557-80) (**Appendix D**). The *bāolī* (*bauṛī*) was named after the famed female warrior Cānda Bībī (aka. Cānda Khātūn and Cānda Sultāna) the daughter of Sultān Hussain Niẓām Shāh I (r. CE 1553-65) of Ahmednagar and wife of Ali 'Adil Shāh I of Bījāpur.

Although not a *bāolī* in the true sense, the Cānda Bauṛī is a bathing tank and exemplifies the grandeur that the 'Adil Shāh Dynasty of Bījāpur espoused. Bījāpur required great amounts of water and luckily there was a plenitude of that element available. Two large reservoirs were constructed to the west of the city — i.e. the Rampling Tālāo and the Torweh Tālāo — and the large Begam Tālāo to the south, all of which more than provided for the city's needs. The Cānda Bauṛī benefitted from water piped from the Rampling Tālāo. In addition, the water table in and around Bījāpur was not inordinately deep allowing for numerous tanks, *bāolī*s and wells to be easily dug within the confines of this royal city. Even, the suburbs of Royal Bījāpur saw construction of elaborate tanks and places of relaxation. The delightful Water Pavilion of Kumatagi is an example — neither *bāolī* (tank), but a pavilion of relaxation midst sprinkling water and cool pools (**Appendix C, Plate 89**).

The tank of the Cānda Bauṛī by being enclosed within high walls takes on the character of a Turkish *hammām* (**Figure 47**). This is not unusual since the origin of the 'Adil Shāhī was the brilliant and corrupt Ottoman court. Although it is open to the sky, bathers were protected from the prying eyes of those who might venture by. A pillared porch leads to a narrow doorway and to several flights of wide steps down to the water level. A terrace-ambulatory surrounds the tank and there are rooms in the centre of the north, south and west sides which open on to the tank. This arrangement became the exemplar employed in many bathing tanks in Bījāpur, especially in the larger and more elaborate Tāj Bauṛī constructed during the reign of Sultān Ibrāhim 'Adil Shāh II.

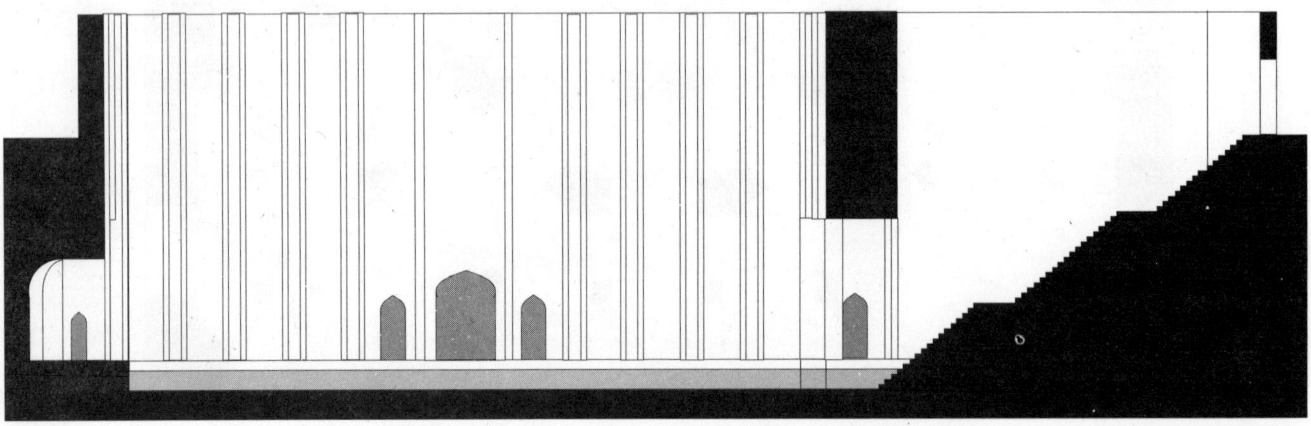

**Figure 47:** Cānda Bauṛī, Bījāpur, sectional view

[65] The plan is redrawn using Adobe PageMaker 6.0 from Henry Cousens, *Bijapur and Its Architectural Remains*, New Delhi: 1996, Plate CXIV and the Internet.

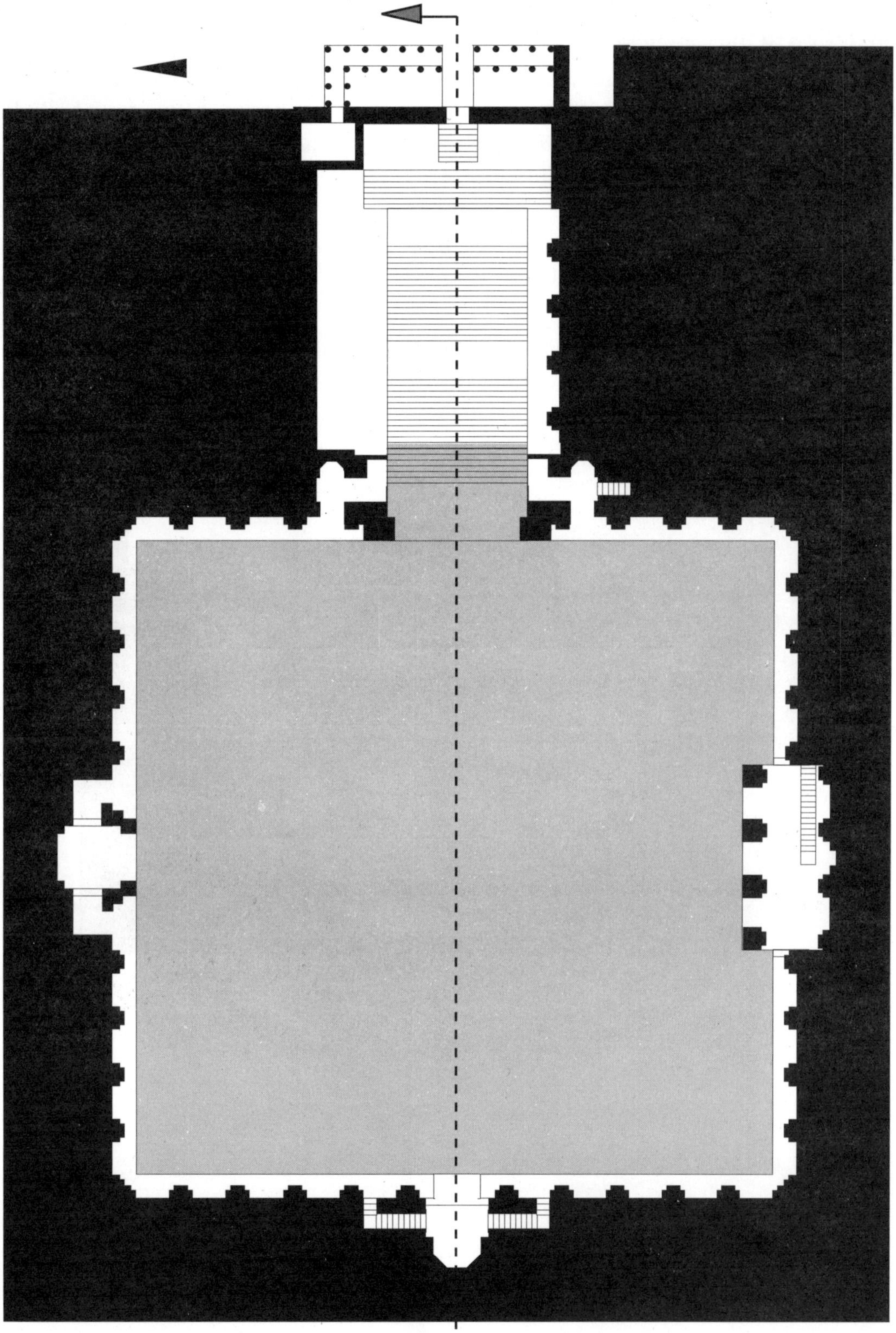

## Amber Tank (Pannā Miā), Amber, Rajasthan (a. CE 1592)

The Pannā Miā or the Amber Tank[66] at Amber in Rajasthan is a *bāolī* set within a semi-arid region. It is similar in design to the Cānda Bāolī of Ābhānerī, also in Rajasthan (**Plate 10**).

The Pannā Miā or the Amber Tank is essentially square, sunk within the surrounding earth and consists of nine levels. Descent is achieved via opposing-step-elements of five steps each. Rather than leading to a narrow walkway as in the Cānda Bāolī, each of the elements are so disposed as to lead directly to the top of the next element. On the second and fifth level, the opposing-step-elements contain a shallow niche, deep enough to allow one to sit and rest, but hardly deep enough to completely shade the occupant (**Figure 48**). On the south side is a multi-storied structure similar to that found in the Cānda Bāolī. It drops all the way to the lower level. On different levels one may enter and get some relief from the sun and heat.

As with the Cānda Bāolī, the Pannā Miā or the Amber Tank is not designed for those with vertigo or acrophobia. Its design and the disposition of the opposing-step-elements is such as to emphasize, even accentuate the depth of the *bāolī*.

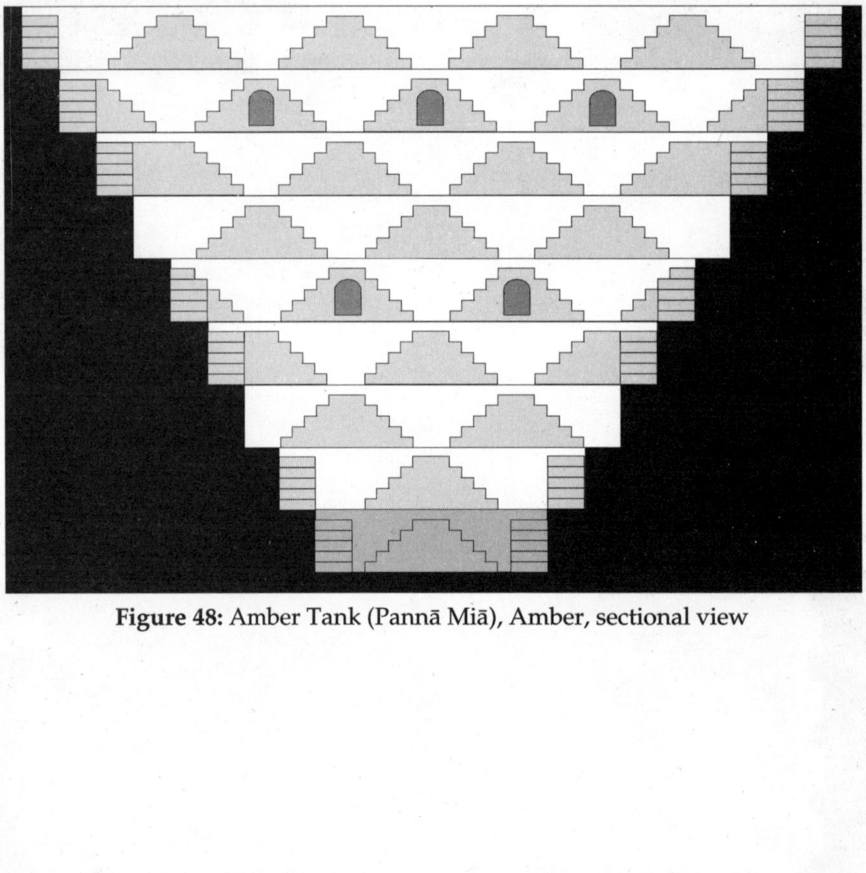

**Figure 48:** Amber Tank (Pannā Miā), Amber, sectional view

[66] The plan is redrawn using Adobe PageMaker 6.0 from ML, p. 110 and from the Internet.

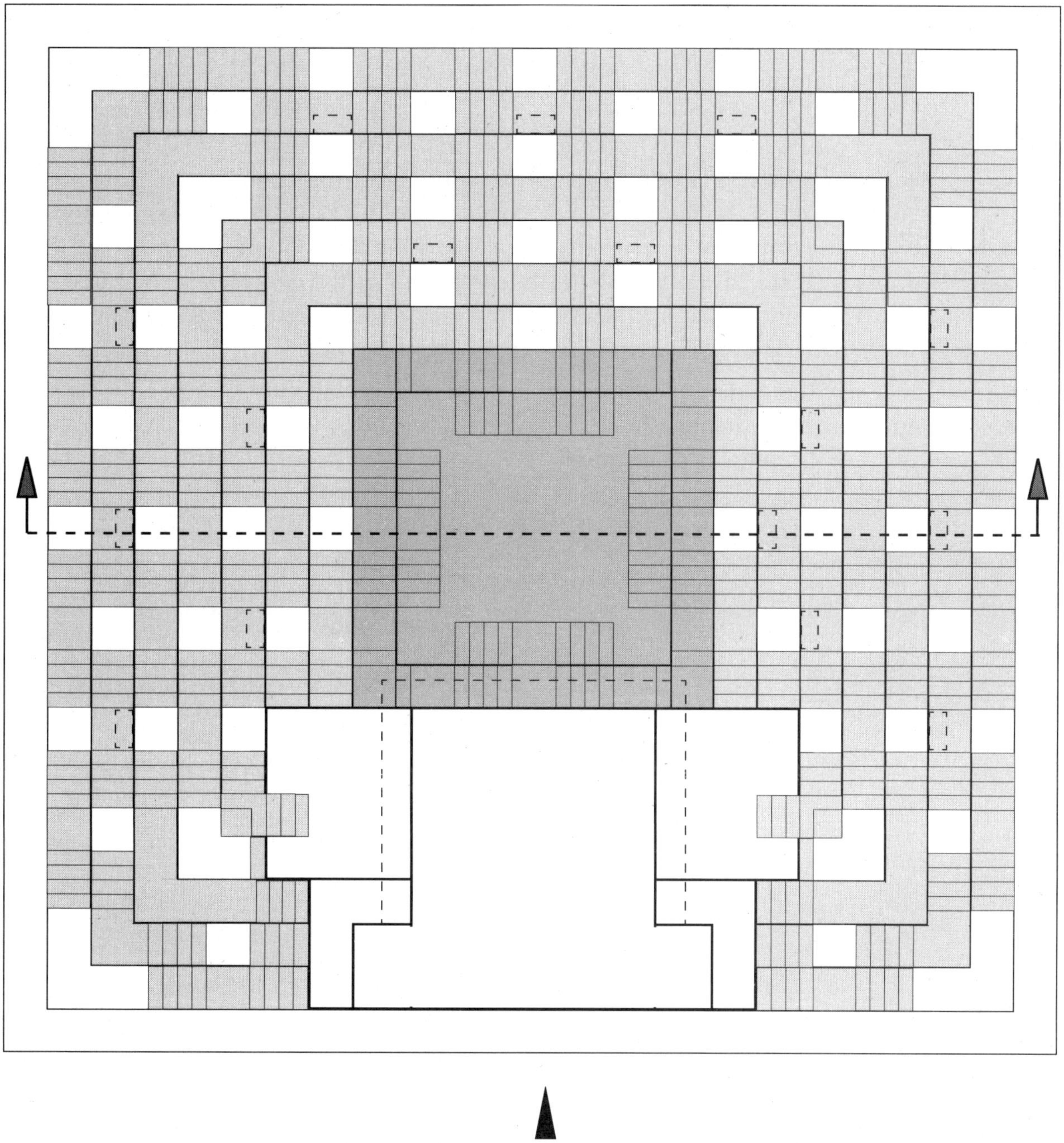

# Mīnākṣī Sundareśvara Temple, Madurai, Tamil Nadu (a. CE 1623)

The Mīnākṣī Sundareśvara Temple[67] of Madurai is one of the most popular and photographed of all the Tamil Nadu Sacred temple cities. It consists of two major shrines in a single complex. The Sundareśvara Shrine (**Plate 56, C**) is in alignment to the towering east and west *gopura* — i.e. the east or Rāya Gopura (**Plate 56, K**) and the west *gopuram* (**Plate 56, L**). Even though this shrine, *en plan*, with five east-west aligned *gopura* appears to be on an axis with the main entrance — i.e. the east *gopura*, it is the South or Mottai Gopuram (**Plate 56, I**) that is considered the main entrance to the complex. The Mīnākṣī Shrine, to the south of the Sundareśvara Shrine (**Plate 56, B**), is aligned to an inner *gopuram* (**Plate 56, N**) as well as an elaborate entrance to the south of the Rāya Gopuram known as the Porch of the Eight Goddesses (**Plate 54, H**). In the north-east corner of the complex is the huge Airakkaḷ Maṇḍapa (Hall of a Thousand Pillars) (**Plate 56, G**). Just inside the South or Mottai Gopuram is the large tank, the Ponthamaraikuḷam (Golden Lotus Tank) (**Plate 56, E**).

Numerous other *maṇḍapa* and minor shrines go to make up this elaborate Temple City. The Mīnākṣī Sundareśvara Temple is a testament to the wealthy, powerful and long reigning Nāyaka Tirumala (r. CE 1623-59) as well as other rulers of this region.

| | |
|---|---|
| **A** | Kulikka Maṇḍapa |
| **B** | Shrine of Mīnākṣī |
| **C** | Shrine of Sundareśvara |
| **D** | Oonjai and Killikoontu Maṇḍapas |
| **E** | Ponthamaraikuḷam (Golden Lotus Tank) |
| **F** | Vīravasantarāya Maṇḍapa |
| **G** | Airakkaḷ Maṇḍapa (Hall of a Thousand Pillars) |
| **H** | Porch of the Eight Goddesses |
| **I** | South or Mottai Gopuram |
| **J** | North Gopuram |
| **K** | East or Rāya Gopuram |
| **L** | West Gopuram |
| **M** | Kambittari Maṇḍapa |
| **N** | Inner Gopuram |
| $\mathbf{O\text{-}O_3}$ | Sundareśvara Gopura |
| $\mathbf{P\text{-}P_1}$ | Mīnākṣī Gopura |

67 The plan is redrawn using Adobe PageMaker 6.0 from Henri Stierlin, *Hindu India: From Khajuraho to the Temple City of Madurai*, 2002, p. 210 and Christopher Tadgell, *The History of Architecture in India*, London: 2002, p. 212.

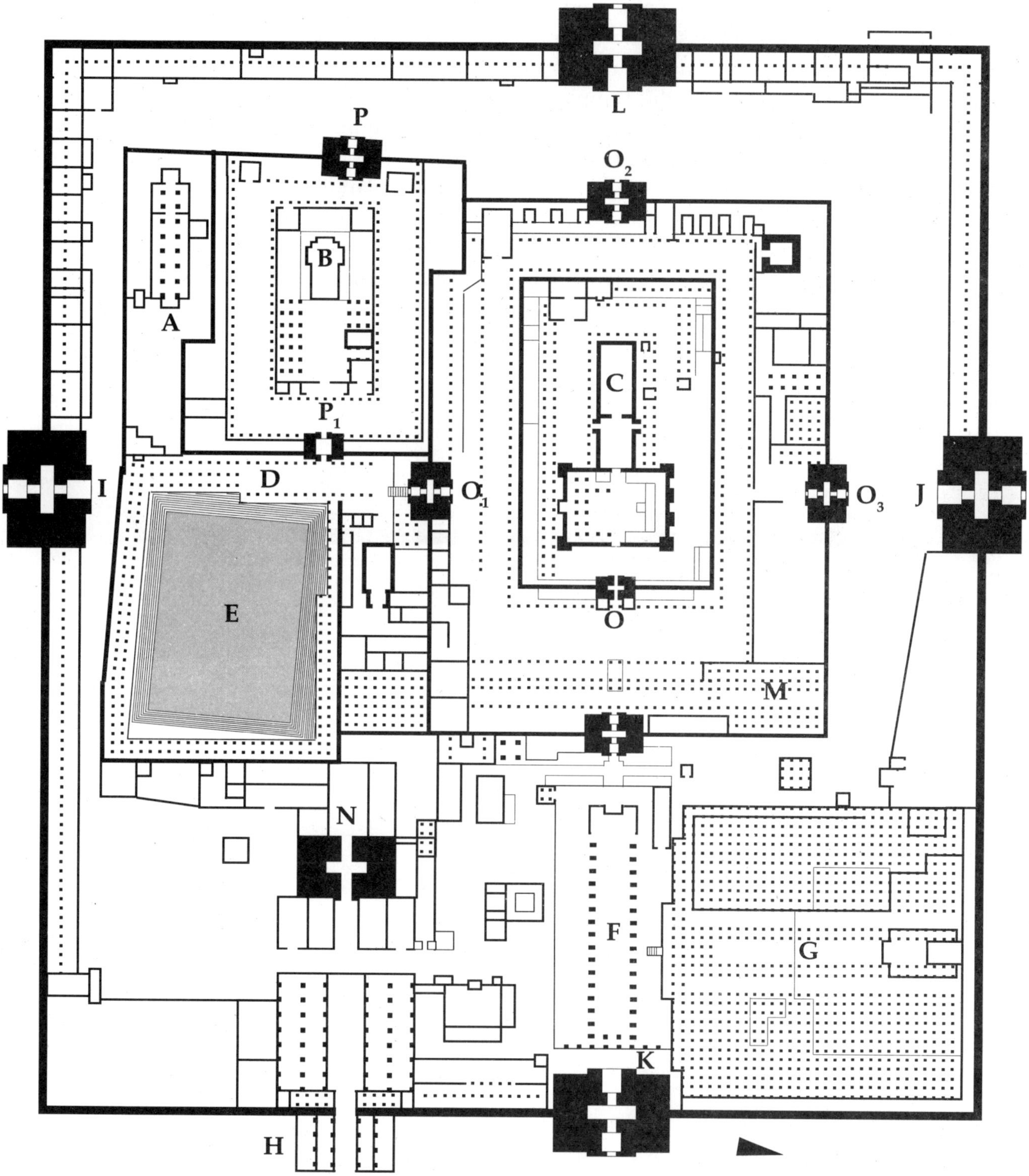
L
P
O2
B
A
C
P1
D
O1
I
O3
J
E
O
M
N
F
G
K
H

## Ponthamaraikuḷam (Golden Lotus Tank)

The Ponthamaraikuḷam[68] or Golden Lotus Tank is a fitting *kuḷam* for the elaborate Mīnākṣī Sundareśvara Temple. Sitting to the east of the southern, main entrance — i.e. the Mottai Gopuram (**Plate 56, I**) — the tank is a relatively large rectangle that is bordered by a covered, double colonnade. As with many of the other tanks associated with the Temple City of Tamil Nadu, this covered colonnade provides the pilgrims a respite from the hot sun. An interesting feature of the Ponthamaraikuḷam (Golden Lotus Tank) is that it is neither aligned to Sundareśvara Shrine nor to the Mīnākṣī Shrine, but to the cardinal points of the compass. This creates a rather eccentric plan to the tank and causes a number of the steps on the north and east sides to become abbreviated as they reach and seemingly melt into the base of the covered colonnade (**Figure 49**). This alignment and the effect it has upon the steps allows uneven access to the tank. The plan of the Ponthamaraikuḷam with its alignment disturbs the otherwise perpendicular axiality of the whole complex.

In the centre of the tank rises an elaborate pillar, a *stambha* set upon a square base. This may refer to the *axis mundi*.

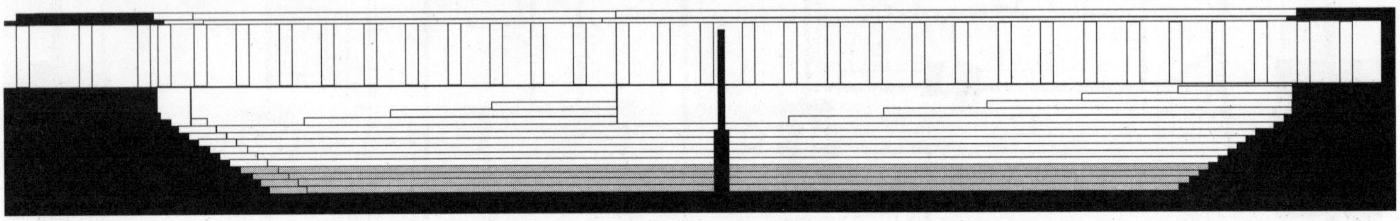

**Figure 49:** Ponthamaraikuḷam, Mīnākṣī Sundareśvara Temple, sectional view

[68] The plan is redrawn using Adobe PageMaker 6.0 from Henri Stierlin, *Hindu India: From Khajuraho to the Temple City of Madurai*, 2002, p. 210 and Christopher Tadgell, *The History of Architecture in India*, London: 2002, p. 212.

## **Plate 57:** Ponthamaraikuḷam (Golden Lotus Tank) plan

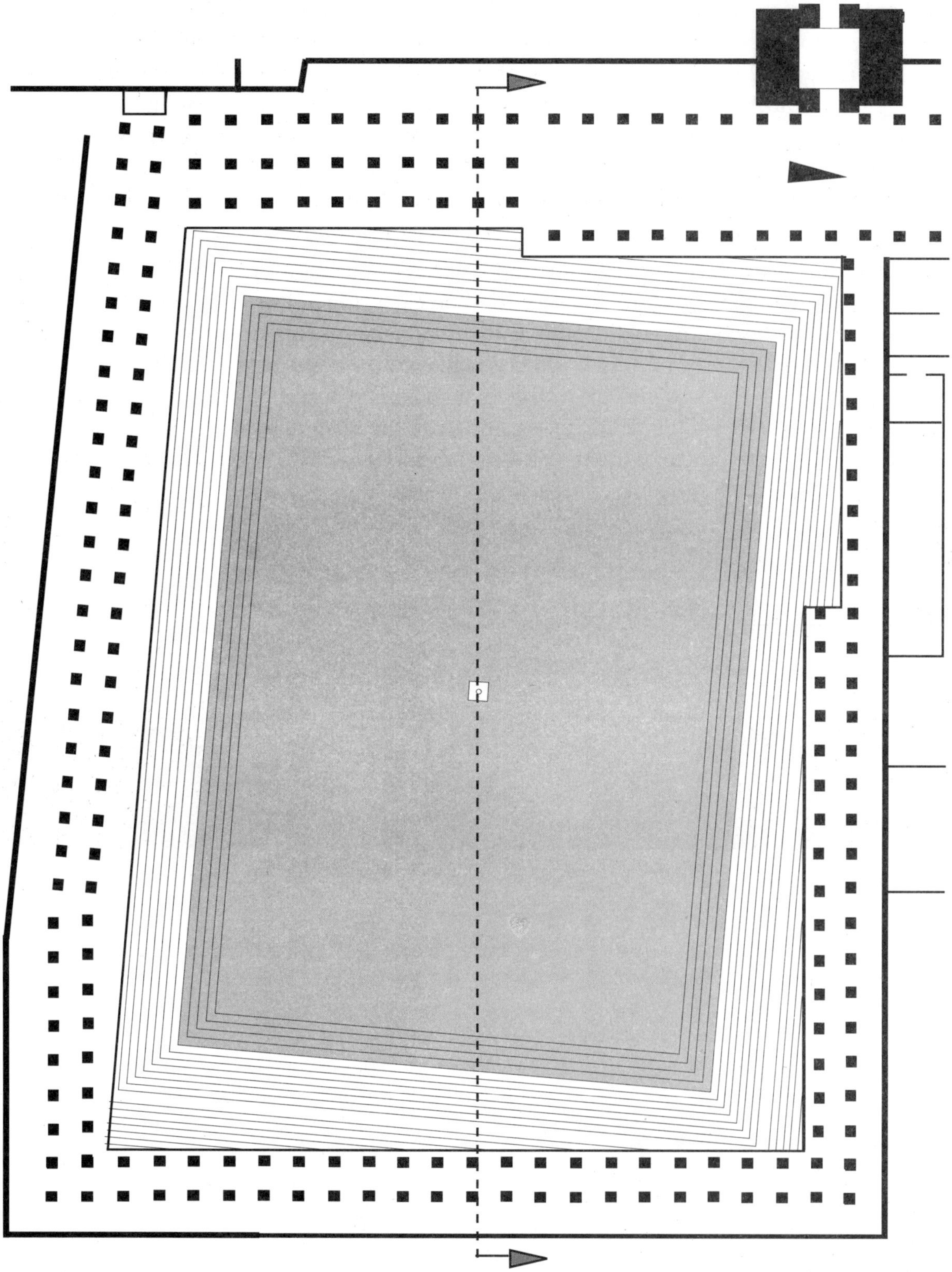

## Fatehpur Bāwḍī, Fatehpur, Rajasthan (a. CE 1700)

The former home of the Nawābs of Fatehpur, members of the Kāimkhānī Dynasty (**Appendix D**), is the site of the impressive and elegant Fatehpur Bāwḍī.[69] The present identification of *bāwḍī* is a bit erroneous as the Fatehpur Bāwḍī is more a tank for bathing than a source of drinking water. Built in the arid region of northern Rajasthan where the availability of adequate drinking water is a problem during much of the year, the *bāwḍī* appears as a luxury for the populous. Luckily, however, the ground water here is not unusually deep, making such a venture feasible. Doubtless, there were *bāolī*s in and around Fatehpur which supplied the populous with sufficient drinking water. In addition, the Fatehpur Bāwḍī is nearly identical to the Māṇḍavā Stepwell (*bāwḍī*) in Māṇḍavā, Rajasthan.

The Fatehpur Bāwḍī is square, *en plan*, and entered via a flight of stairs through an impressive pavilion. Four of these pavilions are found centred on each side of the outer wall, creating a pleasing balance. Once inside the confines of the outer wall, broad terraces step down to the water level (**Figure 50**). Access to the upper storey of the pavilions is achieved by stairs just inside the entrances that lead up to the surrounding terraces and up to the level of the enclosed pavilions. There, visitors to the Fatehpur Bāwḍī may take shelter from the sun and avail themselves of any breezes that pass through these open spaces. In the centre of the tank rises a hexagonal *nirazhi maṇḍapa* with a flight of narrow steps leading up to the columned, roofed *śālā.*

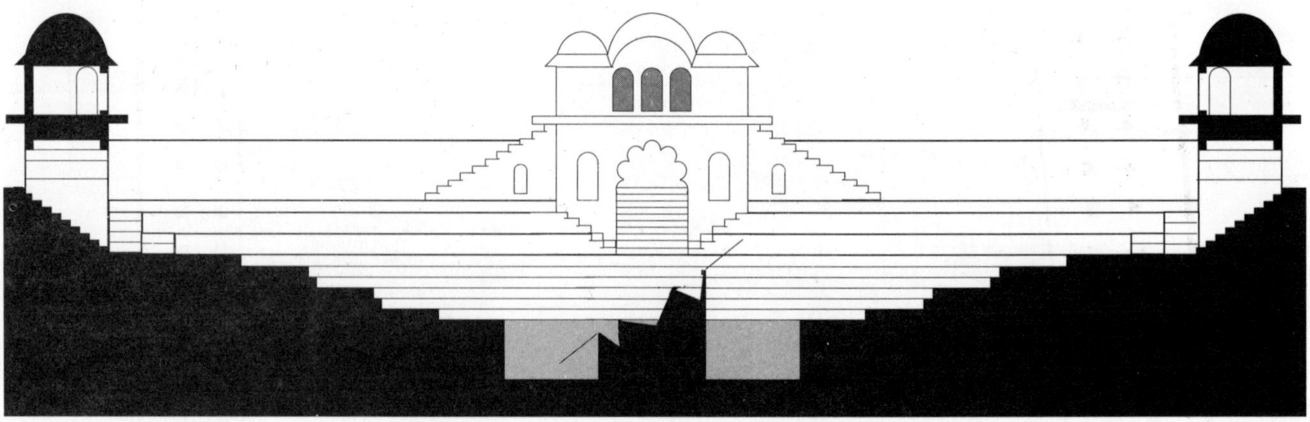

**Figure 50:** Fatehpur Bāwḍī, Fatehpur, sectional view

69 The plan is redrawn using Adobe PageMaker 6.0 from the Internet.

## Plate 58: Fatehpur Bāwḍī plan

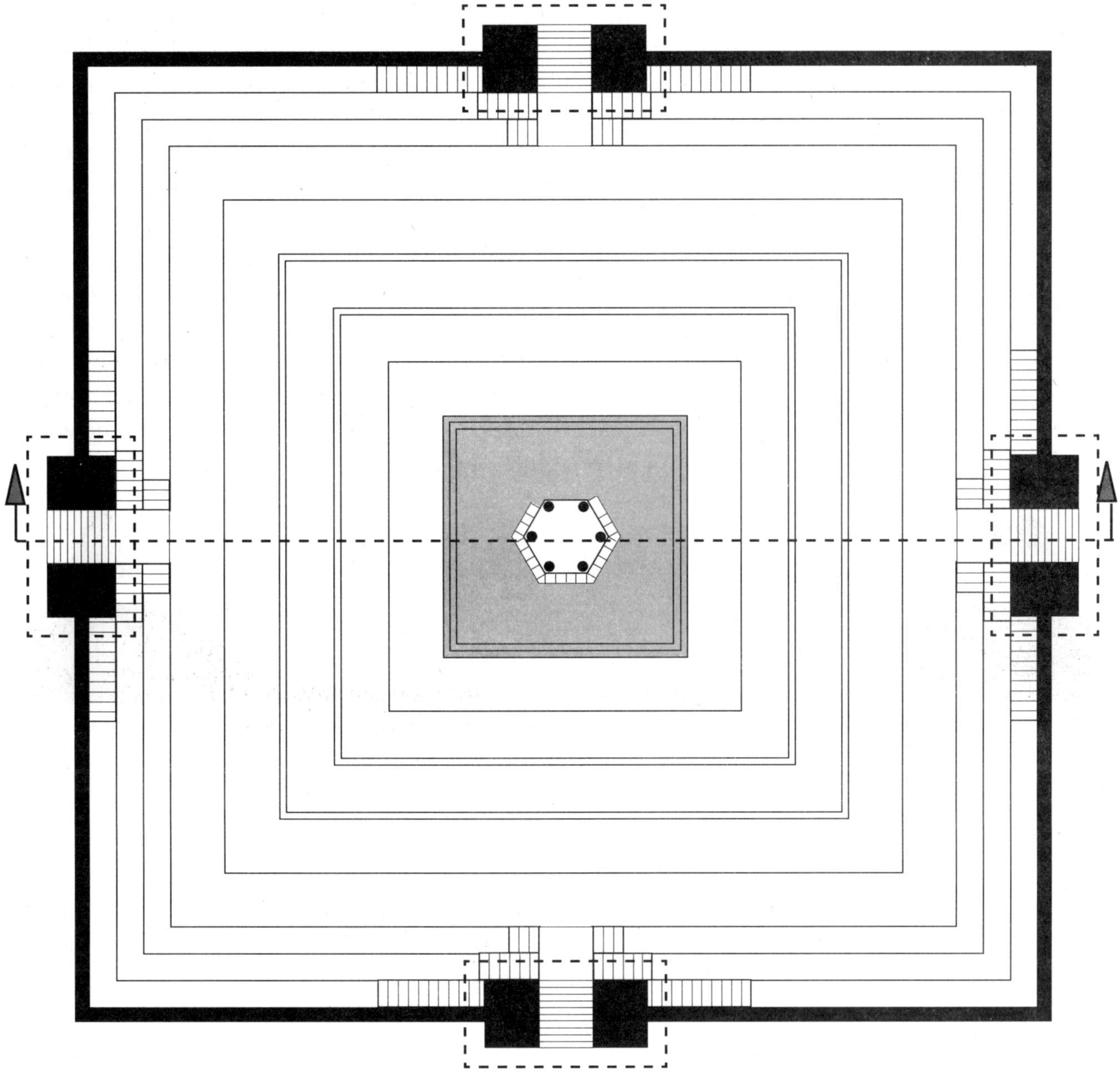

Śravaṇabeḷagoḷa rests between two hills, the Candragiri and the Vindyagiri. It has long been a pilgrimage site in this part of Karnataka, particularly for the Jains. The town is known for the large Jain statue of Gommaṭeśvara which stands over 57 ft tall as well as several noteworthy temples. In the centre of the town is found the substantial tank known as the Kalyāṇī Kuḷam.[70] It is stated that the tank was undertaken by Cikkadevarāja Woḍeyār (r. CE 1673-1704), Mahārājā of Mysore (**Appendix D**). However he died before the vast tank was completed and the completion was accomplished during the reign of Kṛṣṇarāja I (r. CE 1714-32) (**Appendix D**).

The Kalyāṇī Kuḷam is square, *en plan*, and it is surrounded by a high girdle wall. It is fed via the Hemavatī canal and the tank is provided with an outlet a well. Centred in the cardinal points are the colonnaded entrance and opposite a small *gopura* leading to a shrine (**Figure 51**). On the opposite sides are two small shrines. Unlike many tanks of India the Śravaṇabeḷagoḷa tank is not surrounded with a covered, colonnaded cloister. However, a broad terrace surrounds the descending steps and may well have had an awning to protect the devotees from the blazing sun. Steps are laid down to the water's edge and in the centre is a *stambha*, signifying the centre of the world.

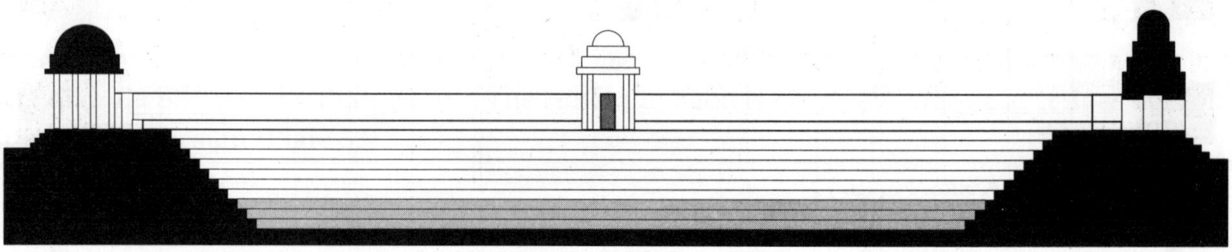

**Figure 51:** Kalyāṇī Kuḷam (Beḷagoḷa), Śravaṇabeḷagoḷa, sectional view

[70] The plan is redrawn using Adobe PageMaker 6.0 from the Internet.

## **Plate 59:** Kalyāṇī Kuḷam (Beḷagoḷa) plan

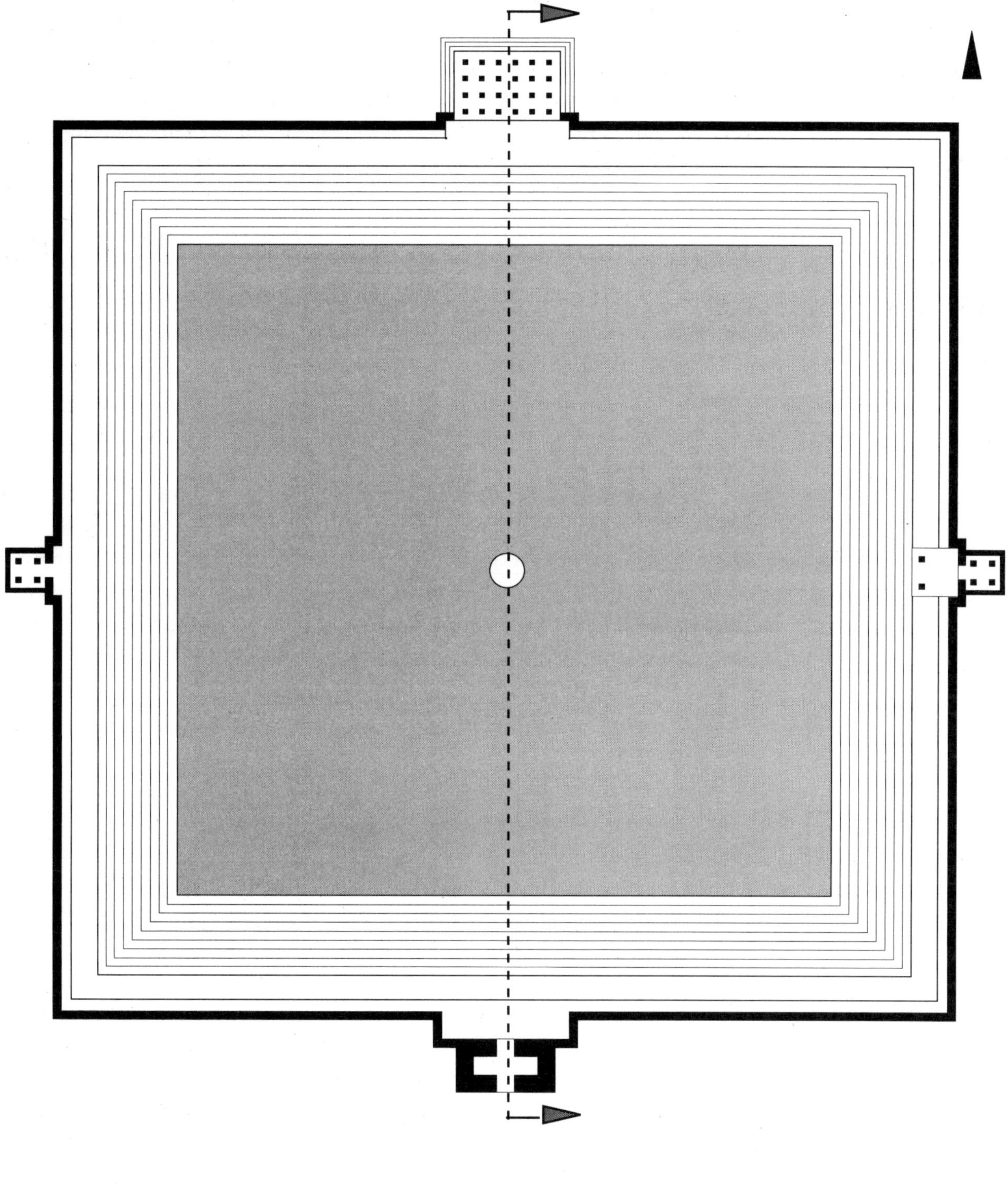

# Shalimar Bagh, Srinagar, Kashmir (a. CE 1610)

The various Muslim rulers, particularly of the Sultanates of Delhi and Gujarat, constructed numerous tanks and step wells for their benefit and the benefit of those they ruled. The Great Mughals were inveterate builders and they too constructed tanks and step wells. However, there was one form which was constructed for the benefit of the Imperial Court alone. Emperor Nūr-ud-Dīn Jahāṅgīr (r. CE 1606-27) built the Shalimar Bagh at Srinagar (**Plate 43**)[71] and the Emperor Shihāb-ud-Dīn Shāh Jahān (r. CE 1628-58), a generation later, constructed the Shalimar Bagh of Lahore (**Plate 44**).[72]

Both the Shalimar Bagh (Garden of the Halls of Love) were crisscrossed with interconnecting waterways, lavish pavilions, commodious bathing pavilions and pools. Here the Imperial Family could relax away from the stifling heat of Agra and Delhi.

Not only were the two Shalimar Baghs lavish locales for relaxing and bathing, they were celestial gardens on earth. The celestial garden with connecting waterways was an Islamic device that the Mughals inherited and took to the epitome of design.

Although the earthly garden is not an exclusive Islamic device, it has been utilized throughout the world as a place of refuge from the maddening crowds and the rigors of daily life. Unfortunately, until recent history, gardens had been the exclusive domain of the royal, the noble and the wealthy — e.g. the Hanging Gardens of Babylon, the Gardens of the Summer Palace outside of Beijing, the Topkapi Gardens of Istanbul, the Gardens of San Souci, the Gardens of Versailles, the Rose Garden of the White House, to mention a few. The nineteenth century saw the establishment of great public gardens — e.g. the Bois de Bologne, Paris; the Central Park, New York; Connaught Place, New Delhi; and Hyde Park, London, again, to mention a few. Here, the common folk could partake of the pleasure formerly confined to the higher classes and meander through bowers, flower beds, shrub-lined walkways, by gurgling brooks and placid ponds.

The *cahār-Bagh*, as we know it, had been employed and refined in Persia and Afghanistan before the Mughals imported it into India where it flourished and was developed even further. The garden was a device that was particularly desirable to Persia, Afghanistan and parts of north-west India. Here, there were and are areas of unyielding desert, the hot season is sweltering and unremitting, and it is an area where dust storms can inundate all. These gardens, these *Bagh*s offered respite from the harshness of the locale. With the exceptions of areas of northern Gujarat and much of Rajastan, India possessed one advantage over Persia and Afghanistan — i.e. the monsoons that regularly and seasonally dumped ample amounts of rain upon the parched soil.[73]

These Garden(s) of Earthly Delights, these *cahār* (fourfold) *Bagh*s (gardens), were but one more way in which wealth and power were manifested in a tangible way by the Mughals. They were and are exemplified in the Shalimar Bagh of Srinagar, the Shalimar Bagh of Lahore, the Bakhsh Bagh and Mahtab Bagh of Delhi's Red Fort, among many. These gardens became one of the hallmarks of the Mughals and ultimately were employed in the ensembles of many major tombs — e.g., the Tomb of Humāyūṅ, the Tomb of Akbar, the Tomb of Jahāṅgīr and the Taj Mahal.

---

71 The plan (**Plate 58**) is redrawn using Adobe PageMaker 6.0 from Henri Stierlin, ed., *Architecture of the World: Islamic India*, Lausanne, n.d., p. 94 and Satish Grover, *Islamic Architecture in India*, New Delhi: 1996, p. 154.

72 The plan (**Plate 59**) is redrawn using Adobe PageMaker 6.0 from ibid. and ibid., p. 155.

73 F.W. Bunce, *Islamic Tombs in India: The Iconography and Genesis of Their Design*, New Delhi: 2004, p. 240.

# **Plate 60:** Shalimar Bagh plan

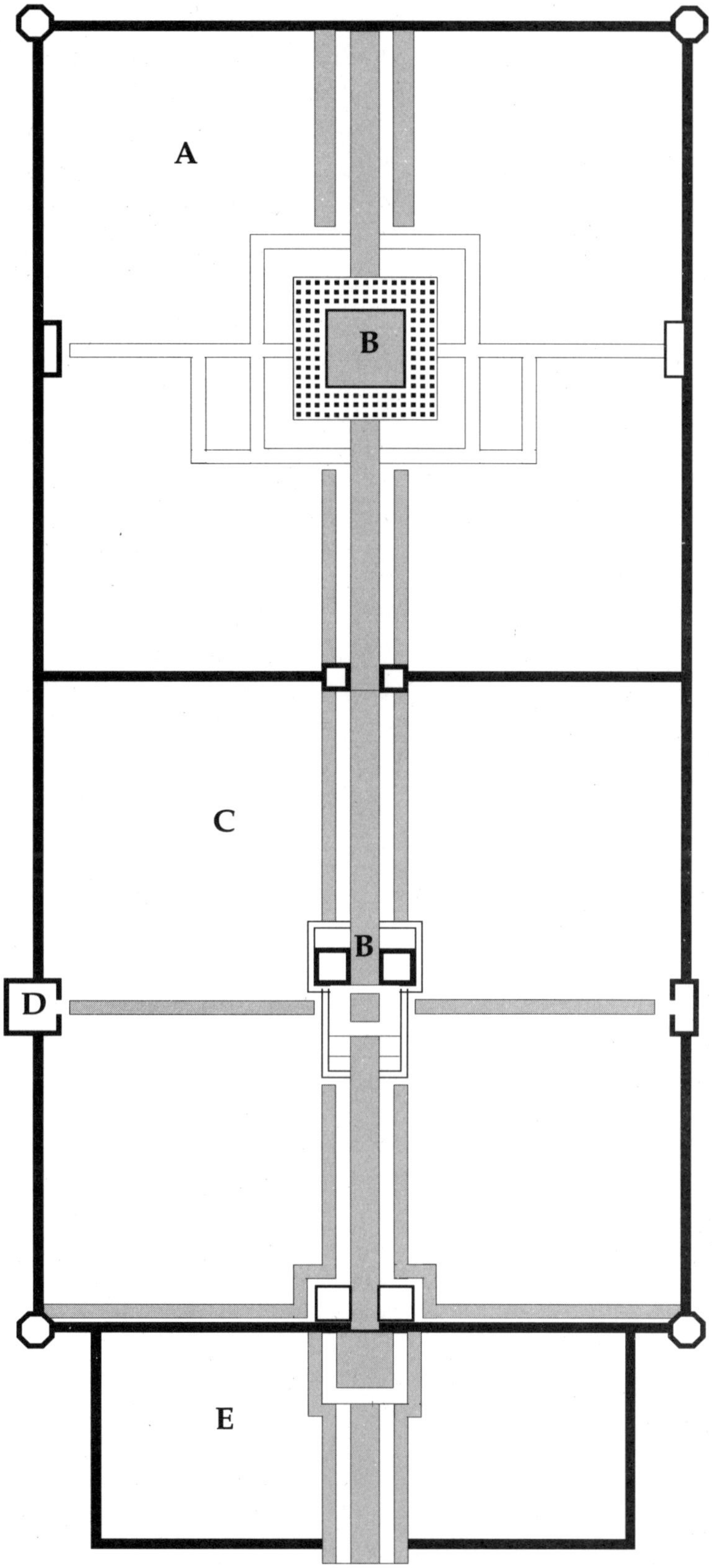


**A** Women's Garden
**B** Bathing Pavilion
**C** Emperor's Garden
**D** Baths
**E** Public Garden

## Shalimar Bagh, Lahore, Punjab, Pakistan (CE 1634)

The acknowledged paragons of the Indo-Islamic mausolea *cahār-Bagh* are the Shalimar Bagh of Srinagar (**Plate 43**) and the smaller Shalimar Bagh of Lahore (**Plate 44**) built by Shāh Jahān. The Lahore pleasure garden consists of three sections — i.e. two squares, one above the other on a terraced hillside, and a lower rectangular section. On the other hand, the Lahore garden, also consisting of the same three elements, but finds the rectangular section bracketed by the square elements. The Srinagar Shalimar Bagh's smaller, rectangular garden is at the foot of the descending terraces of the two squares, and designated as a "public garden". No such designation is attributed to any part of the Lahore *Bagh.* In these gardens are to be found pavilions and *kiosk*s for the delight, pleasure and privacy of the emperor and his party.

The Shalimar Bagh of Srinagar is the simplest in design, although it is larger than its Lahore counterpart. Each of the two square gardens are bisected, cross-axially, creating a Pecaka Maṇḍala of four *pada*s (*cahār*). The Shalimar Bagh of Lahore, on the other hand, is more complex. It, like its predecessor, consists of two square areas that are initially bisected cross-axially forming a primary *maṇḍala* of four *pada*s — i.e. a Pecaka Maṇḍala. Each of these *pada*s is again bisected cross-axially creating a Mahāpīṭha Maṇḍala of sixteen *pada*s each.[74]

[74] F.W. Bunce, op. cit., p. 241.

## **Plate 61:** Shalimar Bagh plan

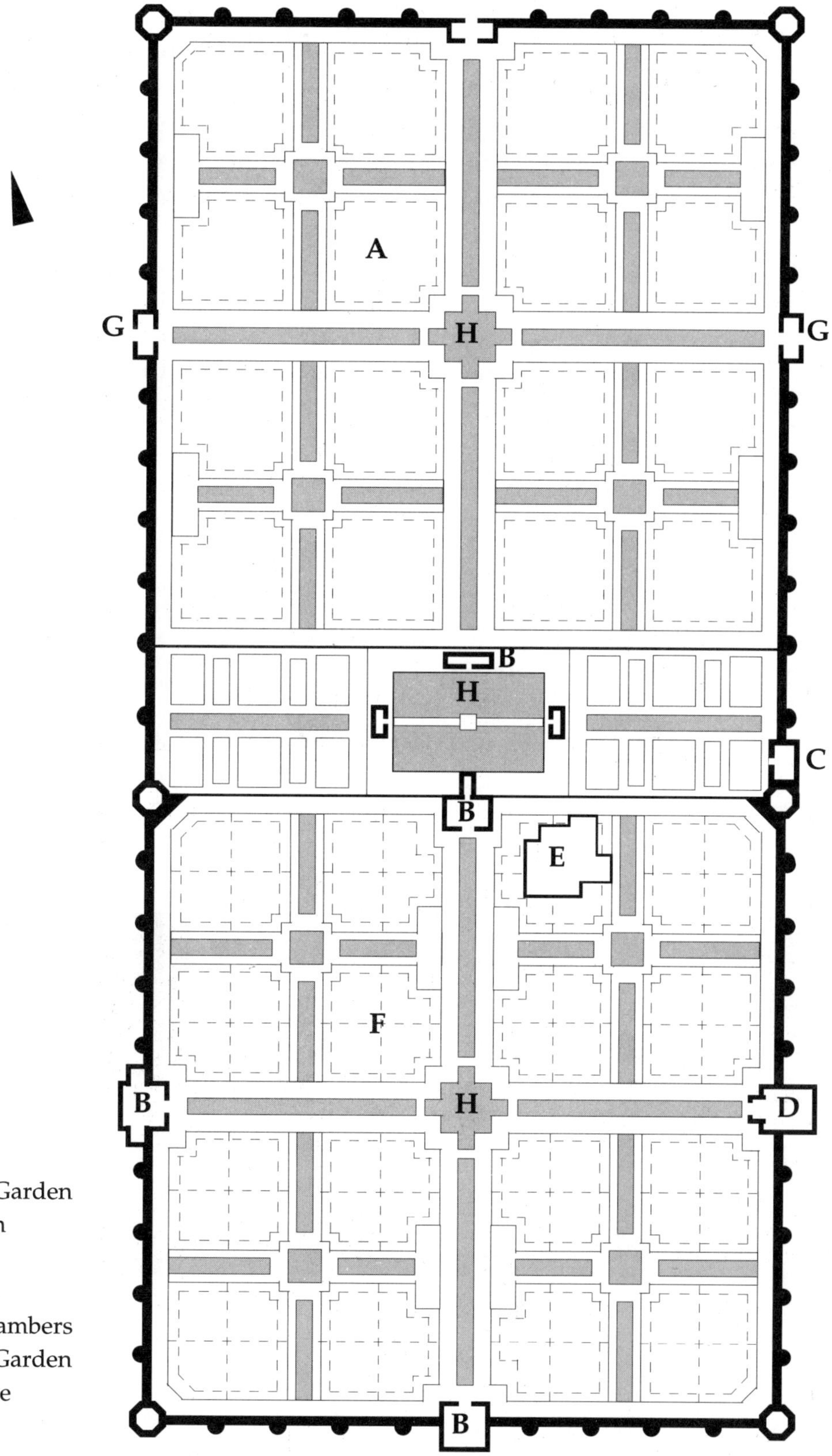


- **A** Upper Garden
- **B** Pavilion
- **C** Bath
- **D** Zenānā
- **E** Bed Chambers
- **F** Lower Garden
- **G** Entrance
- **H** Pool

# Jagadīśa Temple Tank, Udaipur, Rajasthan (a. CE 1651)

The Jagadīśa (aka. Jagadish, Jagdish) Temple Tank[75] is appended to the elaborate temple built by Jagat Singh I (r. CE 1628-52), Mahārāṇa of Udaipur (**Appendix D**). Compared to the lavish temple to which it serves, the tank is relatively simple, even plain in detail.

The Jagadīśa Temple Tank is square, *en plan*, and descends in four levels to the lowest catchment area. The stairways are doubled or opposing as they move downward (**Figure 52**). The first group is centred on each side, the second element descends from the corners, the third centred again on the sides and the last two are single flights which begin in opposite corners and lead down to the lowest basin. The uppermost set of steps numbers nine as they descend to the surrounding terrace-like level from the centre of each side. The next level, the first and only double corner element number ten steps, while the third level, centre element numbers six steps. Although relatively deep, the ratio of rise to tread is ergonomically more acceptable than seen in other tanks (*bāolīs)* of India (**Figure 53**). Its repeat symmetry is reminiscent of the Geometric Tank of Vijayanagara (**Plate 21**).

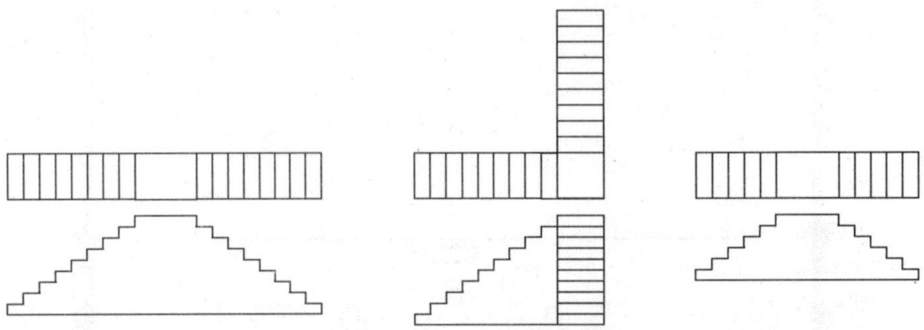

**Figure 52:** Jagadīśa Temple Tank, upper level element (left), second level element (middle) and lower level element (right)

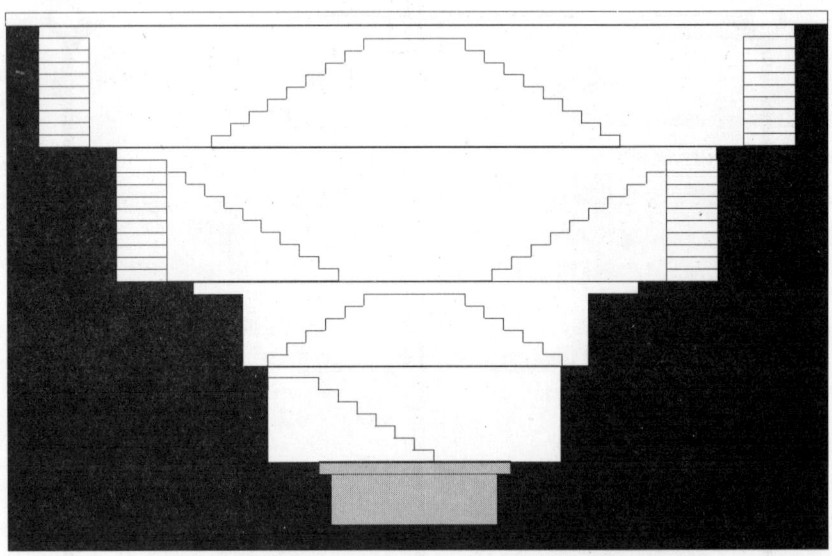

**Figure 53:** Jagadīśa Temple Tank, sectional view

[75] The plan is redrawn using Adobe PageMaker 6.0 from the Internet.

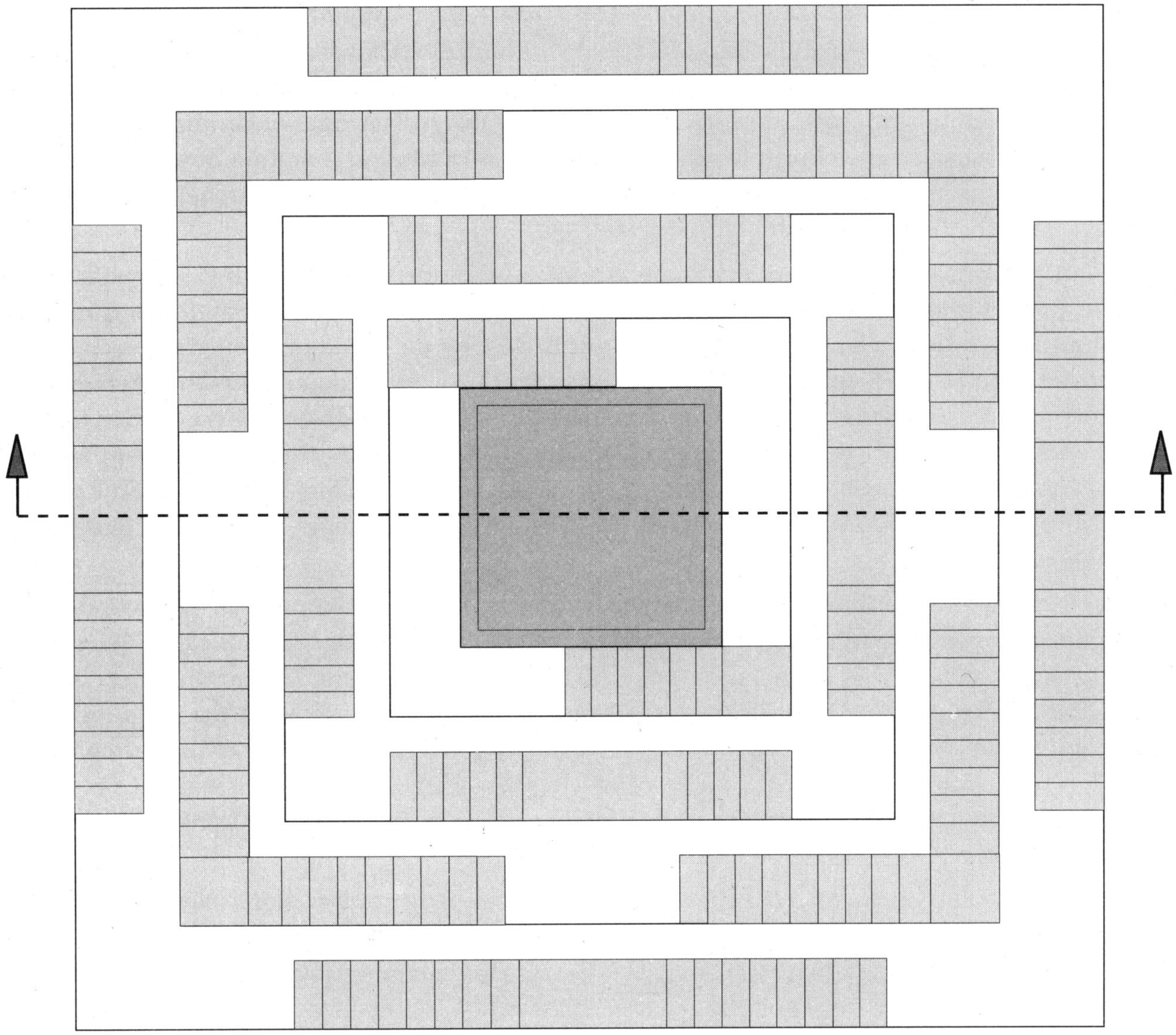

# Garden Palace, Deeg, Rajasthan (CE 1722-63)

The intolerant bigot, Muhiy-ud-Dīn Aurangzeb (r. CE 1658-1707) (**Appendix D**) forced his father Shāh Jahān into retirement and seized the imperial throne for himself. Almost at once the might and glory of the Mughals began to decline at a perceptive rate. Not only was Aurangzeb a bigot, he was also inept in leadership and administration. In his attempt to destroy all vestiges of Hinduism in his realm, he seized and destroyed the Hindu temple of Kesava Deo in Mathurā, among other temples. As a result, the Jāṭs, a group from the Hindu agricultural caste, joined together and rebelled against their overlords. Soon the Jāṭs grew in numbers and strength.

The Jāṭs comprised a number of disparate groups. The Emperor along with the Rājpūts tried to capitalize upon an enmity that had developed between two leaders — i.e. Badan Singh, a Sinsini Jāṭ and Churamun, a Thun Jāṭ. The imperial coalition hoped to destroy the tenuous alliance between the Jāṭs. That was not to be! Badan Singh (r. CE 1722-33) (**Appendix D**) of Sinsini emerged as the overwhelming leader and quickly absorbed his Thun rivals. To placate him, Badan Singh was then awarded the cities of Thun and Deeg. Soon thereafter, the Śrī Sawāī Mahārājā Jai Singh II (r. CE 1727-43) (**Appendix D**) ruler of Amber, wanted a peaceful neighbour to the north-east and declared Sūraj Mal, Rājā of Deeg. The Rājā Sūraj Mal moved his seat of government from Deeg to Bharatapur where he constructed a formidable citadel.

Thus, in the centre of Rājpūt territory, a territory ruled by the proud Rājpūt kṣatriya, a vibrant state arose which was neither part of any of the Rājpūt clans, nor of the ruler/warrior caste. The states of Thun and Deeg were ruled by farmers, vaiśyas who first assumed the title Ṭhākur then were elevated to Rājā, and finally to Mahārājā of Bharatpur. Over the years they constructed the magnificent and lavish Bagh Mahal or Garden Palace[76] from the relatively modest beginnings of the Purānā Mahal (**Plate 61, N**) of Badan Singh. The Garden Palace was in a sense revolutionary as it was not a *gaṛh mahal* (fortified residence). It was enclosed by a wall. That was all! As such it was significant as it proclaimed their strength and fearlessness of the Jāṭs.

The Garden Palace included a number of apparently single storey pavilions (*bhavans*), elaborate, multi-room residences (*mahals*), extensive formal gardens with waterways and pools, a *zenānā* with its own tank (**Plate 61, L**) and two *tālāo* or *sāgara*. The two large tanks (*tālāos*) were incorporated into the design and became the focal point to the east and west — i.e. the large, eccentric shaped Gopāla Sāgara (**Plate 61, G**) to the west and the square Rūpa Sāgara (**Plate 61, H**) to the east. Between these two bodies of water are mirrored the palaces, pavilions and gardens of incomparable delight for the Jāṭ Mahārājās.

All was surrounded by a high girdle wall (**Plate 61, S**), as noted above. This provided visual security for the Royal Court as they visited this retreat from time to time. The Bagh Mahal or Garden Palace could not have fended off any but the smallest and most ill-equipped attack by an equally inept force. It was merely a substantially walled garden holding a number of small palaces and pavilions.

---

[76] The plan is redrawn using Adobe PageMaker 6.0 from George Michell, *The Royal Palaces of India*, London: 1994, p. 190 and G.H.R. Tillotson, *The Rajput Palaces: The Develpoment of an Architectural Style, 1450-1750*, New Delhi: 1999, p. 189.

# Plate 63: Garden Palace plan

**A** Singh Pol
**B** Temple
**C** Nanda Bhavan
**D** Sāvan Bhavan
**E** Gopāl Bhavan
**F** Keśava Badon
**G** Gopāla Sāgara
**H** Rūpa Sāgara
**I** Bhādon Bhavan
**J** Kiśan Bhavan
**K** Sūraj Mahal
**L** Tank
**M** Sūraj Pol
**N** Purānā Mahal
**O** Sūraj Bagh
**P** Hardeva Bhavan
**Q** Śīśa Mahal
**S** Girdle wall

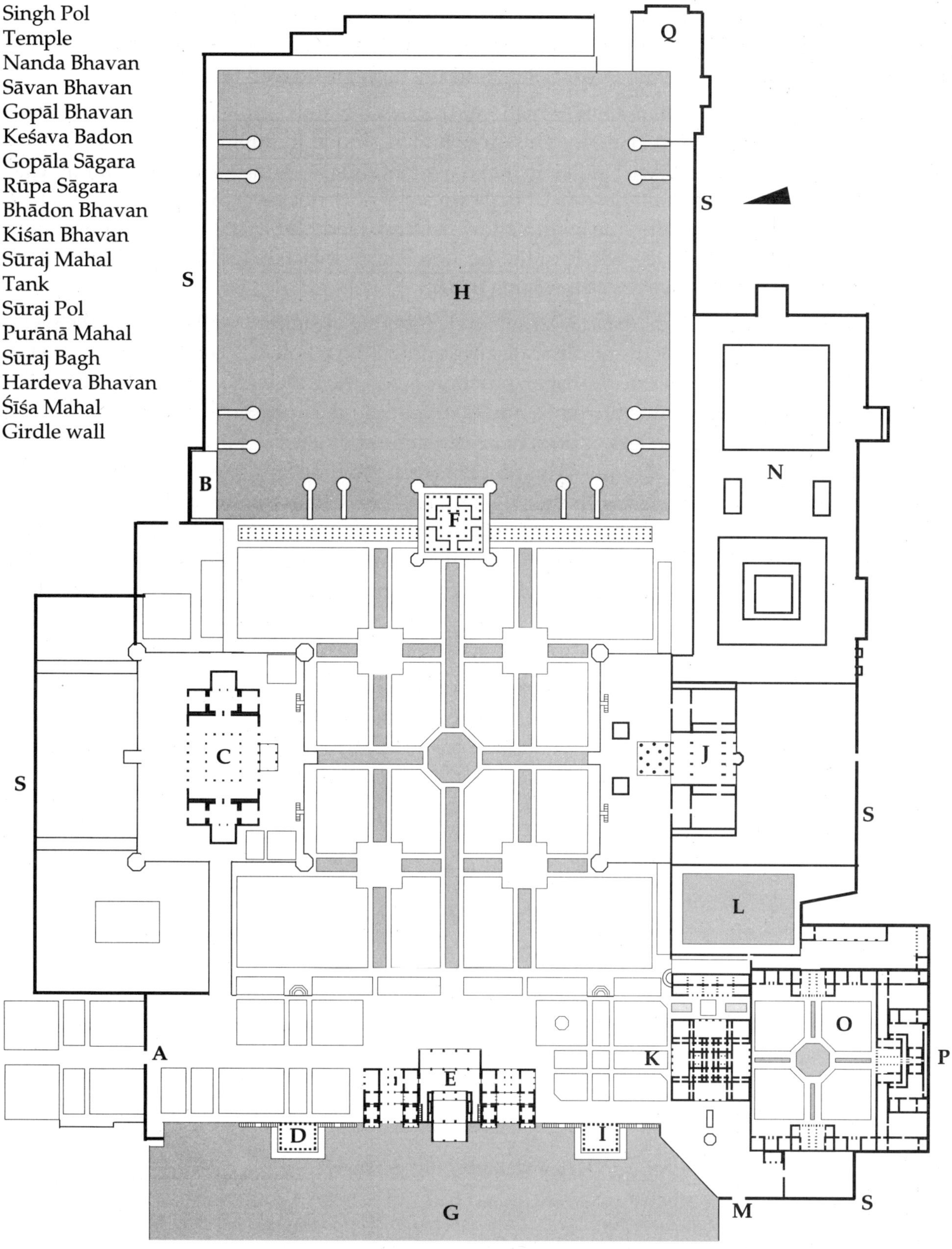

# Rūpa Sāgara, Garden Palace

The Mahārājās of Bharatpur could retire from the hurly-burly of the confined Bharatapur *gaṛh mahal* and find delight in the palaces, pavilions and gardens of their incomparable Garden Palace. The prying eyes of the general public were here avoided behind the high girdle wall (**Plate 61, S**).

The square Rūpa Sāgara[77] (**Plate 61, H**) which was to the east of the complex, provided incomparable vistas and numerous areas for relaxation. This large tank is similar in intent to the Anūpa Tālāo (Cār Caman) (**Plate 50**) of Fatehpur Sīkrī. Similar, in that it was not a place for the general public but reserved for the delight of the Mahārājā and his court. Furthermore, it does not seem to be exclusively a bathing tank, although the steps along the side would allow such activity. Nor, would it be the only source of water for the complex. Indeed, the whole of the Garden Palace was a delightful, lavish retreat with a water theme and one of the jewels was the Rūpa Sāgara.

The Rūpa Sāgara was a sizeable *tālāo* (large tank). A high, substantial wall enclosed the *sāgara* on the north, east and south; part of the girdle wall surrounding the whole Garden Palace. Centred along the west edge of this *sāgara* is a beautiful open pavilion called the Keśava Badon (**Plate 61, F & Figure 54**). This pavilion has been characterized as a *bārādarī*, although it possesses five openings on each side rather than three. None the less, it does have the same character — i.e. an open pavilion with a roof component supported by columns. This elegant open pavilion provides vistas in all directions, including the gardens to the west with their waterways, the Gopāla Bhavan (**Plate 61, E**) and the setting sun. Along the north, south and west borders of the Rūpa Sāgara were narrow walkways that extended out into the water, terminating at circular platforms. These platforms would have been especially private at night and provided a suitable locale for trysting. An entrance to the south-east, with wide steps leading down into the *sāgara*, might well have been employed to bring barges into the tank for the use of the court, or, possibly, even to allow elephants to bathe there.

The water theme, as noted above, of the Garden Palace was but one more way in which the Jāṭs established themselves. Water was and is an icon of power and wealth and the Jāṭs employed it lavishly. The rulers of Jodhpur might have sat high on their rocky hill in a fortified stony isolation, but the Jāṭ Mahārājās of Bharatpur luxuriated in cool splendour among the green gardens of their Garden Palace.

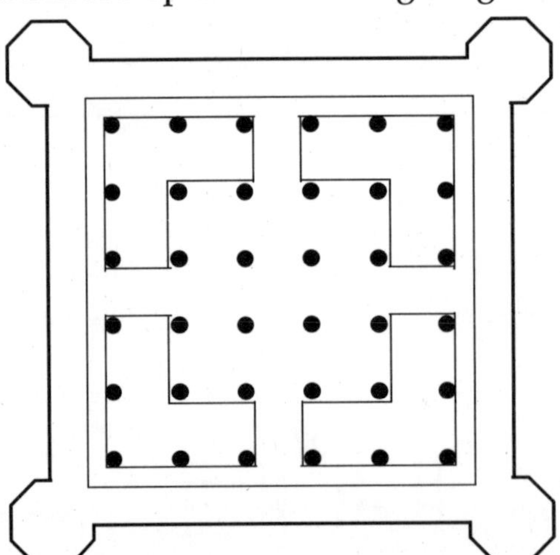

**Figure 54:** Keśava Badon plan

[77] The plan is redrawn using Adobe PageMaker 6.0 from George Michell, *The Royal Palaces of India*, London: 1994, p. 190 and G.H.R. Tillotson, *The Rajput Palaces: The Develpoment of an Architectural Style, 1450-1750*, New Delhi: 1999, p. 189.

## **Plate 64:** Rūpa Sāgara, Garden Palace, plan

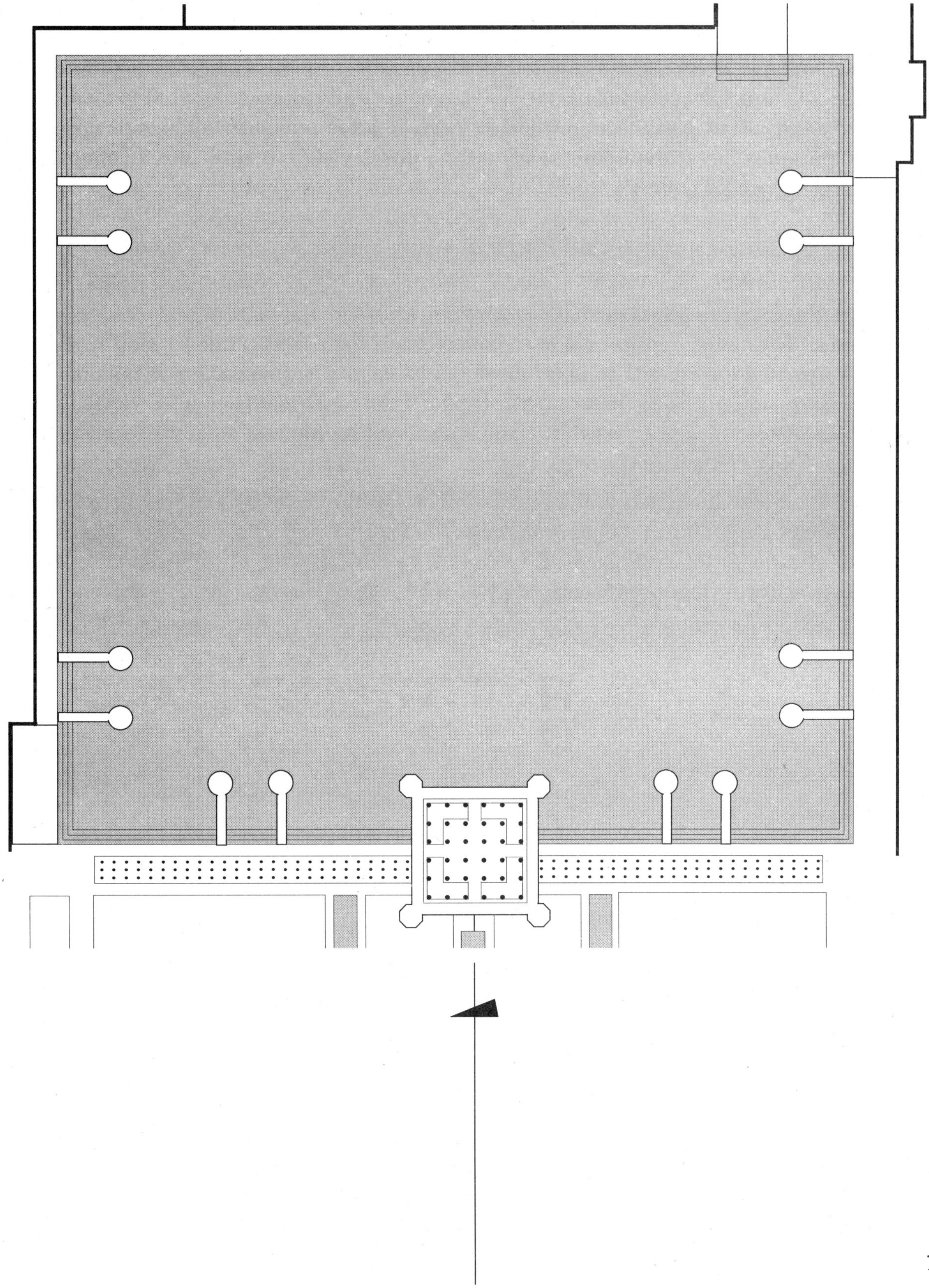

# Padmanābhapuram, Kanyākumārī, Tamil Nadu (CE 1729-58)

Far to the south, far from the madding crowds and intrigues of the Deccan, Tamil Nadu and the Mughals, at the southern tip of the subcontinent rose and flourished the Travancore Mahārājās. This area possessed one distinction, it had, along with Kerala, not suffured the domination of the Muslim north nor the overlordship of Vijayanagara. In this region, the Mahārājā Mārtaṇḍa Varmā III (r. CE 1729-58), of the Cheramān (Kulaśekhara) Dynasty (**Appendix D**), constructed a vast palace complex, Padmanābhapuram[78] (City of the Lotus Born) on the southern edge of the Western Ghāṭs.

It differed considerably from the palaces in the north, primarily in the fact that the majority of the construction material was wood which was in plentiful supply. Padmanābhapuram boasted several tanks. The palace complex was divided into four zones or compounds. The smallest of the zones (**Plate 63, I**) held the Pūmukham hall for public audience. To the north of this compound is a larger area in which the Navarātri Maṇḍapam is found (**Plate 63, II**). To the east of these two zones is a large residential area holding the chambers of the Mahārājā, and the *zenānā* (**Plate 63, III**). To the north of the Lakṣmī Vilāsam is the Zenānā Tank (**Plate 63, K**). Finally, a large tank, the Kalkuḷam is situated to the east of the residential zone (**Plate 63, M**). The Kalkuḷam is fronted by pavilions which look out onto this tank on the south and west sides. There is an access to this tank from the zone of offices and residences of visiting dignitaries (**Plate 63, IV**).

The whole complex is aligned to the cardinal points of the compass and, unlike many palaces, especially of the Rājpūts, Padmanābhapuram palace indicates a logical progression. One zone flows into its neighbour and the internal spaces also possess a design logic that is not frequently encountered. The one exception to this rule of confusion seen in so many other palaces is the famed Govinda Mandir of Datiā.

| | |
|---|---|
| **A** | Main Gate (Pol) |
| **B** | Maidān |
| **C** | Paḍipura |
| **D** | Pūmukham |
| **E** | Navarātri Maṇḍapam |
| **F** | Lakṣmī Vilāsam |
| **G** | Ūppurika Mālikā |
| **H** | Pilamuttu Koṭṭaram |
| **I** | Ūttupura |
| **J** | Homapura |
| **K** | Zenānā Tank |
| **L** | Thai Koṭṭāram |
| **M** | Kalkuḷam |
| **N** | Temple |
| **O** | Indra Vilāsam |
| **I** | 1st compound |
| **II** | 2nd compound |
| **III** | 3rd compound |
| **IV** | 4th compound |

[78] The plan is redrawn using Adobe PageMaker 6.0, from George Michell, *The Royal Palaces of India*, London: 1994, p. 190 and Satish Grover, *Masterpieces of Traditional Indian Architecture*, New Delhi: 2004, pp. 120-21.

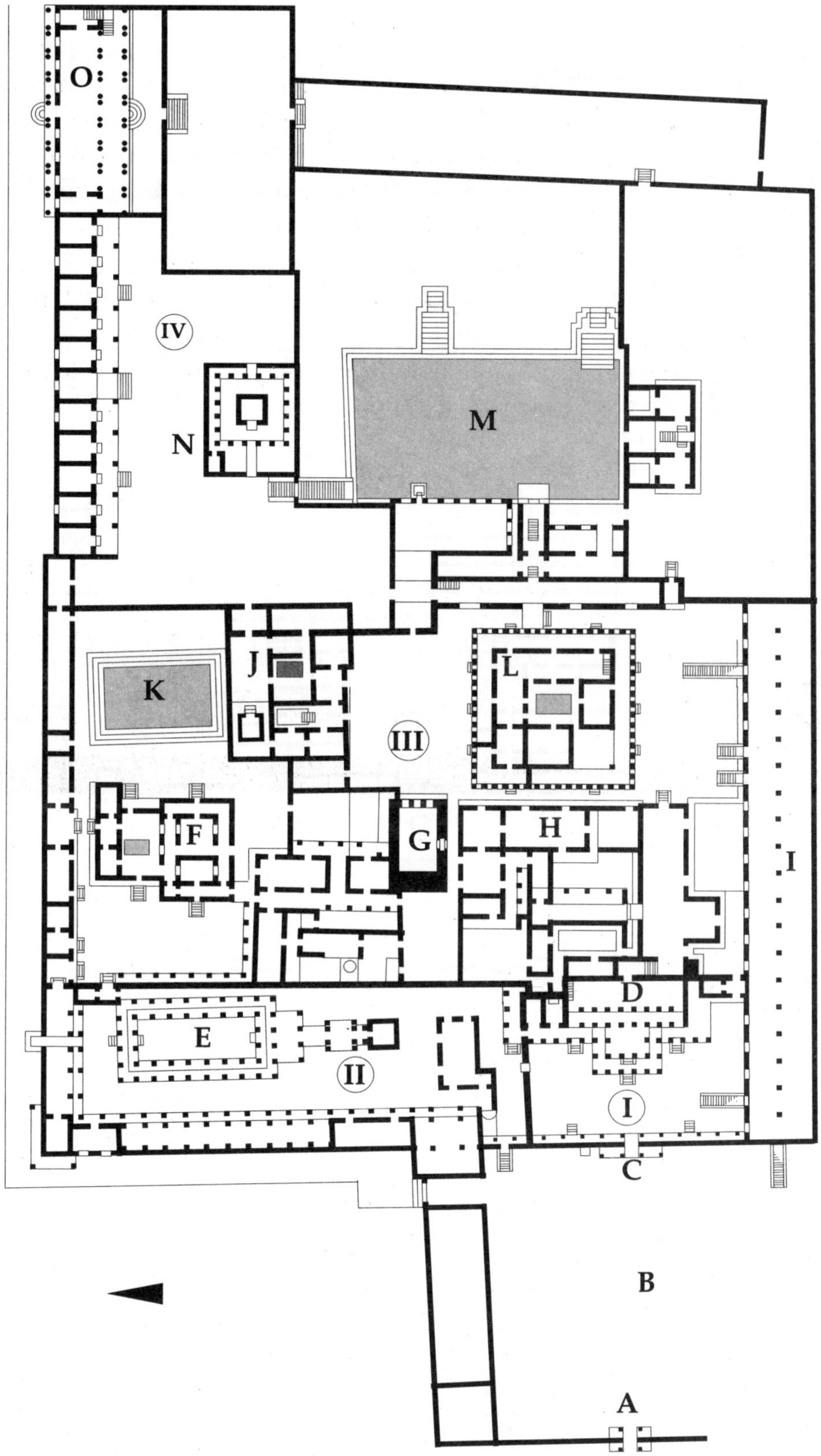
O
IV
N
M
K
J
III
L
F
G
H
I
D
E
II
I
C
B
A

The Kalkuḷam[79] of Padmanābhapuram is rectangular, but not orthogonal. Bounded on the south and west by pavilions which look out over the tank, it occupies a large area in the third compound. The western pavilion is attached to the Thai Koṭṭaram, the residence of the Queen Mother and her demi-court. From here the women of the *zenānā* could observe the tank and its occupants without being observed themselves through the *jālīs* that looked out over the Kalkuḷam. The women of the court had their own tank in Compound III (**Plate 63, K**), east of the Lakṣmī Vilāsam and the Thai Koṭṭaram (**Plate 63, L**) also had a small inner tank for the Queen Mother's use.

Three stairways lead down from the surrounding level to the water's edge. Two of the stairways descend from the east and the third from the north — i.e. from Compound IV. It would be via the northern stairs that the guests, both foreign and domestic, would have access to the tank from their Compound IV (**Plate 63**). The depth of the tank indicates that the water level was not close to the surface (**Figure 55**).

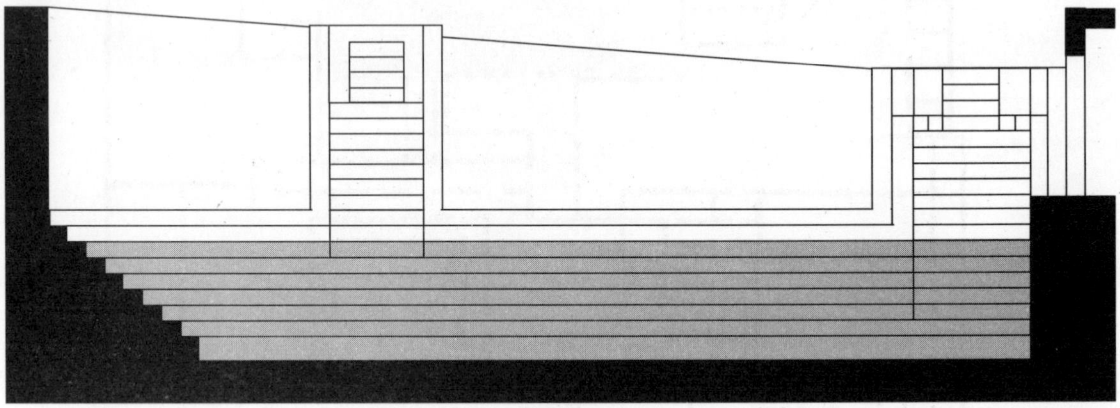

**Figure 55:** Kalkuḷam, sectional view

[79] The plan is redrawn using Adobe PageMaker 6.0 from George Michell, *The Royal Palaces of India*, London: 1994, p. 190 and Satish Grover, *Masterpieces of Traditional Indian Architecture*, New Delhi: 2004, pp. 120-21.

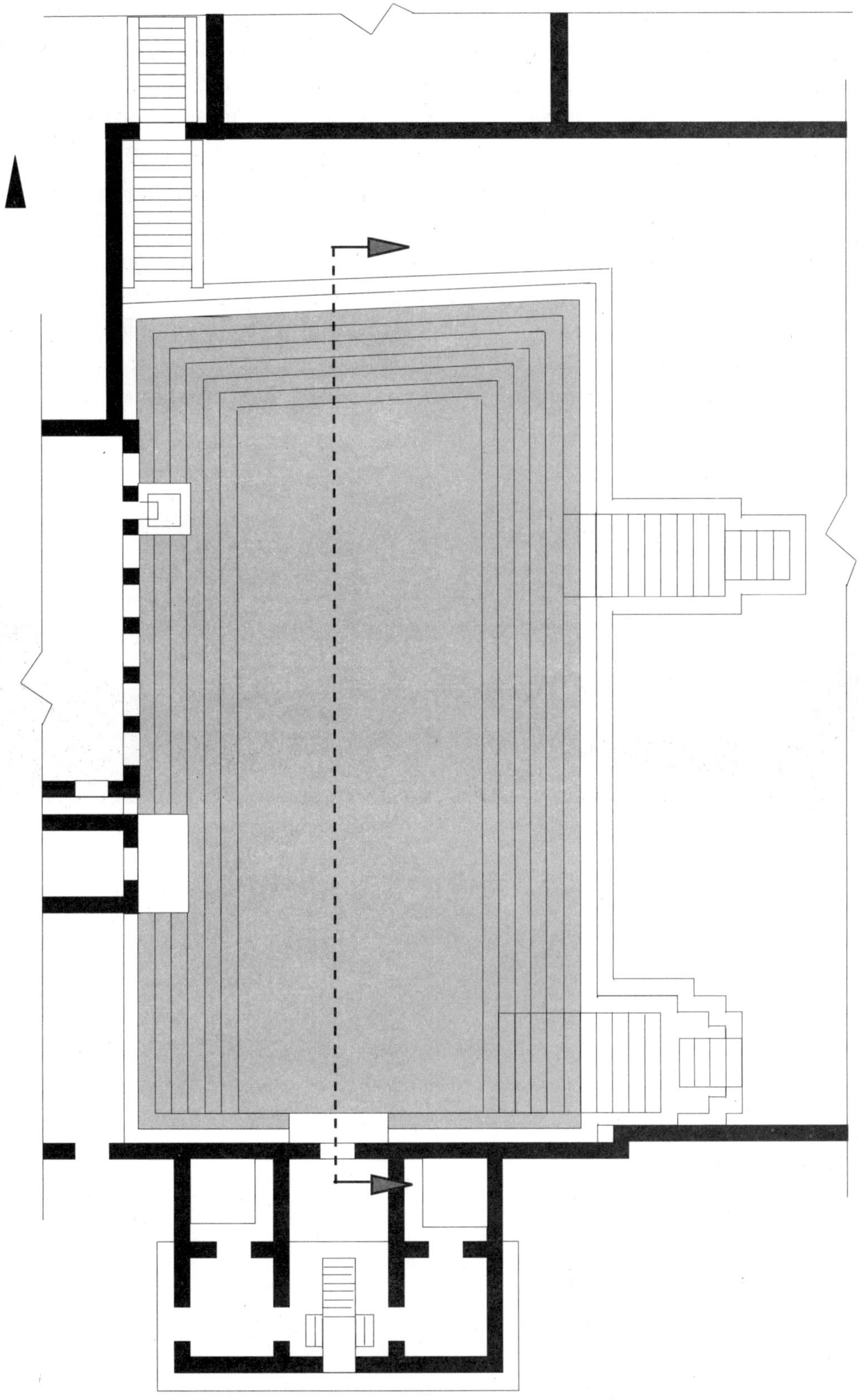

## Guru Nānak Dev-jī Tank, Ṭillā Jogiān, Jhelum, Pakistan (*c.* eighteenth century CE)

Near Sher Shāh Sūri's Rohtās Fort, in Pakistan, is the town of Jhelum and Mount Ṭillā Jogiān. Near the mountain and some kilometers from Jhelum is a tank named after the Sikh founder, Guru (Bābā) Nānak Dev-jī, who had visited this area to complete his *cillā*. It is said that the Guru Nānak Dev-jī Tank[80] was constructed by the Mughals, possibly even Akbar. The pine-shrouded Mount Ṭillā Jogiān provides a calm, cool and picturesque stopover between Lahore and Kashmir.

The Guru Nānak Dev-jī Tank is square, *en plan*, and descends to the water level in a series of ten, surrounding steps. The entrance to the tank is in the north and a flight of steps drops to the water at a shallower angle than the surrounding tank steps (**Figure 56**). A simple wall brackets these steps and terminates in two towers with small, non-functional *chattrīs* on the top. In the north-eastern part of the tank is a low wall. It appears to be a foundation, a foundation that some feel might have risen to a height that could have provided privacy for the Muslim women to bathe in the waters of the Guru Nānak Dev-jī Tank. Replenishment of the tank's water supply is via normal rainfall and from the small mountain streams that flow from Mount Ṭillā Jogiān.

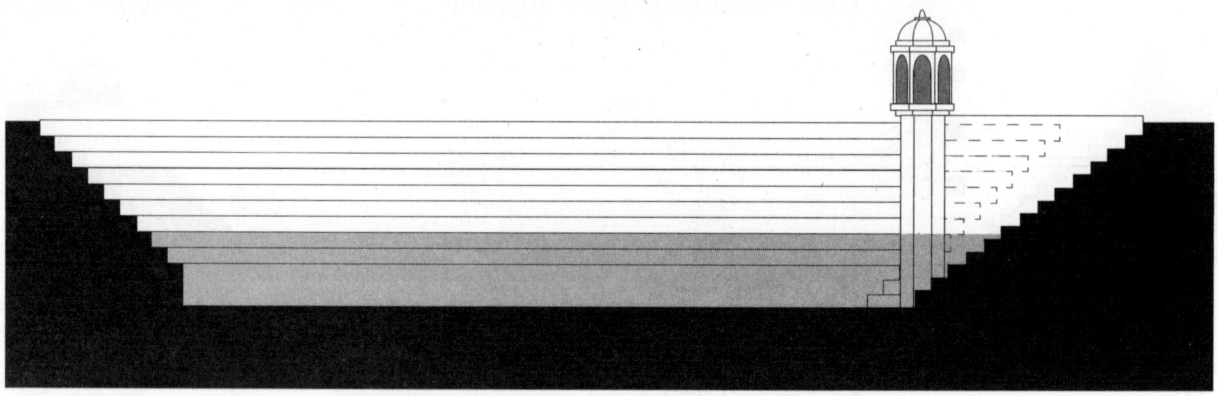

**Figure 56:** Guru Nānak Dev-jī Tank, sectional view

[80] The plan is redrawn using Adobe PageMaker 6.0 from the Internet.

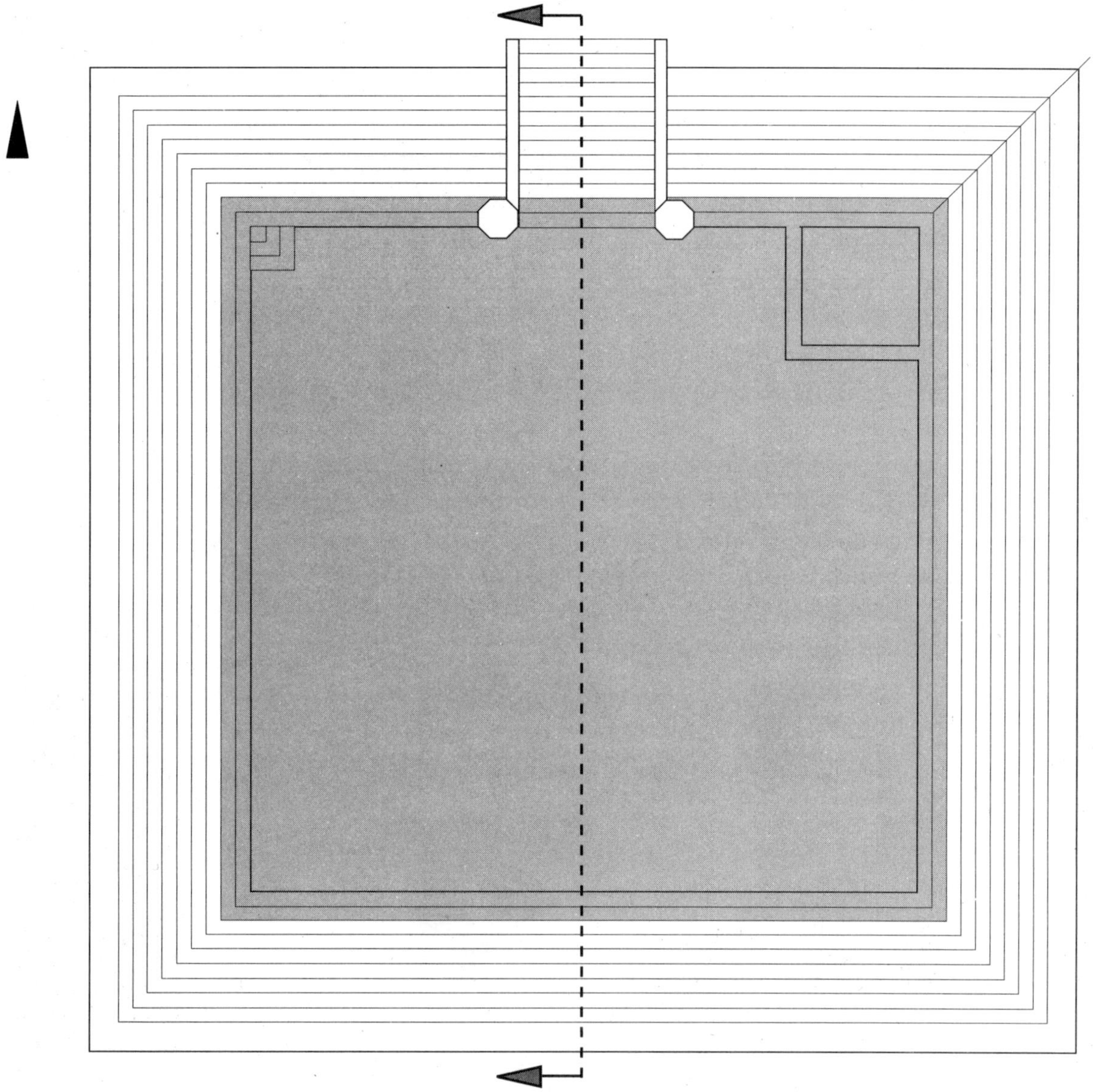

# Nandanvana Bāolī, Nagpur, Maharasthra (a. CE 1788)

The Bhoslās of Nagpur, a minor ruling dynasty, were instrumental in patronizing a number of noteworthy constructions. Specifically during the reign of the Rājā Raghujī II (r. 1788–1816) (**Appendix D**). The Nandanvana Bāolī[81] was one of the more utilitarian monuments undertaken by this ruler and is located some 3 km east of the royal Rajawāḍā.

In many ways the Nandanvana Bāolī is typical of the step wells of north-western India. It consists of a circular shaft (*kūpa*) and a trench with flights of stairs leading down to the water level. In this plan, it is rather typical of *bāolī*s in the region. There are three landings as the stairs descend in this rather narrow trench.

However, there are two elements which make this *bāolī* significantly different from the standard. First, at ground level and extending a short distance over the eastern lip of the *kūpa* is a water storage tank. This ground-level storage tank is a feature encountered on other *bāolī* and simple wells of Nagpur (**Plate 67**). The second differing element, and in some ways the more unique addition to the plan, is the inclusion of a space over the descending staircase and adjacent to the *kūpa*. This is designated as a Śiva Temple. Access to it is gained via a narrow corridor just off the second landing. The interior of this temple would have been nearly dark as there are but two small windows, one towards the *kūpa* and one towards the descending stairs to illuminate the interior (**Figure 57**).

Figure 57: Nandanvana Bāolī, sectional view

[81] The plan is redrawn using Adobe PageMaker 6.0 from K.M. Girhe, *Architecture of Bhoslas of Nagpur*, 2 vols., Delhi: 2004, vol. II, Figure 21.

# Plate 68: Nandanvana Bāolī plan

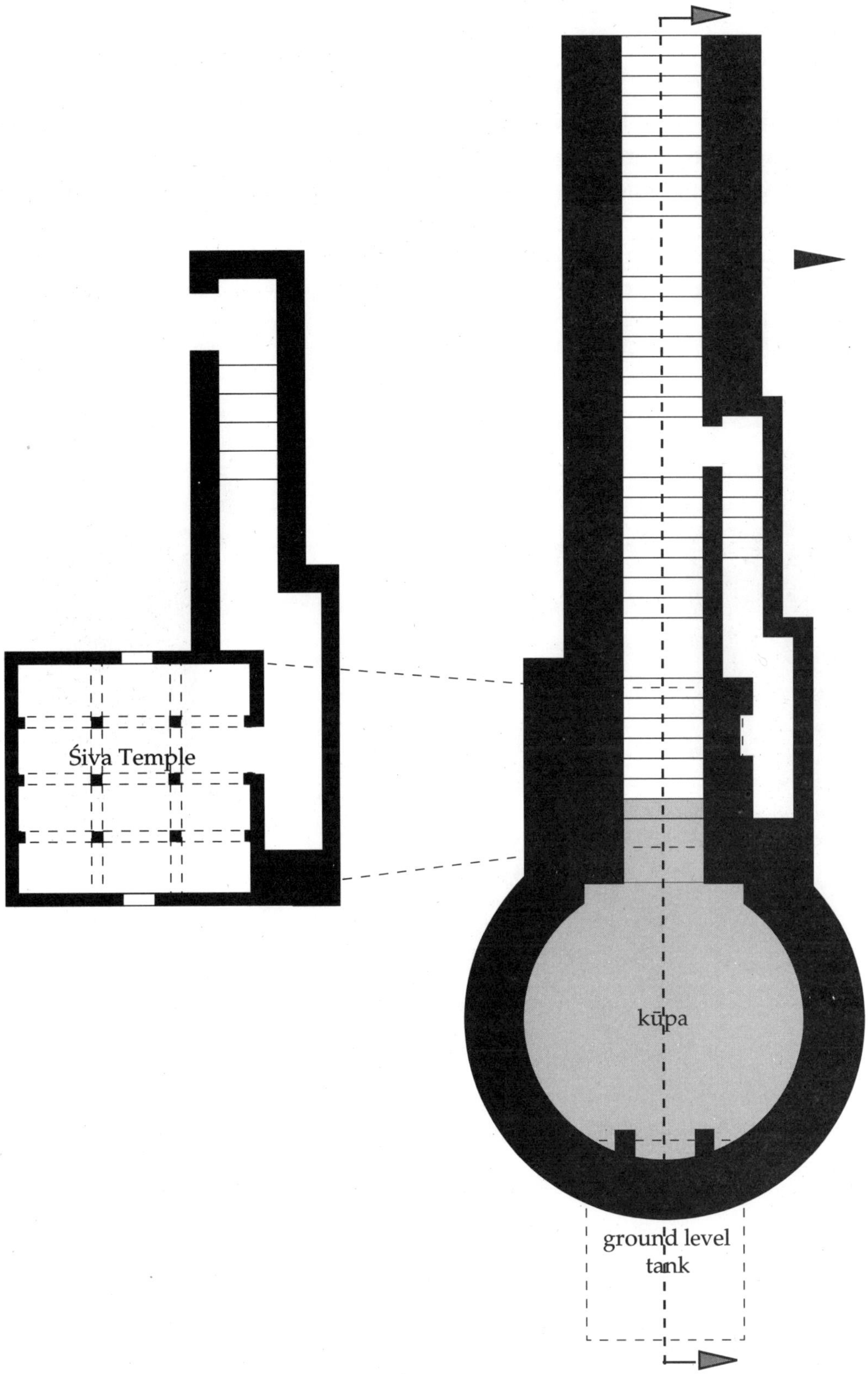

# Sakkardara Bāolī, Nagpur, Maharashtra (a. CE 1788)

The Sakkardara Bāolī[82] is in the south-eastern part of the city of Nagpur. A few kilometers from the Raja's palace, it sets within a small compound, resembling a garden. Its purpose was simply to provide much needed drinking water for the surrounding populous.

It is essentially a simple well (*kūpa*) dodecathedral in shape with two shallow storage tanks overlapping the opening at ground level to the east and west (**Figure 58**). The dodecagon shape provided the essential strength needed to keep the walls from collapsing (**Appendix C, Plate 74**). A platform extends from the south and it too extends over the opening of the *kūpa*. It is surmised that these overlapping forms served a dual purpose. Other than their normal utilitarian *raison d'être,* they provided shade so that the water level in the well might be protected during the hottest part of the day. The aquifer is sufficiently close to the ground level so only a rope and bucket were ever employed to draw water.

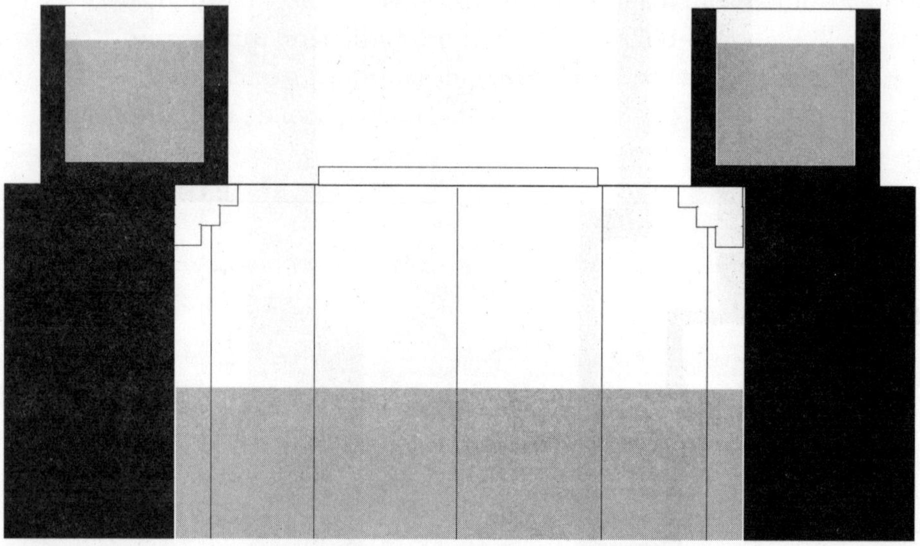

**Figure 58:** Sakkardara Bāolī, sectional view

[82] The plan is redrawn using Adobe PageMaker 6.0 from K.M. Girhe, *Architecture of Bhoslas of Nagpur*, 2 vols., Delhi: 2004, vol. II, Figure 22.

# **Plate 69:** Sakkardara Bāolī plan

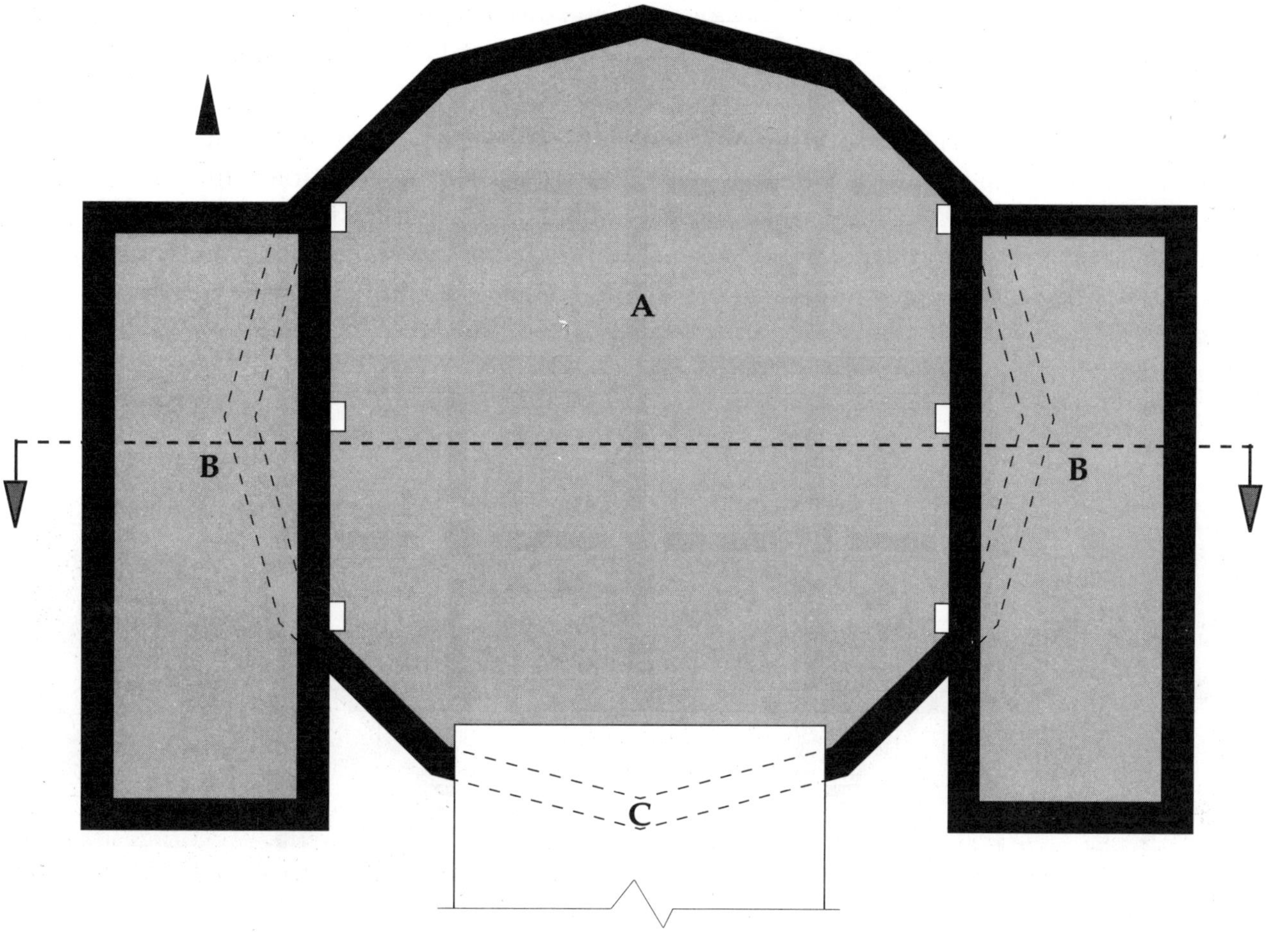


**A** *Kūpa*
**B** Ground level tanks
**C** Platform

## Thathawata-Johaḍ, Thathawata, Rajasthan (twentieth century CE)

To the east of Bikaner, bordering the Great Indian Desert, is a small town of Thathawata. In an area where rainfall is minimal, at best, the need for a *bāolī* is paramount. Here is found the small Thathawata-Johaḍ.[83]

The *bāolī* is unusual, *en plan*, as it is circular and without a trench. Many step wells and other *bāolīs* of Rajasthan and Gujarat possess a circular or octagonal shaft in conjunction with the descending trench. Little is recorded regarding the Thathawata-Johaḍ. Its source of water seems unlikely to come from an aquifer as the ground water would need to be rather close to the surface even though the *bāolī* is sunk into the surrounding ground. Neither is Thathawata in proximity to the mountains with their many glacial streams. Two, opposing stair ways lead from ground level down to a circular terrace that is stepped down to the water level (**Figure 59**).

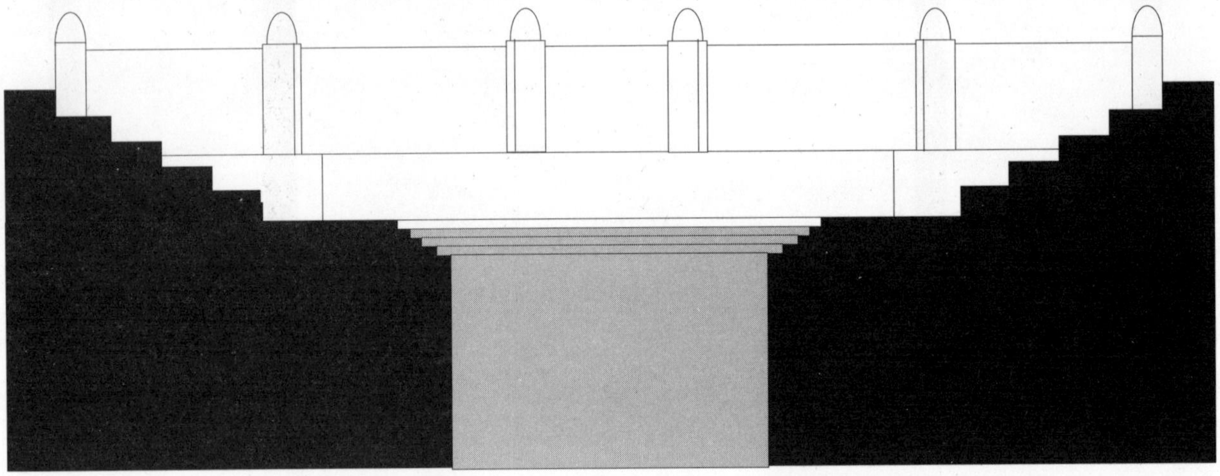

**Figure 59:** Thathawata-Johaḍ, Thathawata, sectional view

[83] The plan is redrawn using Adobe PageMaker 6.0 from the Internet.

# **Plate 69:** Thathawata-Johaḍ plan

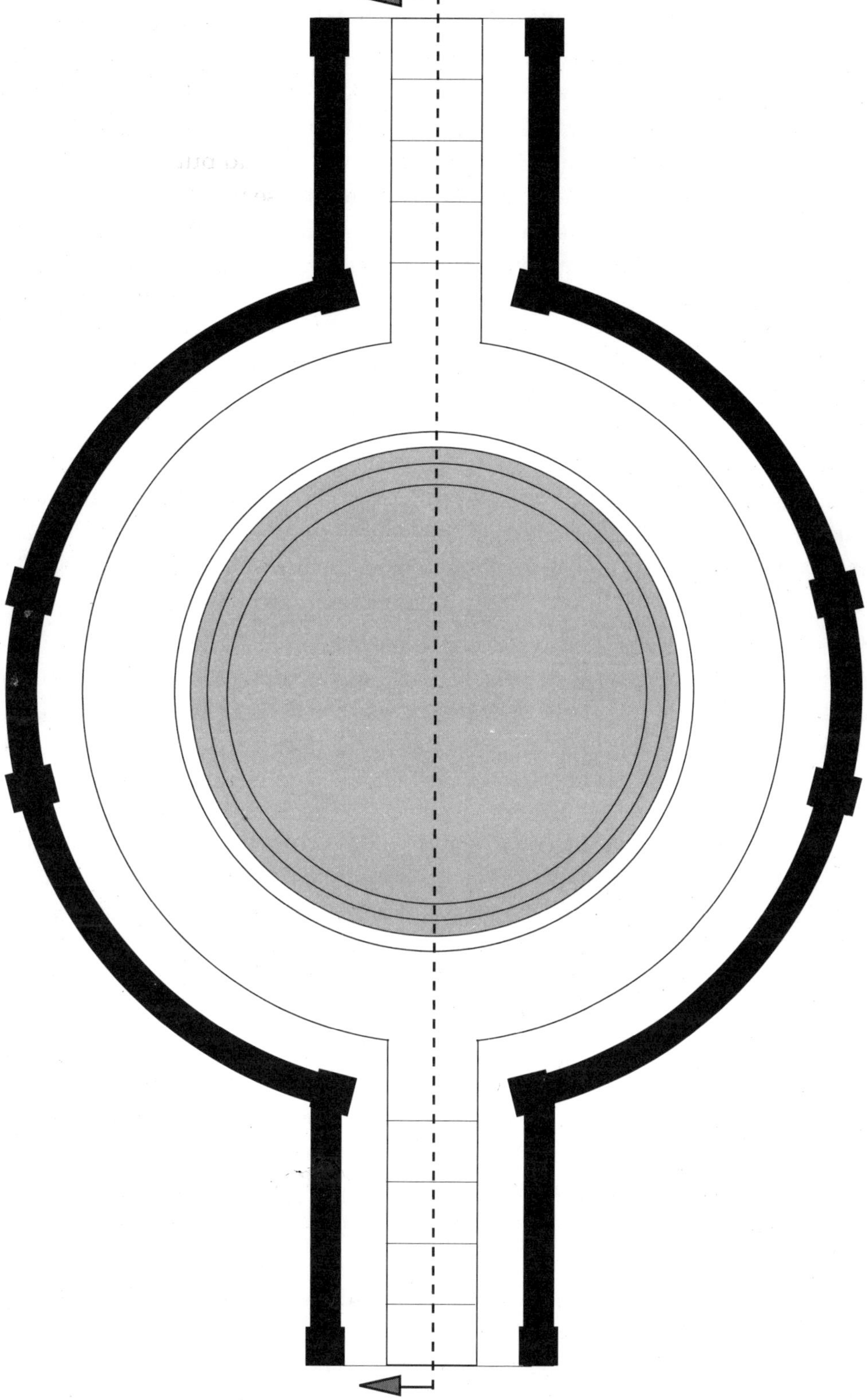

# Final Considerations

India had, for millennia, massive problems of supplying its population with water. Means of water catchment had developed which underline the importance of this culture.

Hundreds of thousands of wells, large and small, had been excavated in India to supply individual families, neighbourhoods and towns with drinking water. This was the primary reason for their existence in the arid and semi-arid regions of India. The simple wells which consist of only a shaft dug into the earth, down to the aquifer — i.e. the *kūpa* — are myriad but not altogether noteworthy, architecturally — e.g. the Sakkardara Bāolī of Nagpur. Later, these wells and tanks became more and more elaborate and extensive. Artificial catchment areas — i.e. *wāv*, *bāolī*, *kuṇḍa*s, *kuḷam*s and *tālāo*s — not only provided drinking water, but in the cases of *kuḷam*, they provided a suitable place for the ritual ablution that the Hindus performed before *pūjā*. Also, Muslims required ritual bathing (*woduk*) before prayers, but generally, the *haud* found in the *masjid*s of India were small and insignificant. There were exceptions — i.e. the Jāmī Masjid Haud of Campāner (**Plate 29**). *Tālāo*s were also a vital element in many of the *gaṛh mahals*, towns and cities.

### Wāv/Bāolī/Kuṇḍa/Kuḷam

Sacred ablution tanks associated with temples — i.e. the Viṣṇu-Puṣkariṇī at the Mahākūṭeśvara Temple in Bādāmī (**Plate 2**), the Roḍā Kuṇḍa (**Plate 3**) or the Maṇikarṇikā Kuṇḍa (Cakra-Puṣkariṇī Kuṇḍa) of Varanasi (**Plate 6**) — were so designed as to allow for easy access to the water. Generally, they tended to be rectangular and indicated a variety of step elements to reach the water.

By the eleventh century, one begins to see step wells (*wāv* or *bāolī*) and step tanks (*kuṇḍa* or *kuḷam*) of not only increasing size, exuberant details, but also of differing forms. A number of *bāolī*s (step wells) were square, *en plan*, and elaborate in design — e.g. the Pañcamahal Kuṇḍa (**Plate 9**) in Gujarat and the Cānda Bāolī (**Plate 10**) at Ābhānerī and later the Amber Tank (Pannā Miā) (**Plate 55**). These square *bāolī*s consisted of progressively smaller terraces as they descended into the earth, but they also included a more or less elaborate structure to shade the patrons from the glaring sun and searing heat. The resting rooms were generally constructed on the south side of the *bāolī*. Often these rooms reached to the deepest part of the well. However, those *bāolī*s which consist of a *kūpa* and inclined trench — e.g. the Gandhak-kī Bāolī (**Plate 16**) of Delhi or the Dādā Harīr Wāv (**Plate 20**) of Asarwa — are more prominent, more noteworthy due to their unique form; singular to the Indian subcontinent. A number of these *wāv*s or *bāolī*s with a *kūpa* and inclined trench are spectacular in their architectural achievement — e.g. the Rudabāī-nī Wāv (**Plate 19**) or Rānī-kī Wāv (**Plate 11**) of Pāṭan.

Tanks (*kuṇḍa*) constructed for the purpose of ritual ablution may also be square, *en plan* — e.g. the Pañcamahal Kuṇḍa (**Plate 9**) or the Kalayni Tank (**Plate 12**). However, a majority of them tended to be rectangular — e.g. Viṭṭhalapura Tank (**Plate 33**), the Kr̥ṣṇa Temple Tank (**Plate 35**) and the Ponthamaraikuḷam (Golden Lotus Tank) (**Plate 57**). Tanks (*kuṇḍa*s) were constructed neither for ritual bathing or providing drinking water, but purely for bathing pleasure and relaxation — e.g. Geometric Tank (**Plate 23**), the Square Water Pavilion (Queen's Bath) (**Plate 24**), the Octagonal Bath (**Plate 23**) and the Rewā Kuṇḍa (**Plate 39**).

The display of water not for any utilitarian value at all, but for purely aesthetic reasons, can also be seen in the Anūpa Tālāo (Cār Chaman) (**Plate 52**) as well as in the reflecting pools and channels at the Garden Palace (**Plate 63**). These two examples were, indeed, examples of conspicuous expenditure.

### Tālāo

The building of *wāv*s, *bāolī*s, *kuṇḍa*s or *kuḷam*s might have entailed varying degrees of expense, time and manpower. But, they pale in the expenditure involved to form and excavate a *tālāo*. Hundreds, even thousands would have been employed to manually dig and then carry the baskets of earth from the excavation. The outlay involved in the construction of the Mūñja Tālāo and the Kāphūra Tālāo at Shādiābād (**Plate 36**) would have been enormous. Similarly, the *tālāo*s constructed at Fatehpur Sīkrī dim in comparison to the expenditure of the whole, but the cost of these reservoirs would not have been insignificant! They just are not as visually spectacular as the Ekastambha Prāsāda (Dīwān-ī-Khās) (**Plate 51, E**) or other monuments of this Imperial city.

### Merit/Charity/Power

Many studies state that the step wells for the use by the general public were often constructed as an act of merit. Merit in that they provided for those whose convenient access to water was severely limited by distance or supply. In this sense merit is connected to the concept of charity. Charity is one of the noble concepts of Buddhism, Hinduism, Christianity and Islam. However, its application, in general, appears to be relevant to the extended family unless one possesses a plethora of riches and then one may observe it being applied to benefit those of the immediate area or region. True altruistic motives were rare. The concept of merit as a means of gaining paradise or *nirvāṇa* is present in all the world religions except Christianity.[1] It is especially important in Buddhism where the raising of temples and *stūpa*s are considered as highly meritorious. The same holds true for Hinduism and to a lesser degree in Islam. Therefore, the construction of a *wāv* or *bāolī* in a region where water may be scarce certainly has meritorious connotations.

But, is there something else involved? The Mañjuśrī Wāv (**Plate 1**) or the Gandhak-kī Bāolī (**Plate 16**) or the Niẓāmuddīn Bāolī (**Plate 17**) may clearly be seen to be efficient architectural-engineering works in which concept, utility and efficacy are united. They simply provide means by which one may access water with minimum effort and time. A second group including the Rānī-kī Wāv (**Plate 11**), the Rudabāī-nī Wāv (**Plate 19**) and the Dādā Harīr Wāv (**Plate 20**) also provide a means by which one may access water with minimum effort and time. However, in comparing the second three with the first three, there is a massive difference! If we use a standard art historical approach of analysis — i.e. subject, matter, form and content — we may say that the "subject" of all six monuments is to provide access to water, a well. The "matter", again in all six cases, is similar — i.e. an excavated space and stone. In all six *wāv*s or *bāolī*s, the "form" is similar — i.e. a shaft (*kūpa*) and an inclined trench. Content? Here we find divergence. The person(s) who were responsible for the first three — i.e. Mañjuśrī Wāv, Gandhak-kī Bāolī, Niẓāmuddin Bāolī — might certainly have gained some merit and respect from those who benefited from their utilitarian construction, but the "content" was clearly utilitarian. On the other hand, the person(s) responsible for the Rānī-kī Wāv, the Rudabāī-nī Wāv and the Dādā Harīr Wāv might also have gained some merit and respect from those who benefited from their utilitarian construction, but the "content" expressed through the ostentation and design was clearly meant to impress. The ostentation and design employed were clearly an indication of extreme wealth and prosperity. And, the expenditure of wealth in these three cases underlines the power that the benefactor

1 Ephesians 2:8-9 states that "works" or "merit" may be acceptable, but they do not guarantee paradise.

possessed.[2] The modern concept of "conspicuous consumption" may be applicable, but mutated to "conspicuous expenditure". In both instances, consumption and expenditure are indications, icons of wealth; indications, icons of power. Since water or the lack thereof is an issue in the subcontinent, it is felt that the supplying and/or the possession of water is more closely connected with power or the exercise of power than an act of merit or charity.

**Power Icon**

A ruler may consolidate his hold on power by a number of ways: furnishing suitable sources of drinking water in a land where it is in short supply is one of these means. Not only was water an absolute necessity for a ruler's personal sustenance in the citadel that protected him, it was vital for his court and retainers as well. In the construction of a *gaṛh mahal,* a suitable and sustainable source of water was a primary consideration. It may be further stated that a source or the construction of a suitable water storage facility was a primary, even, the primary consideration in the choice of location of a *gaṛh mahal*, a defensive citadel. Once this was accomplished or provided, then the ruler's power base could be expanded. This expansion not only included conquest, but once conquest was accomplished, it was incumbent on the ruler to keep his subject in a state of peaceful acquiescence. Providing a source of water was one of them.

The Mālwā Sultāns exercised their power and dominance through the construction of two large *tālāo*s within their citadel of Shādiābād (**Plate 36**) at Māṇḍu. The *tālāo*s not only augmented cool evening breezes and places to bathe, they were cogent icons of the enormous power they possessed and wielded within the region. The power of the Mārwāṛs, the Rāṭhor Rājpūts, was under attack at Mandore. In CE 1459, the Rāo Jodh of Mandore moved his capital to the impregnable heights of Jodhpur. There he constructed the invincible Meherāngaṛh, but not before deep wells and water storage cisterns were incorporated to sustain life during possible sieges. The Bundellā Rājpūts of Orchhā built their *gaṛh mahal*s on an island in the Betwā. Not only did the island provide a primary defensible position, the river supplied the various *gaṛh mahal*s with necessary water. The command of the island and the water that surrounded it was primary to their power.

As the technology of offensive warfare and siege developed, it became evident that few of the *gaṛh mahal*s could withstand bombardment by cannon. Only huge, manoeuvrable forces that were under the command of competent, creative leadership could protect and sustain the power that the various rulers had amassed. Imperial Vijayanagara (**Appendix C, Plate 78**) had grown and developed to such an extent. The Royal Centre was not a citadel as seen in Shādiābād or Meherāngaṛh, rather it was an extended residential, ceremonial and religious complex that was visually protected by high walls. A large, well-equipped and trained army provided the needed defence for the Emperor and his Court. In CE 1505, the Rāya Vīra Narasiṁha (Naresh Tuluva) (r. CE 1505-09) (**Appendix D**) wrested control of the Vijayanagara throne from the Saluvas. It was under the Tuluva Dynasty that the Empire of Vijayanagara ultimately reached its zenith. Based upon the stability and careful expansion of the Saṅgamas, the incomparable supremacy of Vijayanagara was accomplished during the reigns of Rāya Kṛṣṇadeva

[2] These three wells were ostensibly constructed by important women. However, in India at this time, both Hindu and Muslim India, women did not inherit in their own right, nor were they able to independently undertake such constructions. They did so with the understanding of their husbands or sons, as the case may be.

(r. CE 1509-30) and his successor, Rāya Acyutadeva (r. CE 1530-42), the second and third members of the Tuluva line. It wasn't until that the combined forces of the bellicose Deccan Sultanates — i.e. the 'Adil Shāhī of Bījāpur, Qutb Shāhī of Golconda, 'Imād Shāhī of Berār, Niẓām Shāhī of Ahmednagar and Barīd Shāhī of Bīdar — successfully challenged the Vijayanagara hegemony in CE 1565. None the less, the power style of the Saṅgama–Saluva–Tuluva had been established. In this power style, water played a prominent role. Imperial Vijayanagara exhibited an inordinate use of water — i.e. the Great Tank (**Plate 22**), the Geometric Tank (**Plate 23**), the Square Water Pavilion (Queen's Bath) (**Plate 24**), Octagonal Bath and the various Residence Tanks (**Plate 25**). All were icons of power. Later, the Sinsini Jāṭs, neither brāhmaṇa or kṣatriya, gained power and built the beauteous Garden Palace (**Plate 63**) which was replete with all manner of water usage — i.e. *tālāo*s (tanks), reflecting pools and fountains.

The power of the Timurid, the Great Mughals, was such that the use of water became an important feature of their palaces and citadels. For them, water as an icon of power reached it apogee in the Anūpa Tālāo (aka Cār Chaman) (**Plate 52**) of Fatehpur Sīkrī. In this monument water is utilized almost exclusively as an aesthetic element. It was not, primarily, a bathing pool, although people might have bathed there. It was not utilized as a source of drinking water and it was too small to have provided cooling breezes for the person of the Emperor. Similarly, the Nahr-ī-Bahist (**Figure 60**) of Shāhjahānābād, which physically connected the Hammām with the Mumtāẓ Mahal and running through the Diwan-ī-Khās with its fabled "Peacock Throne", the Khās Mahal and the Raṅga Mahal, was purely iconographic and an exceptional display of Imperial power.

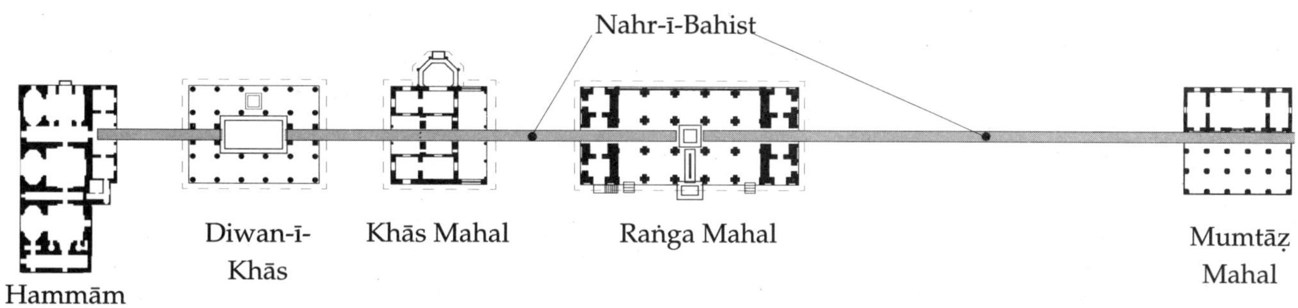


**Figure 60:** Nahr-ī-Bahist and its relationship to other monuments at Shāhjahānābād

Whether *bāolī* or *wāv* or *tālāo* or *kuṇḍa* or *kuḷam*, these architectural-engineering monuments of India ostensibly provided water for drinking or bathing or irrigation. In a semi-arid land teeming with people, this was most important. The *bāolī* or *wāv* or *tālāo* or *kuṇḍa* or *kuḷam*, more importantly, developed into indications, icons of power.

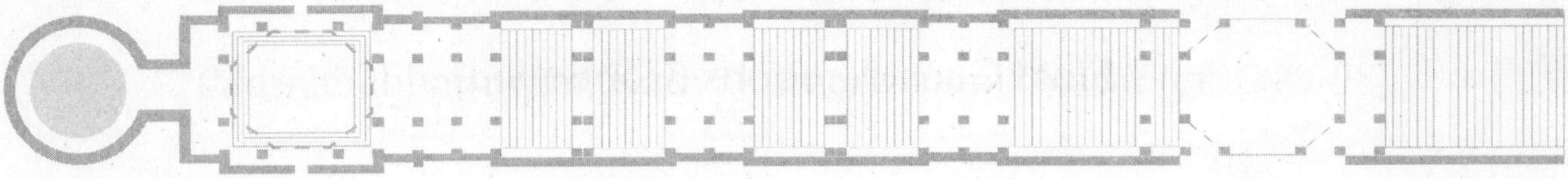

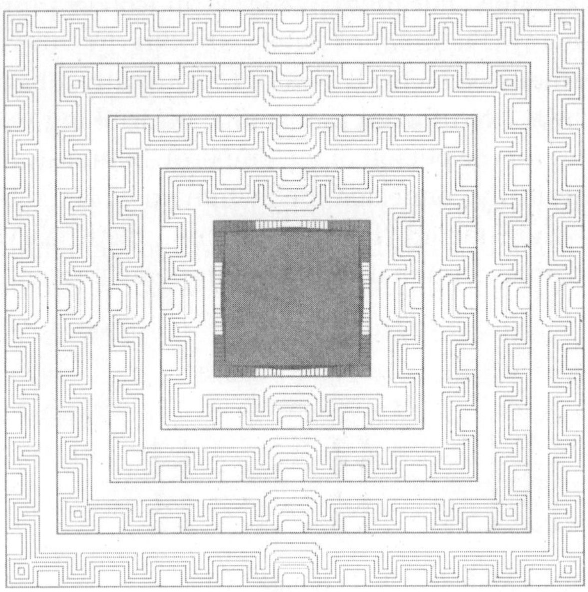

# Appendix A

## Location of Important Tanks and Stepped Wells in the Subcontinent

## List of Notable Stepped Wells and Tanks in the Subcontinent

# List of Notable Stepped Wells and Tanks in the Subcontinent

The following list is but a small number of the thousands of *kūpas*, *bāolīs*, tanks, *tālāo*s and other water catchment devices found in the subcontinent.

### Andhra Pradesh

Alaghumallāri Kr̥ṣṇa Temple Tank, Alaghumallāri Kr̥ṣṇa Temple, Mannarpolur.

Bhavanāśī Varāha Narasiṁha Tank, Bhavanāśī Varāha Narasiṁha Temple, Singarayakoṇḍa.

Brahma Tīrtham, Bhavanārāyaṇa Temple, Ponnur.

Hanumān Guṇṭa Temple Tank, Hanumān Guṇṭa, Warangal.

Kalyāṇī Kuḷam, Siddheśvara Temple, Hemavati.

Kalyāṇī Puṣkariṇī, Panaka Narasiṁha Temple, Maṅgalagiri.

Mahendragaṛh Bāolī, Mahendragaṛh.

Nāgeśvara Temple Tank, Cheprolu.

Pasavakkarai, Śiva Temple, Penukoṇḍa.

Puṣkariṇī, Narasiṁha Temple, Yadagirigatta.

Puṣkariṇī, Śiva Temple, Śrīsailam.

Rāma Tīrtham, Rāmathīrtha Devālayam Temple, Puram.

Saṅgameśvara Temple Tank, Saṅgameśvara Temple, Chinnekothapalle.

Sarasvatī Tīrtham (Śrī Kalahastīśvara Temple Tank), Kalahasti.

Śiva Temple Tank, Śiva Temple, Warangal.

Śiva Temple Tank, Śiva Temple, Śrīsailam.

Swāmi Puṣkariṇī, Veṅkaṭeśvara Temple, Tirumala.

Swāmi Puṣkariṇī, Śrī Veṅkaṭeśvara Temple, Tirupati.

Tiruchanur Temple Tank, Padmāvatī Temple, Tiruchanur.

Vīrabhadra Swāmi Temple Tank, Vīrabhadra Swāmi Temple, Lepakshi.

### Assam

Joysagar Tank, Jeydol (Viṣṇu) Temple, Śiva Temple and Devī Ghara and Ghanaśyāma (Naṭī Gosāiṅ) Temple, Rangpur.

Sibsāgar Tank, Śiva Temple, Viṣṇu Temple and Devīdol Temple, Sibsāgar.

### Bihar

Abhiṣeka Puṣkariṇī Tank, Vaiśālī.

Dakṣiṇa Mānas, Dakṣiṇārka Sun Temple, Gayā.

Sūrya Kuṇḍa, Dakṣiṇārka Sun Temple, Gayā.

### Delhi

Agar Sain kī Bāolī.

Bāolī Haranand, Palam.

Bāolī near Mūbarak Shāh's tomb.

Bāolī near Sikandar Lodhi's tomb.

Bāolī-Surai, Khizrabad.

Dargāh of Kākī Sāheb Bāolī.

Feroz Shāh Koṭlā Bāolī.

Gandhak-kā Bāolī, Mehrauli.

Haus-ī-Alai Tālāo.

Hauz-ī-Sultani Tank (Hauz-ī-Iltumish).

Hauz Khās Bāolī.

Khārī Bāolī.

Niẓāmuddīn Bāolī, Dargāh.

Qutb-Badarpur Road Bāolī.

Rajoṅ-kī-Bāolī (Rajoṅ-kī-Bain), Mehrauli.

Tughlaqābād Fort Jheel.

### Goa

Mangueshi Temple Tank, Mangueshi-Priol.

Lakṣmī-Narasiṁha Temple Tank, Nagueshi.

Pañcagaṅgā Tīrtha (Saptakoṭeśvara Temple Tank), Narva-Diwar.

### Gujarat

Ādi Cādī Bāolī, Jūnāgaḍh Hills.

Ādi Kaḍi Well, Jūnāgaḍh.

Akhaj Stepped Kuṇḍa, Akhaj.

Ambāpur Wāv, Buḍthal.

Aṅkol Mātā Stepwell, Davad.

Āsāpurī Stepwell, Ahmedabad.

Bahādur Singh-kī Wāv, Pāṭan.

Bāī Harīr Wāv, Asarwa.

Batris Koṭha, Kapadvanj.

Bharmaja Stepwell, Mehmūdābād.

Brahma Kuṇḍa, Sirhore.

Chatral Stepwell, Chatral.

Caumukhī Stepwell, Chobari.

Dabhoi Stepwell, Dabhoi.

Dādā Harīr Wāv (Bāolī Bāī Harīr), Asarwa.

Dāmodara Kuṇḍa, Dāmodarrāyajī Temple, Mount Girnār.

Dhandhalapur Stepwell, Dhandhalapur.

Gaṅgā Stepwell, Dedadara.
Gaṅgā Stepwell, Vadhvan.
Ghaṇṭī Stepwell, Idar.
Halisa Vāv, Halisa.
Hanpur Stepwell, Hampur.
Helical Wāv, Pāvāgaṛh.
Holemata Wāv, Holemata.
Īśānpur Stepwell, Ahmedabad.
Jeṭhā Stepwell, Ghumlī.
Jeṭhābhāī Stepwell, Īśānpur.
Jhilanī Wāv, Dhaṅk.
Jñān Stepwell, Visvada.
Kāleśvarī-ī-Nal Kuṇḍa, Kāleśvarī.
Kaṅkura Tank, Ahmednagar.
Kapadvañj Stepwell, Kapadvañj.
Khuṭal Stepwell, Zhinzhuvāḍā.
Koṭaḍī Stepwell (ruin), Koṭaḍī.
Limbhoi Stepwell, Limbhoi.
Madha Stepwell, Vadhvan.
Mañjuśrī Stepwell, Dhank.
Mañjuśrī Wāv, Jūnāgaḍh Hills.
Mātā Bhavinī Wāv, Asarwa.
Mātṛ Stepwell, Kaṅkavatī.
Minai Stepwell, Vīrpur.
Minmal Stepwell, Bhalej.
Nāgā Bāva Stepwell, Dhrangadhrā.
Nārāyaṇa Rāo Stepwell, Idar.
Navaghan Wāv, Jūnāgaḍh Hills.
Navalakhī Stepwell, Baroda.
Pandoli Stepwell, Pandoli.
Rāigaḍh Stepwell, Rāigaḍh.
Ra Kheṅgar Stepwell, Vanthali.
Rāma Stepwell, Bhuj.
Rānī-kī Wāv (Rānī Udayamatī), Pāṭan.
Rataba Stepwell, Rāmpura.
Roḍā Stepped Pond, Roḍā.

Rudabāī-nī Wāv, Adalaj.
Sabeli Stepwell, Sabeli.
Sahasarliṅga Tank, Pāṭan.
Sampa Stepwell, Sampa.
Sāsū Stepwell, Kāleśvarī.
Sathamba Stepwell, Sathamba.
Sevasi Stepwell, Sevasi.
Sūrya Temple Tank, Moḍherā.
Tārāpur Stepwell, Tārāpur.
Tintoi Stepwell, Tintoi.
Trikambarot Stepwell, Jūnāgaḍh.
Umta Wāv, Umta.
Utvali Mātā Stepwell, Dholkā.
Vaḍtal Wāv, Vaḍtal.
Vahu Stepwell, Kāleśvarī.
Vethiya Wāv, Mukhana, Kāṭhiāwāṛ.
Vikia Stepwell, Ghumlī.
Wankaner Stepwell, Wankaner.
Zhinzhuvāḍā Stepwell, Zhinzhuvāḍā.

**Haryana**

Alijān-kī Bāolī, Narnaul.
Bādshāpur Bāolī, Bādshāpur.
Bhīṣma Kuṇḍa (Bāṇa Gaṅgā), Kurukṣetra.
Brahma Sarovara, Sarveśvara Mahādeva Temple, Kurukṣetra.
Farrukhnagar Bāolī, Farrukhnagar.
Ghaus Khan-kī Bāolī, Farrukhnagar.
Hoḍal Bāolī, Hoḍal.
Kaithal Bāolī, Kaithal.
Kapāla Mocana Sarovara, Kapāla Mocana.
Meham Bāolī, Meham.
Nābhi Kamala Tank, Kurukṣetra.
Nāgpurīoṅ-kī Bāolī, Narnaul.
Narnaul Bāolī, Narnaul.
Pinnaṅghave Bāolī, Pinnaṅghave.
Rohtak Bāolī, Rohtak.
Saidu Kalāl Bāolī, Meham.

Sannihit Sarovara, Dhruva Nārāyaṇa Temple, Kurukṣetra.

Silānā Bāolī, Silānā.

Sirsā Bāolī, Sirsā.

Sūraj Kuṇḍa Sarovara, Kapāla Mocana.

Thānesar Bāolī, Thānesar.

**Jharkhand**

Kamal Sarobar, Jala Mandir, Pāwāpurī.

**Karnataka**

Akka Thankai Kuḷam, Nārāyaṇaswāmi Temple, Melkoṭe.

Ambu Tīrtham, Rāmaeśvara Temple, Ambutīrtham.

Antharagaṅgā, Nandīśvara Temple, Kolar.

Bhasvarajan Kuḷam, Nārāyaṇaswāmi Temple, Melkoṭe.

Brahmacārī Kuḷam, Nārāyaṇaswāmi Temple, Melkoṭe.

Brahma Kuṇḍa, Kāverī Temple, Talakāverī.

Candra Puṣkariṇī, Śrī Vināyaka and Śrī Mahāliṅgeśvara Temple, Kumbākṣī.

Chokkaṭṭan Kuḷam, Nārāyaṇaswāmi Temple, Melkoṭe.

Dattātreya Temple Tank, Dattātreya Temple and Dādā Hayāt Dargāh, Bababudan Giri.

Devī Karai, Dattātreya Cāmuṇḍeśvarī Temple, Cāmuṇḍī Hills.

Devī Karai, Gaṅgādāreśvara Temple, Hassan.

Geometric Tank, Vijayanagara.

Grand Tank, Vijayanagara.

Kadri Mañjunāth Temple Tanks (9), Mangalore.

Kalyāṇī Kuḷam (Beḷagoḷa), Śrī Gomaṭeśvara Temple, Śravaṇabeḷagoḷa.

Kalayni Tank, Hulikere.

Koṭi Tīrtham and Rāma Tīrtham, Mahābaleśvara Temple, Gokarṇa.

Kṛṣṇa Temple Tank, Hampī (Vijayanagara).

Mādhva Tīrtham, Śrī Kṛṣṇa Temple, Udupi.

Mahāliṅgeśvara Temple Tank, Mahāliṅgeśvara Temple, Kota.

Manmatha Tank, Virūpākṣa-Pampī Temple Complex, Hampī (Vijayanagara).

Narasiṁha Tīrtham, Vīra Āñjaneyaswāmi Tank and Viṭṭhala Kuḷam, Śrī Yoga Narasiṁhar Temple, Taḷa.

Octagonal Bath, Vijayanagara.

Octagonal Well, Malpannaguḍi (Vijayanagara).

Padmasarovara, Mahiṣamardinī Temple, Udupi.

Residence Tank (north-west of Rāmacandra Temple [$W_1$]), Vijayanagara.

Skanda Puṣkariṇī, Śrī Vadapanda Balarāma Temple, Malpe.

Square Water Pavilion (Queen's Bath), Vijayanagara.
Śrī Janārdana Temple Tank, Śrī Janārdana and Śrī Mahālakṣmī Temple, Ambalapadi.
Sūrya Puṣkariṇī, Śrī Vināyaka and Śrī Mahāliṅgeśvara Temple, Kumbākṣī.
Thirunārāyaṇaswāmi Kuḷam, Nārāyaṇaswāmi Temple, Melkoṭe.
Vaidīśvarar Temple Tank, Vaidīśvarar Temple, Talakadu.
Vāsudeva Kalyāṇī Tīrtham, Cennakeśava Temple, Belūr.
Viṭṭhalapura Tank, Viṭṭhalapura, Hampī (Vijayanagara).
Yoga Narasiṁhaswāmi Kuḷam, Nārāyaṇaswāmi Temple, Melkoṭe.

**Kashmir**

Shalimar Bagh, Śrīnagar, Kashmir.

**Kerala**

Ambala Kuḷam, Kr̥ṣṇa Temple, Alappuzhā.
Ambala Kuḷam, Mahādeva Temple, Koṭṭayam.
Ambala Kuḷam, Mullakkal Devī Temple, Alappuzhā.
Ambala Kuḷam, Nāgarājan Temple, Alappuzhā.
Ambala Kuḷam, Rāma Temple, Koṭṭayam.
Ambala Kuḷam, Sarasvatī Temple, Koṭṭayam.
Ambala Kuḷam, Śiva Temple, Alappuzhā.
Ambala Kuḷam, Śrī Subrahmaṇya Swāmī Temple, Koṭṭayam.
Ambala Kuḷam and Kunnathu Kuḷam, Alappuzhā.
Ambala Kuḷam, Sun Temple, Ādityapuram.
Bhīsma Kuḷam, Ayyappan Temple, Pathanamthiṭṭa.
Cakra Tīrtham, Janārdana (Viṣṇu) Temple, Thiruvananthapuram.
Gaṅgā Tīrtham, Vaikathappan Temple, Koṭṭayam.
Jalavanthi (Ghaṇṭakarṇa Tīrtha) and Puthen Kuḷam, Vallabha (Viṣṇu) Temple, Tiruvalla.
Kulpinī Tīrtham and Kuṭṭtan Tīrtham, Kūdalmaṇikyam Temple, Iriñjalakuḍa.
Padiñjārr Cira, Vaḍakkunāthan Temple, Thrissur.
Padmanābha Tīrtham, Padmanābha Temple, Thiruvananthapuram.
Rudra Tīrtham, Guruvāyūrappan Temple, Guruvāyur.
Śaktikuṇḍa Tīrtham and Gaṅgā Tīrtham, Bagavathi Ambalam, Chenganoor.
Śivānanda Tīrtham, Vaikathappan Temple, Koṭṭayam.
Subrahmaṇya Ambala Kuḷam and Raktha Kuḷam, Cottānikkara Bhagavatī Temple Tank, Ernakulam.
Swāmi/Pārvatī Devī Temple, Koṭṭayam.
Thali Kuḷam, Śiva Temple, Kaduthuruthy.
Varāha Ambala Kuḷam, Lakṣmī Varāha Mūrty Temple, Thiruvananthapuram.
Vaṭṭa Kuḷam, Bhagavatī (Kaṇṇagi) Temple, Kodungallur.

Villu Kuḷam, Śiva Temple, Ettumānur.

**Madhya Pradesh**

Campā Bāolī, Shādiābād Mandir, Māṇḍu.

Kāphūra Tālāo, Shādiābād Mandir, Māṇḍu.

Mūñja Tālāo, Shādiābād Mandir, Māṇḍu.

Rewa Kuṇḍa, Shādiābād Mandir, Māṇḍu.

**Maharashtra**

Ambāgaḍa Fort Tank, Ambāgaḍa Fort (Bhaṇḍārā).

Ambājharā Tālāv, Nagpur.

Bhiwānī Bāolī, Bhiwānī.

Carkhā Bāolī, Carkhi Village, Bhiwānī.

Corbahuli Bāolī, Corbahuli, Nagpur.

Jaganātha Budhawari Bāolī, Nagpur.

Nandanvan Bāolī, Nandanvan, Nagpur.

Rāmakuṇḍa, Nashik.

Sakkardara Bāolī, Sakkardara, Nagpur.

Śukrawāra (Jumā) Tālāv, Nagpur.

Sonegāon Tālāv, Sanegāon, Nagpur.

Telaṅgkheḍī Tālāv, Nagpur.

Telaṅgkheḍī Garden Tank, Nagpur.

**Odisha**

Bindu Sarovara, Bhubaneshwar.

Indradyumna Sarovara, Purī.

Narendra Sarovara, Jagannātha Temple, Purī.

Pārbatī Sarovara, Loknātha Temple, Purī.

Rohiṇī Kuṇḍa, Jagannātha Temple, Purī.

Śvetagaṅgā Kuṇḍa, Jagannātha Temple, Purī.

**Pakistan**

East Well (Bāolī), Rohtās.

Lakho Dehar Bāolī, Lahore.

Shāhī Bāolī, Rohtās.

Shalimar Bagh, Lahore.

South Gate Bāolī, Rohtās.

**Punjab**

Aṭṭārī Tank, Aṭṭārī.

Baniye Ke Tank, Baniye Ke.
Bhaknā Kalān Tank, Bhaknā Kalān.
Boparāī Kalān Tank, Boparāī Kalān.
Chamiyārī Tank, Chamiyārī.
Chanet Bāolī, Chanet, Ludhiana.
Chawinda Devī Tank, Cḥawinda Devī.
Daraha Bāolī, Daraha, Ludhiana.
Dayālgaṛh Bāolī, Dayālgaṛh.
Dhand Kasel Tank, Dhand Kasel.
Fatehābād Tank, Fatehābād.
Gandiwind Tank, Gandiwind.
Gurdwārā Bāolī Sāhib, Goindwāl (Pehowā), Amritsar.
Gurudwārā Dukh Nivāraṇ Sāhib, Patiala.
Khāngaṛh Tank, Khāngaṛh.
Koṭlā Vasawa Singh Tank.
Nauraṅgābād Tank, Nauraṅgābād.
Preet Nagar Tank, Preet Nagar.
Pul Kañjirī Tank, Pul Kañjirī.
Rajatal Tank, Rajatal.
Rāmatīrath (Sarovara) Tank, Rāmatīrath.
Serāī Imānet Khan Tank, Serāī Imānet Khan.
Tarn Tāran Sarovara, Gurudwārā Tarn Tāran Sāhib.

**Rajasthan**

Abhinava Kuṇḍa Vāv, Bhinmali.
Ablā Kuṇḍa Vāv, Bhinmali.
Alwar Palace Tank.
Amber Bāolī (Pannā Miā), Amber.
Baggar Tank, Sekhāvaṭī.
Bālsamand Garden Palace Bāolī, Bālsamand.
Bandarej Stepwell, Bandarej.
Baṛī Bāolī, Bandarej.
Cānda Bāolī, Ābhānerī.
Caṇḍānāth-kī Wāv, Bhinmal.
Chirava Bāoṛī, Sekhāvaṭī.
Churu Bāoṛī, Sekhāvaṭī.
Daedali Kuṇḍa Vāv, Bhinmal.

Fatehpur Bāwḍī, Fatehpur.

Gadsisar Tank, Jaisalmer.

Galta Temple Tank, Galta.

Ghāṭ-kī-Goony, Jaipur.

Hādī Rānī Vāv, Toda Raisingh.

Jhālarā Bāoṛī, Jhālarā.

Katan Vāv, Ossian.

Mandākinī Kuṇḍa, Mt. Ābu.

Mandavan Stepwell, Sekhāvaṭī.

Nahagaṛh Cistern (Tālāo), Jaipur.

Nimrāṇā Stepwell, Nimrāṇā.

Pannā Miāṅ-kī Bāolī (Pannā Miāṅ-kī Kuṇḍa), Amber.

Rānī Stepwell, Būndī.

Rānī Stepwell, Nadol.

Sarāī Bāolī, Amber.

Śiva Vāḍī Temple Tank, Bikaner.

Thathawata-Johaḍ, Thathawata.

Vasantgaḍh Stepwell, Vasantgaḍh.

**Tamil Nadu**

Agastīśvarār Kovil Kuḷam, Nuṅgambakkam Agastīśvarār Prasanna Veṅkaṭeśa Perumāḷ Temple, Chennai.

Agastya Tīrtham, Nuṅgambakkam Thirumazhisai Ottandiśwarār Temple, Chennai.

Agni Tīrtham, Jambukeśvarār Temple, Tiruchirappalli.

Agni Tīrtham, Maṅgaleśvarār Temple, Pondicherry.

Agni Tīrtham, Vedhanathar Temple, Madurai.

Airāvata Tīrtham, Kovur Sundareśvarār Temple, Chennai.

Amr̥ta Puṣkariṇī, Koṇeśwarār Temple, Tiruvarur.

Amr̥ta Puṣkariṇī, Villivakkam Sowmiya Dāmodara Perumāl Temple, Chennai.

Amr̥ta Tīrtham, Bhavaouṣadheśvarār Temple, Tiruvarur.

Aṅgala Parameśvarī Kovil Kuḷam, Royapuram Aṅgala Parameśvarī Temple, Chennai.

Aṅgalamman Kuḷam, Choolai Aṅgala Parameśvarī Amman Temple, Chennai.

Anaikuḷam, Inner Fort, Gingee.

Ānanda Tīrtham, Viśvanāthar Temple, Tirunelveli .

Aṅgāraka Tīrtham, Villivakkam Agastīśvarār Temple, Chennai.

Annapūraṇī Kuḷam, Ayanavaram Paraśurāmaliṅgeśvarār Temple, Chennai.

Ardhanārīśvarār Kovil Kuḷam, Naṅganallur Ardhanārīśvarār Temple, Chennai.

Arkīśvarār Temple Tank, Pammal Arkīśvarār Temple, Chennai.

Aruṇācaleśarār Kovil Kuḷam, Thoṇḍaiyarpeṭ Aruṇācaleśvarār Temple, Chennai.
Bhṛgu Puṣkariṇī, Thirumazhisai Jagannātha Perumāḷ Temple, Chennai.
Brahma Tīrtham, Ayanavaram Paraśurāmaliṅgeśvarār Temple, Chennai.
Brahma Tīrtham, Bhavaouṣadheśvarār Temple, Tiruvarur.
Brahma Tīrtham, Kolilināthar Temple, Tiruvarur.
Brahma Tīrtham, Mayūrnāthaswāmi Temple, Mayavaram.
Brahma Tīrtham, Muruganātha Swāmi Temple, Coimbatore.
Brahma Tīrtham, Thiruvotriyur Tyāgarājaswāmi Temple, Chennai.
Brahma Tīrtham, Thrisoolam Thrisoolanāthar Temple, Chennai.
Brahma Tīrtham, Sattaināthar Temple, Nāgapaṭṭinam.
Brahma Tīrtham, Vedanāthar Temple, Madurai.
Cakra Tīrtham, Yoga Narasiṁha Temple, Sholingur.
Cakra Tīrtham, Soundararāja Perumāḷ and Azhagar Temple, Madurai.
Cakra Tīrtham, Śrī Vikrama-Ulagalantha Perumāḷ Temple, Vilupuram.
Candra Puṣkariṇī, Raṅganāthar Temple, Tiruchirapalli.
Candra Tīrtham, Kolilināthar Temple, Tiruvarur.
Chitra Kuḷam, Mylapore Ādi Keśava Perumāḷ Temple, Chennai.
Devarāja Puṣkariṇī, Kāladipeṭ Kalyāṇa Varadarāja Perumāḷ Temple, Chennai.
Easwaran Temple Tank, Easwaran Temple, Tiruchirapalli.
Ekāmbareśvarār Temple Tank, Ekāmbareśvarār Temple, Chennai.
Gaja Puṣkariṇī, Kālīśvarār Temple, Kaalaiyar Koil.
Gaṅgā Tīrtham, Puraṣawalkam Gaṅgādīśvarār Temple, Chennai.
Garbharakṣāmbiga Temple Tank, Thanjavur.
Gomatī Tīrtham, Gomaṭeśvarār Temple, Nāgapaṭṭinam.
Gowmara Tīrtham, Maṅgaleśvarār Temple, Pondicherry.
Gowrī Puṣkariṇī, Kālīśvarār Temple, Kaalaiyar Koil.
Hanumān Guṇḍa Tīrtham, Rāmanāthaswāmi Temple, Rāmanāthapuram.
Hanumān Tīrtham, Mambalam Kothaṇḍarāmar Temple, Chennai.
Indra Tīrtham, Jambukeśvarār Temple, Tiruchirappalli.
Indra Tīrtham, Saidapet Karanishwarar Temple, Chennai.
Indra Tīrtham, Velachery Karaṇīśvarār Temple, Chennai.
Jalakaṇḍeśvarār Temple Tank, Jalakaṇḍeśvarār Temple, Vellore.
Jambukeśvarār Tank, Jambukeśvarār Temple, Tiruchirapalli.
Janmanāśnī Tīrtham, Thiruvanmiyur Maruṇḍīśvarār Temple, Chennai.
Jñāna Tīrtham, Muruganātha Swāmi Temple, Coimbatore.
Jñānadhāyinī Tīrtham, Thiruvanmiyur Maruṇḍīśvarār Temple, Chennai.

Kaccalīśvarār Thiru Kuḷam, Kaccalīśvarār Temple, Chennai.

Kairaveni Tank, Thiruvallikeni Pārthasārathi Temple, Chennai.

Kalyāṇa Tīrtham, Thirumullaivoyal Masilamaṇīśvarār Temple, Chennai.

Kamalakanniamman Kuḷam, Rājagiri, Gingee.

Kamalālayam, Tyāgarājaswāmi Temple, Tiruvarur.

Kāmanāśinī Tīrtham, Thiruvanmiyur Maruṇḍīśvarār Temple, Chennai.

Kannudaya Nāyakī Temple Tank, Kannudaya Nāyakī Temple, Śivagaṅgā.

Kapāli Tīrtham, Mylapore Kabalīśvarār Temple, Chennai.

Kāruṇya Amr̥ta Tīrtham, Mahāliṅgeśvarār Temple, Tañjāvur.

Kāśī Viśvanāthar Theppakuḷam, Kāśī Viśvanāthar Temple, Chennai.

Kavvaikakadal, Bhūmināthar Temple, Virudhunagar.

Kirupa Tīrtham, Pāśupatanāthar Temple, Cuddalore.

Kr̥ṣṇa Puṣkariṇī, Mannady Kr̥ṣṇan Temple, Chennai.

Kṣīra Tank, Mullaivananāthar Temple, Tanjavur.

Lakṣmaṇa Tīrtham, Rāmanāthaswāmi Temple, Rāmanāthapuram.

Lakṣmī Tīrtham, Subrahmaṇya Temple, Madurai.

Lava-Kuśa Tīrtham, Koyambedu Kuruṅgalīśvarār Temple, Chennai.

Liṅgeśvara Temple Tank, Liṅgeśvara Temple, Coimbatore.

Mahāmagam Kuḷam, Kumbheśvarar Temple, Tanjavur.

Mallikuḷam, Rājagiri, Gingee.

Mānasa Puṣkariṇī, Kr̥pā Samudra Perumāḷ Temple, Nāgapaṭṭinam.

Maṇikarṇikā Taṭṭkam, Thiruneermalai Raṅganātha Perumāḷ Temple, Chennai.

Mannarguḍī Temramaple Tank, Raja Temple, Thiruvarur.

Maragatha Tīrtham, Candra Cūḍeśvarār Temple, Kr̥ṣṇagiri.

Mariamman Tank, Muthu Mariamman Temple, Pudukoṭṭai.

Mariamman TheppaKuḷam, Vandiyur.

Marudhapurī Tank, Murugan Temple, Śivagaṅgā.

Melavaiyapurī Tank, Murugan Temple, Śivagaṅgā.

Mokṣadāyinī Tīrtham, Thiruvanmiyur Maruṇḍīśvarār Temple, Chennai.

Muruga Tīrtham, Muruganātha Swāmi Temple, Coimbatore.

Naazhi Kinaru Senthil Aandavar Temple Tank, Naazhi Kinaru Senthil Aandavar Temple, Tuticorin.

Nadappankuḷam, Outer Fort, Gingee.

Nāgāraja Temple Tank, Nāgarāja Temple, Kanyākumārī.

Nandīśvara Tīrtham, Vedagirīśvarār Temple, Kāñcīpuram.

Nitya Puṣkariṇī, Śrīmūṣnam Temple, Cuḍḍalore.

Pakṣī Tīrtham, Vedagirīśvarār Temple, Kāñcīpuram.

Palani Andavar Kovil Kuḷam, Vadapalani Palani Andavar Temple, Chennai.

Pāpanāśinī Tīrtham, Thiruvanmiyur Maruṇḍīśvarār Temple, Chennai.

Pārthasārathi Temple Tank, Triplicane, Chennai.

Periya Udayar Koil Tank, Periya Udayar Koil Temple, Ariyalur.

Pillayarpatti Temple Tank, Karpaga Vinayagar Temple, Śivagaṅgā.

Ponthamaraikuḷam (Golden Lotus Tank), Mīnākṣī Sundareśvara Temple, Madurai.

Potramarai Kuḷam, Mīnākṣī Temple, Madurai.

Prasanna Veṅkaṭa Narasiṁha Perumāḷ Kovil Kuḷam, Saidapeṭ Prasanna Veṅkaṭa Narasiṁha Perumāḷ Temple, Chennai.

Puṣkariṇī, Karmegha Perumāḷ Temple, Madurai.

Rāja Rājeśvarī Amman Temple Tank, Rāja Rājeśvarī Amman Temple, Śivagaṅgā.

Rāma Tīrtham, Yoga Narasiṁha Temple, Sholingur.

Śaṅkaranārāyaṇa Temple Tank, Śaṅkaranārāyaṇa Temple, Tirunelveli.

Śaṅku Tīrtham, Vedagirīśvarār Temple, Kāñcīpuram.

Santāna Puṣkariṇī, Mylapore Mādhava Perumāḷ Temple, Chennai.

Saravana Poigai, Kandaswāmi Temple, Chennai.

Saravana Poigai, Kosapeṭ Kandaswāmi Temple, Chennai.

Saravana Poigai, Kundrathur Murugan Temple, Chennai.

Saravana Poigai, Murugan Temple, Śivagaṅgā.

Saravana Poigai, Subrahmaṇya Temple, Madurai.

Saravana Tīrtham, Valīśvarār Temple, Perambalur.

Satyagiri Tīrtham, Satyamūrti Temple, Pudukoṭṭai.

Setu Mādhava Tīrtham, Rāmanāthaswāmi Temple, Rāmanāthapuram.

Sevagapperumāḷ Temple Tank, Sevagapperumāḷ Temple, Śivagaṅgā.

Śukra Tīrtham, Valasaravakkam Agastīśvarār Temple, Chennai.

Siddha Amṛta Kuḷam, Vaidyanāthaswāmi Temple, Nāgapaṭṭinam.

Śivagaṅgā Kali Tīrtham, Kālīśvarār Temple, Kālaiyar Koil.

Śivagaṅgā Tank, Aruṇācala (Annamalaiar) Temple, Tiruvannamalai.

Śivagaṅgā Tank, Bṛhadīśvara Temple, Tanjavur.

Śivagaṅgā Tank, Naṭarājārām Temple, Cidambaram.

Śivagaṅgā Tīrtham, Kovur Sundareśvarār Temple, Chennai.

Śrī Vedapurīśvarār Temple Tank, Pondicherry.

Sūrya Puṣkariṇī, Kundrathur Vada Nāgeśvarār Temple, Chennai.

Sūrya Puṣkariṇī, Raṅganāthar Temple, Tiruchirapalli.

Sūrya Tīrtham, Vyāsarpādi Raviśvarār Temple, Chennai.

Svastik Tank, Puṇḍarīkāṣa Perumāḷ Temple, Thiruvellarai.

Sveda Puṣkariṇī, Poonamalee Śrī Varadarāja Perumāḷ Temple, Chennai.

Thekkan Kuḷam, Yoga Narasiṁha Temple, Vellore.
Theppa Kuḷam, Raṅganāthar Temple, Tiruchirapalli.
Thirukani Tīrtham, Murugan Temple, Pudukkoṭṭai.
Thiru Kuḷam, Cennakeśava Perumāḷ Temple, Chennai.
Thiru Kuḷam, Cintādripeṭ Ādipurīśvarār Temple, Chennai.
Thiru Kuḷam, Mannady Chengazhunir Pillaiyar Temple, Chennai.
Thiru Kuḷam, Muthialpet Mallikeśvarār Temple, Chennai.
Thiru Kuḷam, Nandīśvarār Temple, Chennai.
Thiru Kuḷam, Teynampeṭ Bālāsubrahmaṇia Swāmi Temple, Chennai.
Thiru Kuḷam, Thiruvattīśvaran Pettai Thiruvattīśvaran Temple, Chennai.
Tiruvarur or Thyāgarājaswāmi Temple Tank, Tiruvarur.
Varuṇa Tīrtham, Thiruninravur Śrī Bhaktavatsala Perumāḷ Temple, Chennai.
Virūpākṣīśvarār Kovil Kuḷam, Mylapore Virūpākṣīśvarār Temple, Chennai.
Yama Tīrtham, Velachery Daṇḍīśvarār Temple, Chennai.

### Uttarakhanda

Bhīmgoḍa Tank, Hardwar.
Gaurī Kuṇḍa, Rudraprayag Dist..
Tapta Kuṇḍa, Badrīnāth.

### Uttar Pradesh

Babur's Bāolī, Fatehpur Sīkrī.
Bāolī, near Baḍā Imāmbāṛā, Lucknow.
Bāolī Buland Darwāẓā, Fatehpur Sīkrī.
Hāthī Pol Bāolī, Fatehpur Sīkrī.
Kañcana Kuṇḍa, Mathurā.
Maṇikarṇikā Kuṇḍa, Varanasi.
Rādhā Kuṇḍa, Mathurā.
Śyāma Kuṇḍa, Mathurā.

# Appendix B

# Glossary

# Glossary

**ahar** (Indic) (aka. *ceruvu, dhora, khadin*): The term refers to a catchment area or basin found in Bihar. See *khadin*.

**ap** (Indic): The term refers to water, one of the five elements (Ind.: *pañca-tanmātras*).

**axiality**: The symmetrical arrangement of the elements of a plan along a given axis. If that axis is the east–west axis, then the northern half is a duplicate of the southern half.

**bandhara** (Indic): The term refers to a check dam or weir that diverts water into *ahar, dhora, ceruvu* and *khadin*.

**bahulī** (Indic) (aka. *bāolī, bauḍī, baulī, bauṛī, bāv, bāvaḍī, bāw, bāwḍī, bāwṛī, boulī, pairayachi vihir, vai, vāpī, vāpikā, vāv, vāvaḍī, wāv*): One of the numerous terms referring to a well, specifically a step well. See *step well*.

**bāolī** (Indic) (aka. *bahulī, bauḍī, baulī, bauṛī, bāv, bāvaḍī, bāw, bāwḍī, bāwṛī, boulī, pairayachi vihir, vai, vāpī, vāpikā, vāv, vāvaḍī, wāv*): One of the numerous terms referring to a well, specifically a step well. See *step well*.

**bārādarī** (Indic): The term strictly refers to an open pavilion with three openings on each side — i.e. supported by twelve columns along the outside perimeter. However, it is often used to describe any square, open pavilion with a roof structure supported by columns.

**bauḍī** (Indic) (aka. *bahulī, bāolī, baulī, bauṛī, bāv, bāvaḍī, bāw, bāwḍī, bāwṛī, boulī, pairayachi vihir, vai, vāpī, vāpikā, vāv, vāvaḍī, wāv*): One of the numerous terms referring to a well, specifically a step well. See *step well*.

**baulī** (aka. *bahulī, bāolī, bauḍī, bauṛī, bāv, bāvaḍī, bāw, bāwḍī, bāwṛī, boulī, pairayachi vihir, vai, vāpī, vāpikā, vāv, vāvaḍī, wāv*): One of the numerous terms referring to a well, specifically a step well. See *step well*.

**bauṛī** (aka. *bahulī, bāolī, bauḍī, baulī, bāv, bāvaḍī, bāw, bāwḍī, bāwṛī, boulī, pairayachi vihir, vai, vāpī, vāpikā, vāv, vāvaḍī, wāv*): One of the numerous terms referring to a well, specifically a step well. See *step well*.

**bāv** (aka. *bahulī, bāolī, bauḍī, baulī, bauṛī, bāvaḍī, bāw, bāwḍī, bāwṛī, boulī, pairayachi vihir, vai, vāpī, vāpikā, vāv, vāvaḍī, wāv*): One of the numerous terms referring to a well, specifically a step well. See *step well*.

**bāvaḍī** (Indic) (aka. *bahulī, bāolī, bauḍī, baulī, bauṛī, bāv, bāw, bāwḍī, bāwṛī, boulī, pairayachi vihir, vai, vāpī, vāpikā, vāv, vāvaḍī, wāv*): One of the numerous terms referring to a well, specifically a step well. See *step well*.

**bāw** (Indic) (aka. *bahulī, bāolī, bauḍī, baulī, bauṛī, bāv, bāvaḍī, bāwḍī, bāwṛī, boulī, pairayachi vihir, vai, vāpī, vāpikā, vāv, vāvaḍī, wāv*): One of the numerous terms referring to a well, specifically a step well. See *step well*.

**bāwḍī** (Indic) (aka. *bahulī, bāolī, bauḍī, baulī, bauṛī, bāv, bāvaḍī, bāw, bāwṛī, boulī, pairayachi vihir, vai, vāpī, vāpikā, vāv, vāvaḍī, wāv*): One of the numerous terms referring to a well, specifically a step well. See *step well*.

**bāwṛī** (Indic) (aka. *bahulī, bāolī, bauḍī, baulī, bauṛī, bāv, bāvaḍī, bāw, bāwḍī, boulī, pairayachi vihir, vai, vāpī, vāpikā, vāv, vāvaḍī, wāv*): One of the numerous terms referring to a well, specifically a step well. See *step well*.

**birka** (Arab.) (aka. *fisqiya, haud, saqaya, sihrij*): The term refers to an ablution tank that is normally found within the confines of a *masjid*.

**boulī** (Indic) (aka. *bahulī, bāolī, bauḍī, baulī, bauṛī, bāv, bāvaḍī, bāw, bāwḍī, bāwṛī, pairayachi vihir, vai, vāpī, vāpikā, vāv, vāvaḍī, wāv*): One of the numerous terms referring to a well, specifically a step well. See *step well*.

**cera** (Indic) (aka. *kalayni, karai* (Mysore), *kohil, kuḷam, kuṇḍa, puṣkariṇī, sāgara, sarobara, sarovara, tank, tīrtham*): One of a number of terms that refers to a pool or tank associated with a temple and used in ritual bathing particularly in Kerala and the south. It may often be a scared site in and of itself. See *tank*.

**ceruvu** (Indic) (aka. *ahar, dhora, khadin*): The term refers to a catchment area or basin found in Chitoor and Cuḍḍapah districts. See *khadin*.

**cross-axiality**: The symmetrical arrangement of the elements of a plan along both axes, precisely duplicating each quadrant. This term is not applied to rectangular plans. The north–south and east-west axes divide the form. The north-west quadrant is a duplicate of the north–east, south-west and south-east quadrants. This quadrant can either be mirrored in any of the other three or rotated 90°, 180° and 270° to achieve the whole.

**curtain wall** (aka. girdle wall): See *girdle wall*.

**dhora** (Indic) (aka. *ahar, ceruvu, khadin*): The term refers to a catchment area or basin utilized for irrigation. See *khadin*.

**dungs** (Indic) (aka. *jampois, pynes*): The term refers to artificial irrigation channels in West Bengal.

**eri** (Indic) (aka. *kere, taḍāka, tālāo, tālāv, yert, zing*): The term refers to a pond, a reservoir for water. See *tālāo*.

**fisqiya** (Arab.) (aka. *birka, haud, saqaya, sihrij*): The term refers to an ablution tank that is normally found within the confines of a *masjid*.

**gaṛh mahal** (Indic): The term refers to a fortified residence, a citadel.

**girdle wall** (aka. curtain wall): The term refers to a wall surrounding a structure or group of structures.

**haud** (Arab.) (aka. *birka, fisqiya, saqaya, sihrij*): The term refers to an ablution tank found in a *sahn*.

**jampois** (Indic) (aka. *dungs, pynes*): The term refers to artificial irrigation channels in West Bengal.

**johara wells**: The term refers to a well for cattle or other livestock. It is constructed as a wide trench with a shallow ramp leading down to the water's edge.

**kalayni** (Indic) (aka. *cera, karai* (Mysore), *kohil, kuḷam, kuṇḍa, puṣkariṇī, sāgara, sarobara, sarovara, tank, tīrtham*): One of a number of terms that refers to a pool or tank associated with a temple and used in ritual bathing. It may often be a scared site in and of itself. See *tank*.

**karai** (Mysore) (aka. *cera, kalayni, kohil, kuḷam, kuṇḍa, puṣkariṇī, sāgara, sarobara, sarovara, tank, tīrtham*): One of a number of terms that refers to a pool or tank associated with a temple and used in ritual bathing. It may often be a scared site in and of itself. See *tank*.

**kere** (Indic) (aka. *eri, taḍāka, tālāo, tālāv, yert, zing*): The term refers to an artificial lake or large tank. See *tālāo*.

**khadin** (Indic) (aka. *ahar, ceruvu, dhora*): The term refers to a catchment area or basin, usually at the base of a shallow slope with an earthen or masonry dam (*khadin bund*) frequently found on three sides which is constructed to hold back the water. Its main use was for irrigation.

**khadin bund** (Indic): The term refers to an earthen or masonry dam built at the base of a shallow slope to hold back water, forming a *khadin*.

**kohil** (Indic) (aka. *cera, kalayni, karai* (Mysore), *kuḷam, kuṇḍa, puṣkariṇī, sāgara, sarobara, sarovara, tank, tīrtham*): One of a number of terms that refers to a pool or tank associated with a temple and used in ritual bathing. It may often be a scared site in and of itself. See *tank*.

**kuḷam** (Indic) (aka. *cera, kalayni, karai* (Mysore), *kohil, kuṇḍa, puṣkariṇī, sāgara, sarobara, sarovara, tank, tīrtham*): One of a number of terms that refers to a pool or tank associated with a temple and used in ritual bathing particularly in Kerala and the south. It may often be a scared site in and of itself. See *tank*.

**kuṇḍa** (Indic) (aka. *cera, karai* (Mysore), *kohil, kuḷam, puṣkariṇī, sāgara, sarobara, sarovara, tank, tīrtham*): One of a number of terms that refers to a pool or tank associated with a temple and used in ritual bathing. *En forme*, it descends to the water level in a series of steps. It may often be a scared site in and of itself and differs from a *wāv* or *bāolī* in that it is essentially a bathing tank rather than a source of drinking water. See *tank*.

**kuṇḍī** (Indic) (aka. *kuṇḍa*): A variation on the term *kuṇḍa*. See *kuṇḍa*.

**kuā** (Indic) (aka. *kuo, kūpa*): Another term referring to a simple well. See *kūpa*.

**kuo** (Indic) (aka. *kuā, kūpa*): Another term referring to a simple well. See *kūpa*.

**kūpa** (Indic) (aka. *kuā, kuo*): A simple well in which a shaft is dug into the ground until the aquefer is reached. Water is extracted via a bucket attached to a long rope. Also, the term employed to refer to the shaft of a stepped well.

**mirror cross-axiality**: The symmetrical arrangement of the elements of a plan along both axes, mirroring each quadrant. This term is applied to rectangular plans. The north–south and east–west axes divide the form. The north-west quadrant is a mirror duplicate of the north-east, south-west and south-east quadrants. This quadrant can be mirrored in any of the other three. It cannot be rotated to achieve the whole as can be seen in the cross-axial plan.

**nirazhi maṇḍapa** (Indic): A term that refers to a small shrine or structure in the centre of a tank.

**pañca-tanmātras** (Indic): A term that refers to the five basic elements. They are: earth (Ind.: *kṣiti*), ether (Ind.: *ākāśa*), wind (Ind.: *marut*), water (Ind.: *ap*) and fire (Ind.: *tejas*).

**prākāra** (Indic): A term that refers to an area within a temple complex that is enclosed by a girdle wall. Many of the southern temples may have a number of enclosing, nested walls and, therefore, a number of *prākāras*.

**pairayachi vihir** (Indic) (aka. *bahulī, bāolī, bauḍī, baulī, baurī, bāv, bāvaḍī, bāw, bāwḍī, bāwṛī, boulī, vai, vāpī, vāpikā, vāv, vāvaḍī, wāv*): One of the numerous terms referring to a well, specifically a step well. See *step well*.

**pūjā** (Indic): The term refers to prayers or worship.

**puṣkariṇī** (Indic) (aka. *cera, kalayni, karai* (Mysore), *kohil, kuḷam, kuṇḍa, sāgara, sarobara, sarovara, tank, tīrtham*): One of a number of terms that refers to a pool or tank associated with a temple and used in ritual bathing particularly in Andhra Pradesh. It may often be a scared site in and of itself. See *tank*.

**pynes** (Indic) (aka. *dungs, jampois*): The term refers to artificial irrigation channels associated with *ahars* in Bihar.

**rise:** The term "rise" refers to the height of a step in a flight of stairs from one "tread" to the next.

**sāgara** (Indic) (aka. *cera, kalayni, karai* (Mysore), *kohil, kuḷam, kuṇḍa, puṣkariṇī, sarovara, tank, tīrtham*): The term refers to a lake and is used to describe the tanks of Dig. See *tank*.

**śālā** (Indic): The term refers to an open pillar supported roofed structure.

**sarobara** (Indic) (aka. *cera, kalayni, karai* (Mysore), *kohil, kuḷam, kuṇḍa, puṣkariṇī, sāgara, sarovara, tank, tīrtham*): One of a number of terms that refers to a pool or tank associated with a temple and used in ritual bathing. It may often be a scared site in and of itself. See *tank*.

**sarovara** (Indic) (aka. *cera, kalayni, karai* (Mysore), *kohil, kuḷam, kuṇḍa, puṣkariṇī, sāgara, sarobara, tank, tīrtham*): The term refers to a *tank* or *kuṇḍa* associated with a temple and used for ritual bathing. It is often considered as a sacred site. See *tank*.

**sihrij** (Arab.) (aka. *saqaya, birka, haud, fisqiya*): The term refers to an ablution tank that is normally found within the confines of a *masjid*.

**śodhana** (Indic): The term refers to ritual bathing, cleansing, especially every morning and before special prayers.

**somparas** (Indic): The term refers to the workers from the lower castes who were generally utilized in the excavation and construction of the step wells and tanks. They were not *sthapati*, but often possessed considerable knowledge in the construction of step wells and deep tanks.

**steptank** See *kuṇḍa*.

**step well** (aka. *bāolī, bauḍī, baulī, bāv, bāvaḍī, bāw, bāwḍī, bāwṛī, boulī*): The term refers to a shaft (*kūpa*) and trench dug into the earth until the water table is reached. The trench is sloped downward to the water source with a series of paved steps. Usually the sides of the trench are lined with rock and/or masonry. The step well is utilized as a source of drinking water. It also may afford spaces as a place to rest out of the blinding glare of the sun.

**sthapati** (Indic.): The term refers to an architect-builder, who in the Indian Hindu tradition was what the "master builder" was to the European Medieval tradition.

**taḍāka** (Indic) (aka. *eri, kere, tālāo, tālāv, yert, zing*): The term refers to a reservoir and is related to a *tālāo*. See *tālāo*.

**tālāo** (Indic) (aka. *eri, kere, taḍāka, tālāv, yert, zing*): The term refers to an artificial lake or large tank, often irregular in shape which may have steps leading down to the surface of the water at one point, but not surrounding the whole. The *tālāo* is often excavated either wholly or partially and lined with an impermeable material, generally clay, to inhibit the seeping of water into the surrounding soil. A *tālāo* is a feature often associated with a fortified city or palace.

**tālāv** (Indic) (aka. *eri, kere, taḍāka, tālāo, yert, zing*): The term refers to an artificial lake or large tank. See *tālāo*.

**tank** (Indic) (aka. *cera, kalayni, karai* (Mysore), *kohil, kuḷam, kuṇḍa, puṣkariṇī, sāgara, sarobara, sarovara, tīrtham*): The term refers to pools associated with temples and are used for ritual bathing. Tanks may be either within the precincts of the temple or, in some cases, beyond the outer wall. Generally, they are excavated from the surrounding soil, lined with clay or other impermeable material, and surrounded by a series of steps that leads down to and below the surface of the water.

**ṭaṅka** (Indic): The term refers to a small tank or reservoir for the storage of water. Generally, found in residences. It is lined with clay, brick and/or tile and covered.

**tīrtham** (Indic) (aka. *kalayni*, *karai* (Mysore), *kohil*, *kuḷam*, *kuṇḍa*, *puṣkariṇī*, *sāgara*, *sarobara*, *sarovara*, *tank*): One of a number of terms that refers to a pool or tank associated with a temple and used in ritual bathing. It may often be a scared site in and of itself. See *tank*.

**tread**: The term refers to the depth of a step in a staircase. It is that area upon which the foot is placed as the stair is mounted and separated one from the other "tread" by a "rise".

**vai** (Indic) (aka. *bahulī*, *bāolī*, *bauḍī*, *baulī*, *bauṛī*, *bāv*, *bāvaḍī*, *bāw*, *bāwḍī*, *bāwṛī*, *boulī*, *pairayachi vihir*, *vāpī*, *vāpikā*, *vāv*, *vāvaḍī*, *wāv*): One of the numerous terms referring to a well, specifically a step well. See *step well*.

**vāpī** (Indic) (aka. *bahulī*, *bāolī*, *bauḍī*, *baulī*, *bauṛī*, *bāv*, *bāvaḍī*, *bāw*, *bāwḍī*, *bāwṛī*, *boulī*, *pairayachi vihir*, *vai*, *vāpikā*, *vāv*, *vāvaḍī*, *wāv*): One of the numerous terms referring to a well, specifically a step well. See *step well*.

**vāpikā** (Indic) (aka. *bahulī*, *bāolī*, *bauḍī*, *baulī*, *bauṛī*, *bāv*, *bāvaḍī*, *bāw*, *bāwḍī*, *bāwṛī*, *boulī*, *pairayachi vihir*, *vai*, *vāpī*, *vāv*, *vāvaḍī*, *wāv*): One of the numerous terms referring to a well, specifically a step well. See *step well*.

**vāv** (Indic) (aka. *bahulī*, *bāolī*, *bauḍī*, *baulī*, *bauṛī*, *bāv*, *bāvaḍī*, *bāw*, *bāwḍī*, *bāwṛī*, *boulī*, *pairayachi vihir*, *vai*, *vāpī*, *vāpikā*, *vāvaḍī*, *wāv*): One of the numerous terms referring to a well, specifically a step well. See *step well*.

**vāvaḍī** (Indic) (aka. *bahulī*, *bāolī*, *bauḍī*, *baulī*, *bauṛī*, *bāv*, *bāvaḍī*, *bāw*, *bāwḍī*, *bāwṛī*, *boulī*, *pairayachi vihir*, *vai*, *vāpī*, *vāpikā*, *vāv*, *wāv*): One of the numerous terms referring to a well, specifically a step well. See *step well*.

**wāv** (Indic) (aka. *bahulī*, *bāolī*, *bauḍī*, *baulī*, *bauṛī*, *bāv*, *bāvaḍī*, *bāw*, *bāwḍī*, *bāwṛī*, *boulī*, *pairayachi vihir*, *vai*, *vāpī*, *vāpikā*, *vāv*, *vāvaḍī*): One of the numerous terms referring to a well, specifically a step well. See *step well*.

**woduk** (Arabic): The term refers to the act of ritual bathing that all Muslims perform before prayer.

**yert** (Indic) (aka. *eri*, *kere*, *taḍāka*, *tālāo*, *tālāv*, *zing*): The term refers to an artificial lake, particularly in Tamil Nadu. See *tālāo*.

**zing** (Indic) (aka. *eri*, *kere*, *taḍāka*, *tālāo*, *tālāv*, *yert*): The term refers to an artificial lake or a large tank in Ladakh. Frequently, they are constructed upon exposed bed-rock. See *tālāo*.

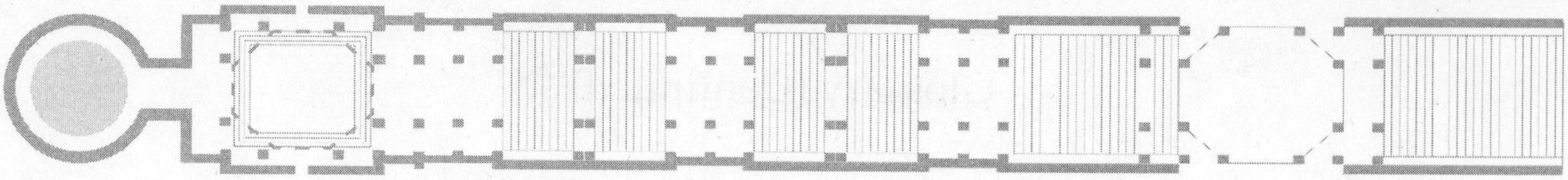

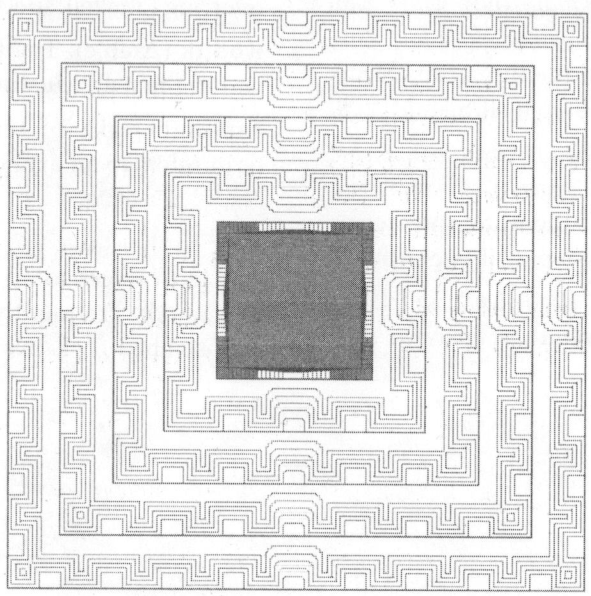

# Appendix C

# Miscellaneous Plans and Elevations

# **Plate 71:** Ta Keo, Angkor half plan (below) and proportional elevation (above)

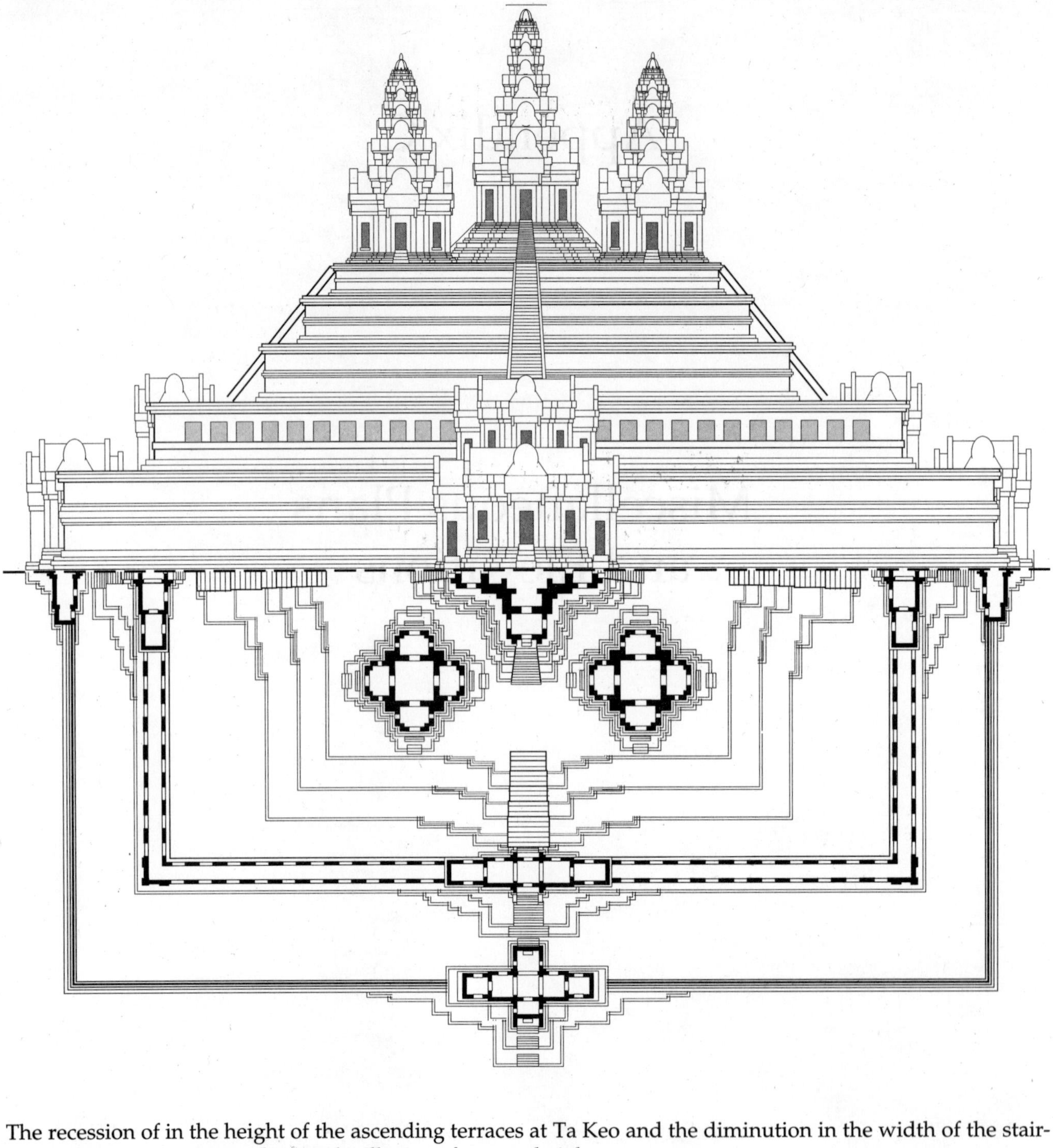

The recession of in the height of the ascending terraces at Ta Keo and the diminution in the width of the stairways as they move upwards give the illusion of greater height.

**Plate 72:** Jambukeśvarār Temple, Southern Gopura, proportional elevation

## Plate 73: Jambukeśvarār Temple, 3rd Gopura from the South, proportional elevation

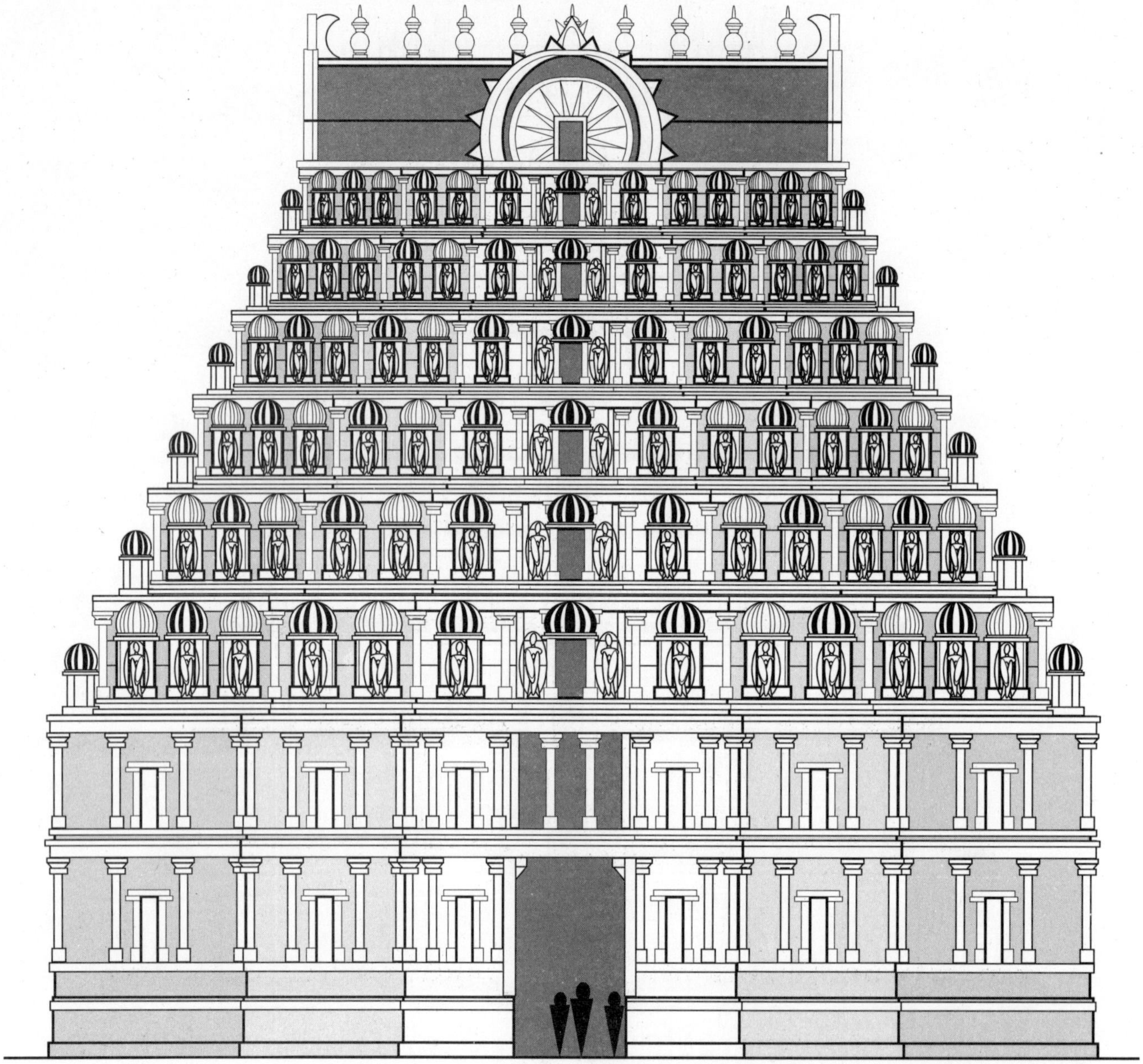

## Plate 74: Gandhak-kī Bāolī elevation towards shaft

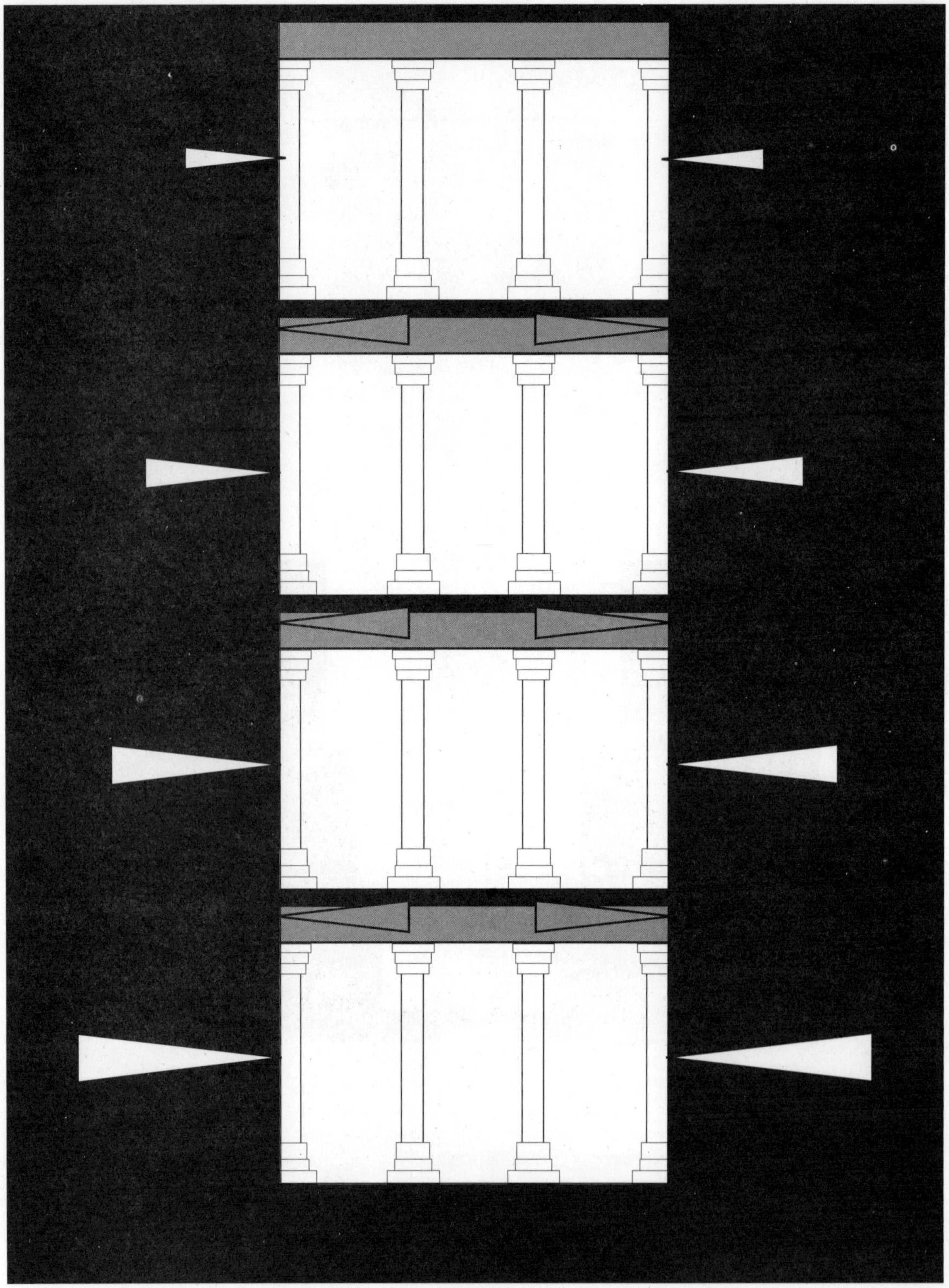

= counterthrust

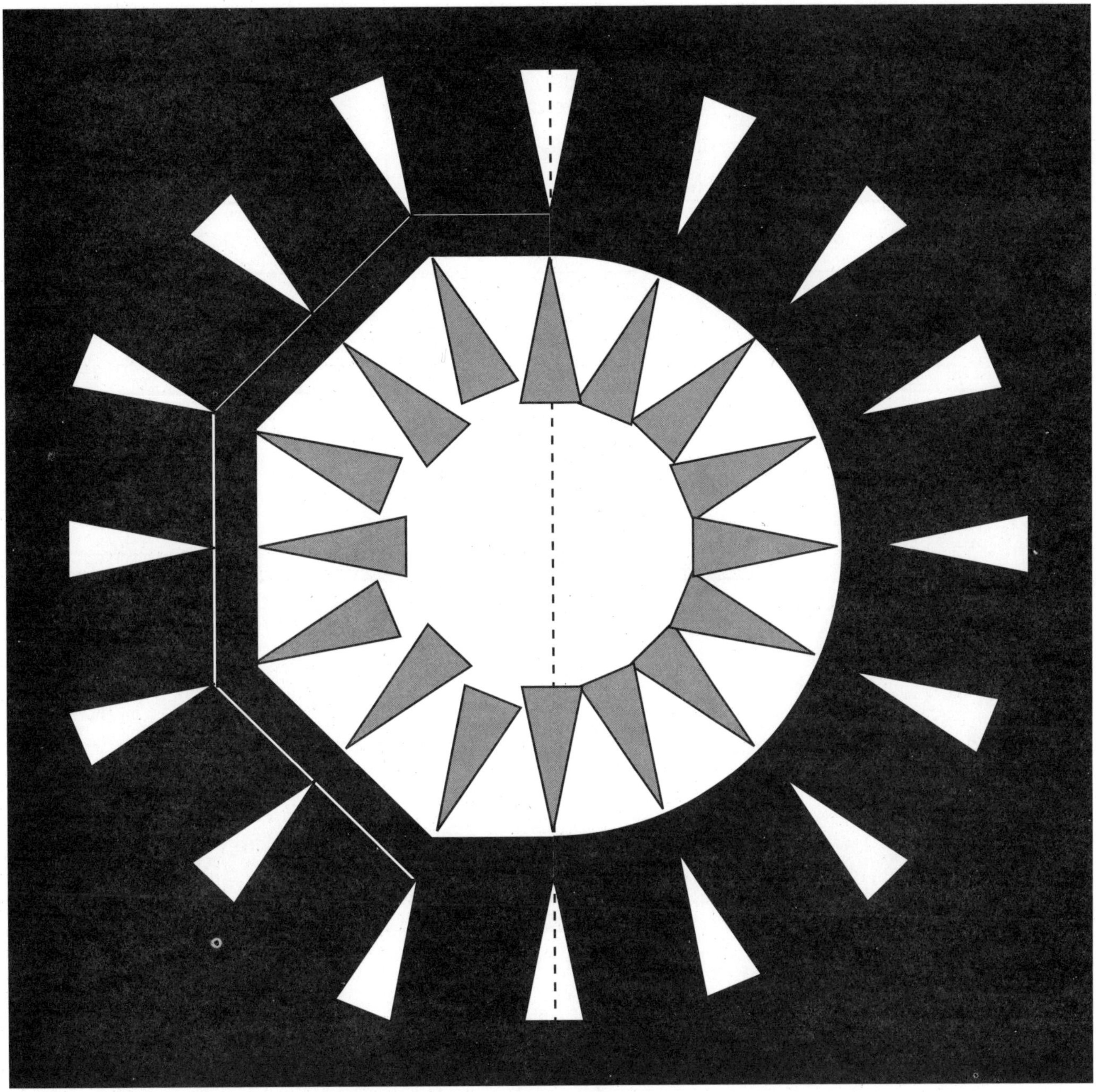

= counterthrust

## Plate 77: Rānī-kī Wāv, Pāṭan, proportional elevation at open colonnade

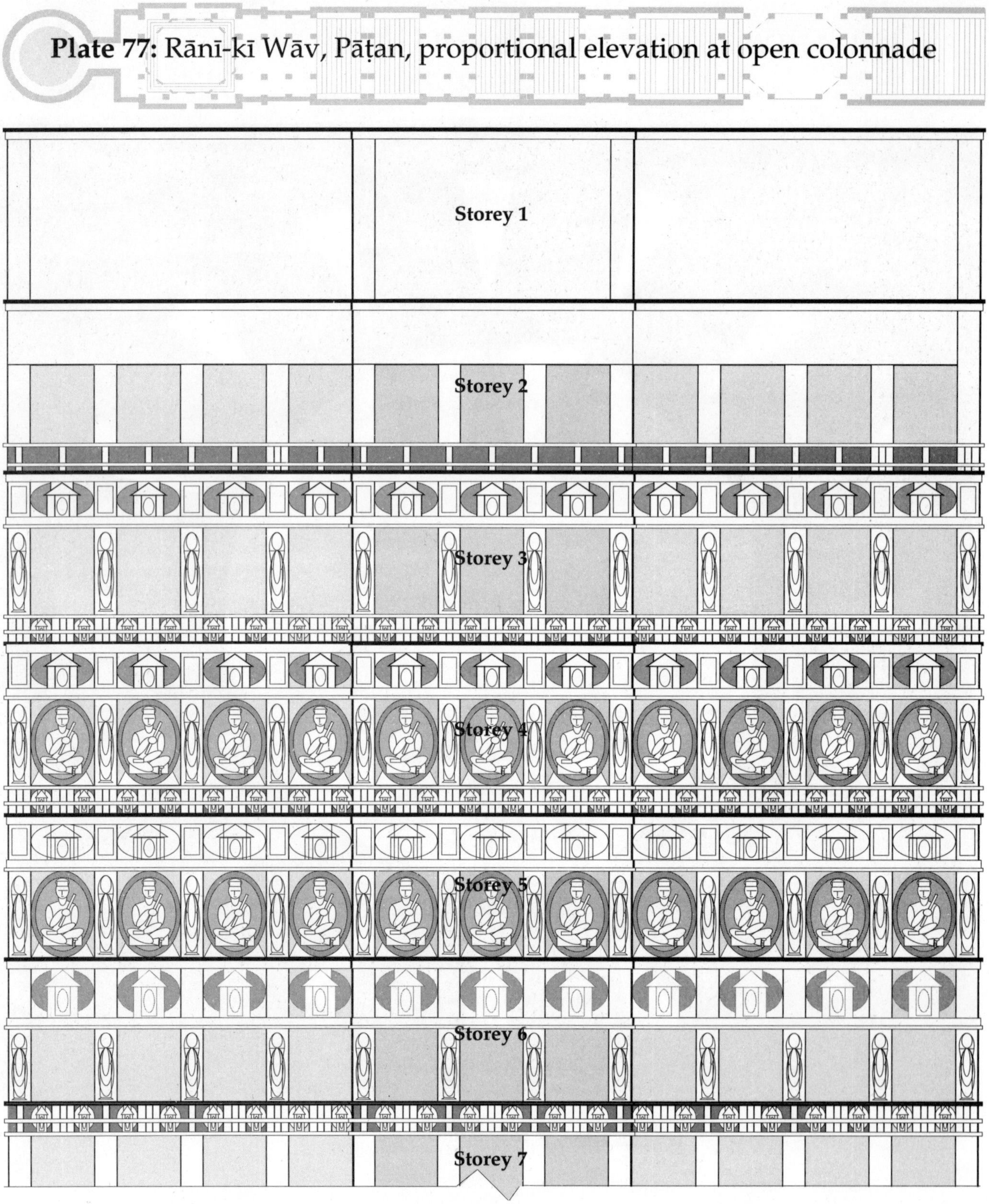

# Plate 78: Dādā Harīr Wāv proportional elevation (partial)

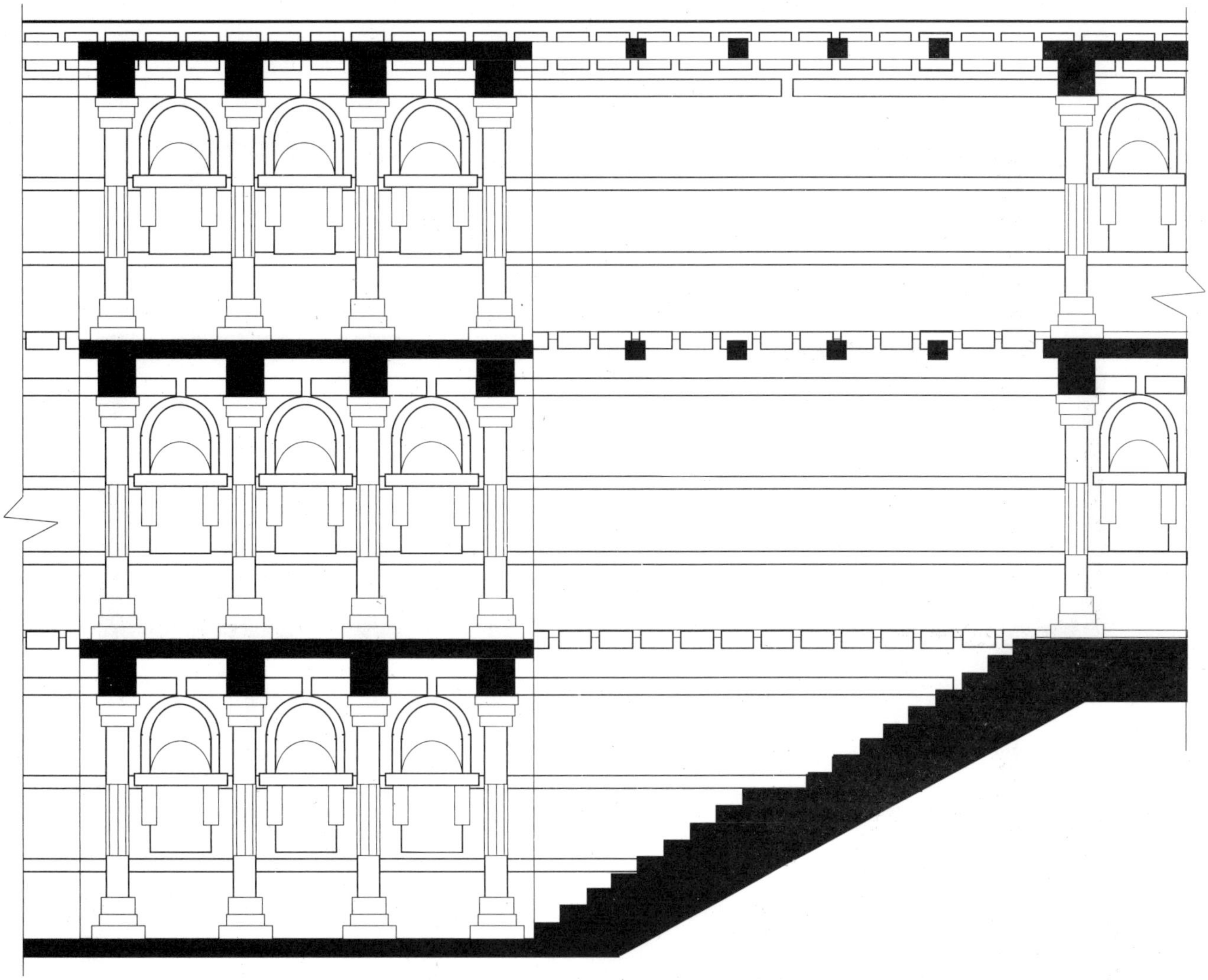

# Plate 79: Noblemen's Quarter with tank, Vijayanagara

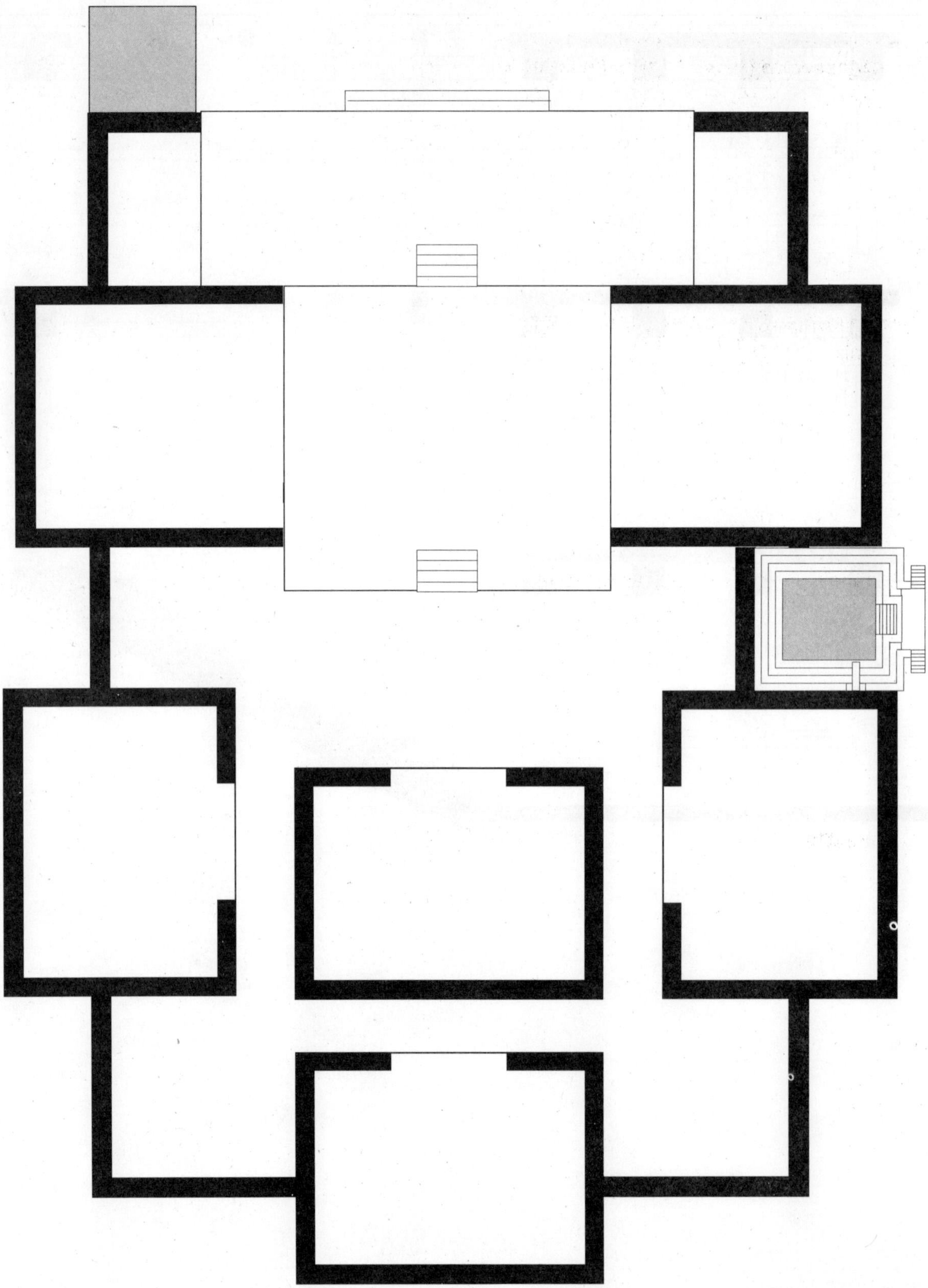

## Plate 80: Imperial Vijayanagara

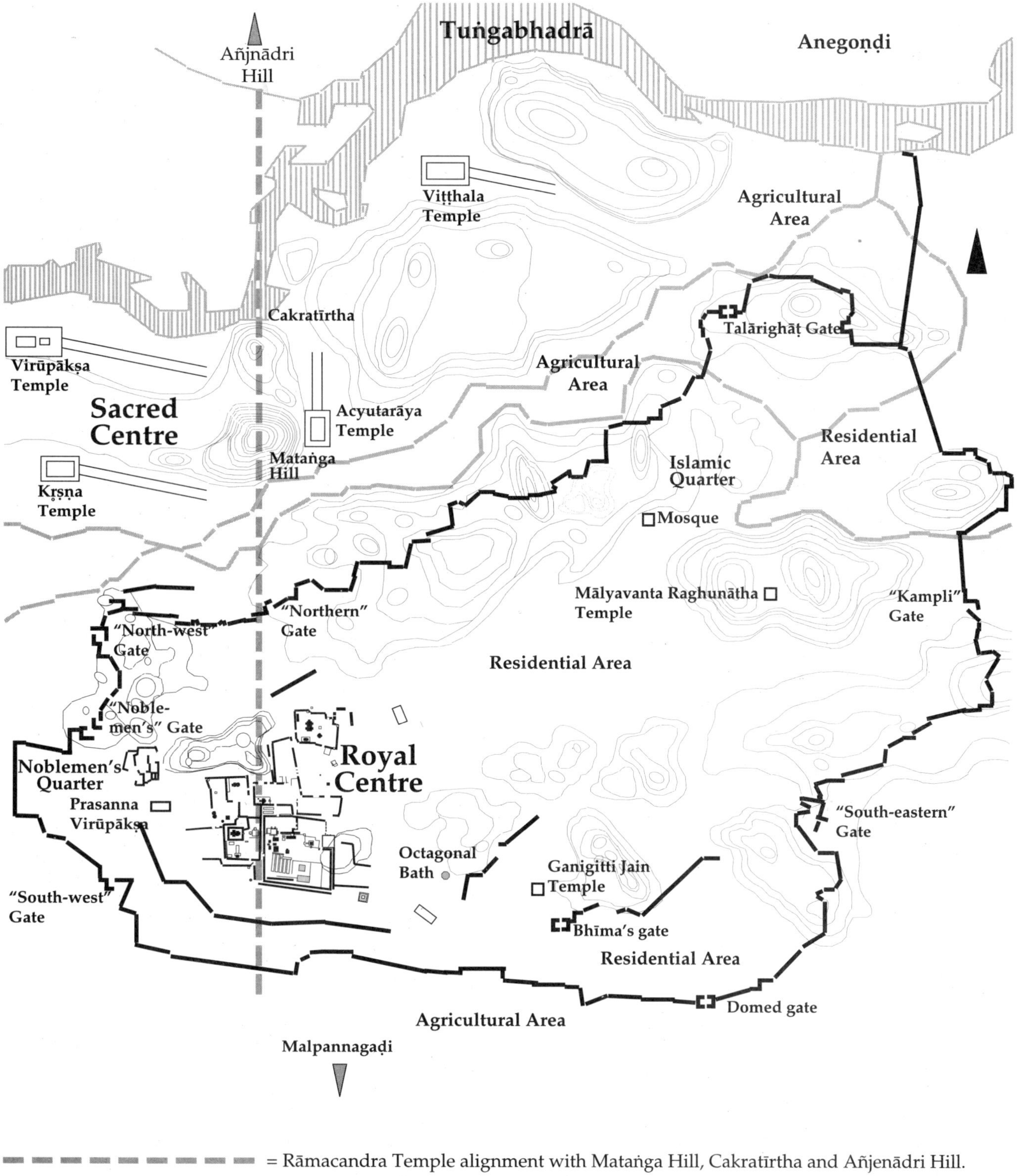


= Rāmacandra Temple alignment with Mataṅga Hill, Cakratīrtha and Añjenādri Hill.

# Plate 81: Sacred Centre, Vijayanagara, Hampī

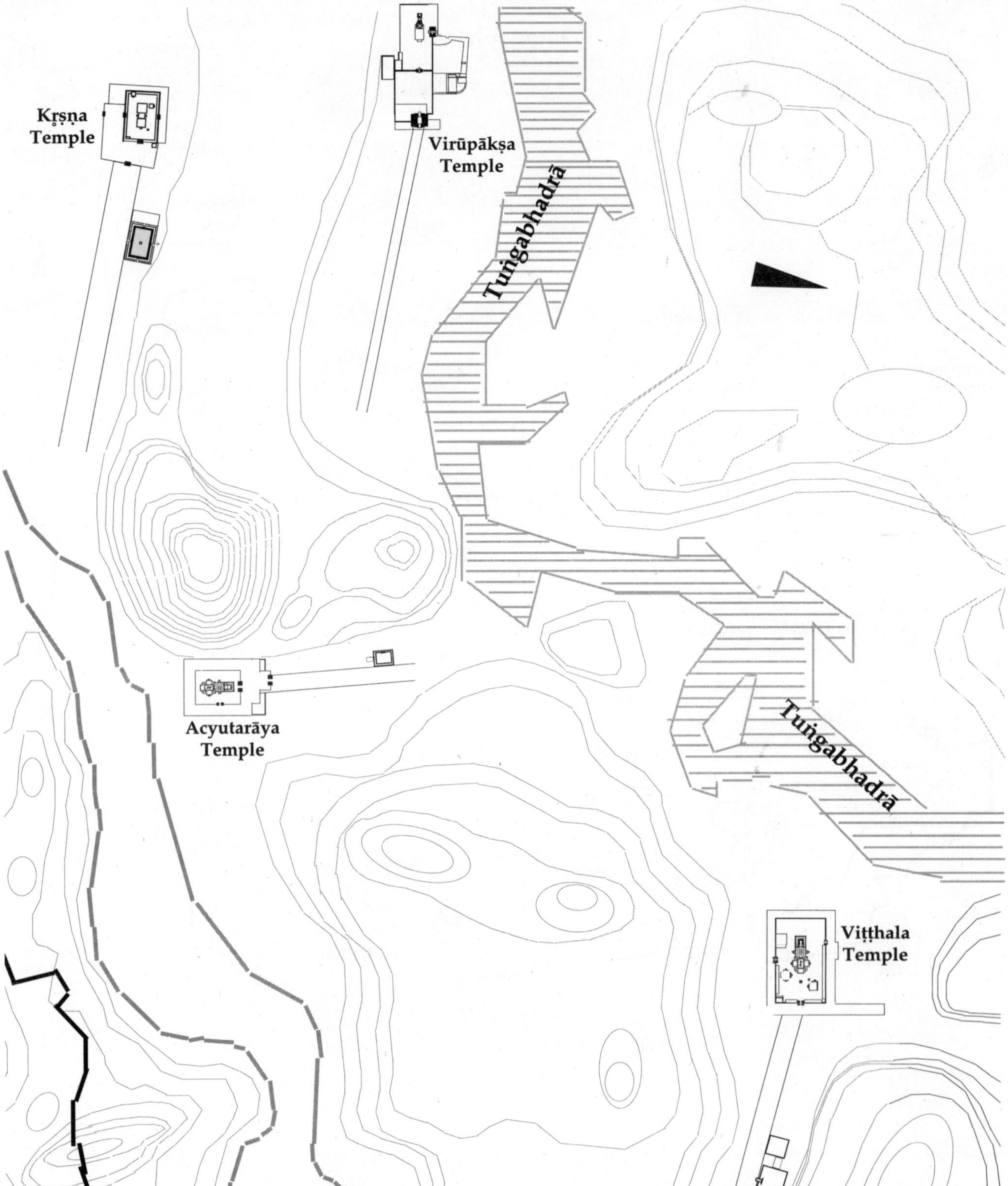

**Plate 82:** Virūpākṣapura, showing Processional Avenue, Hampī

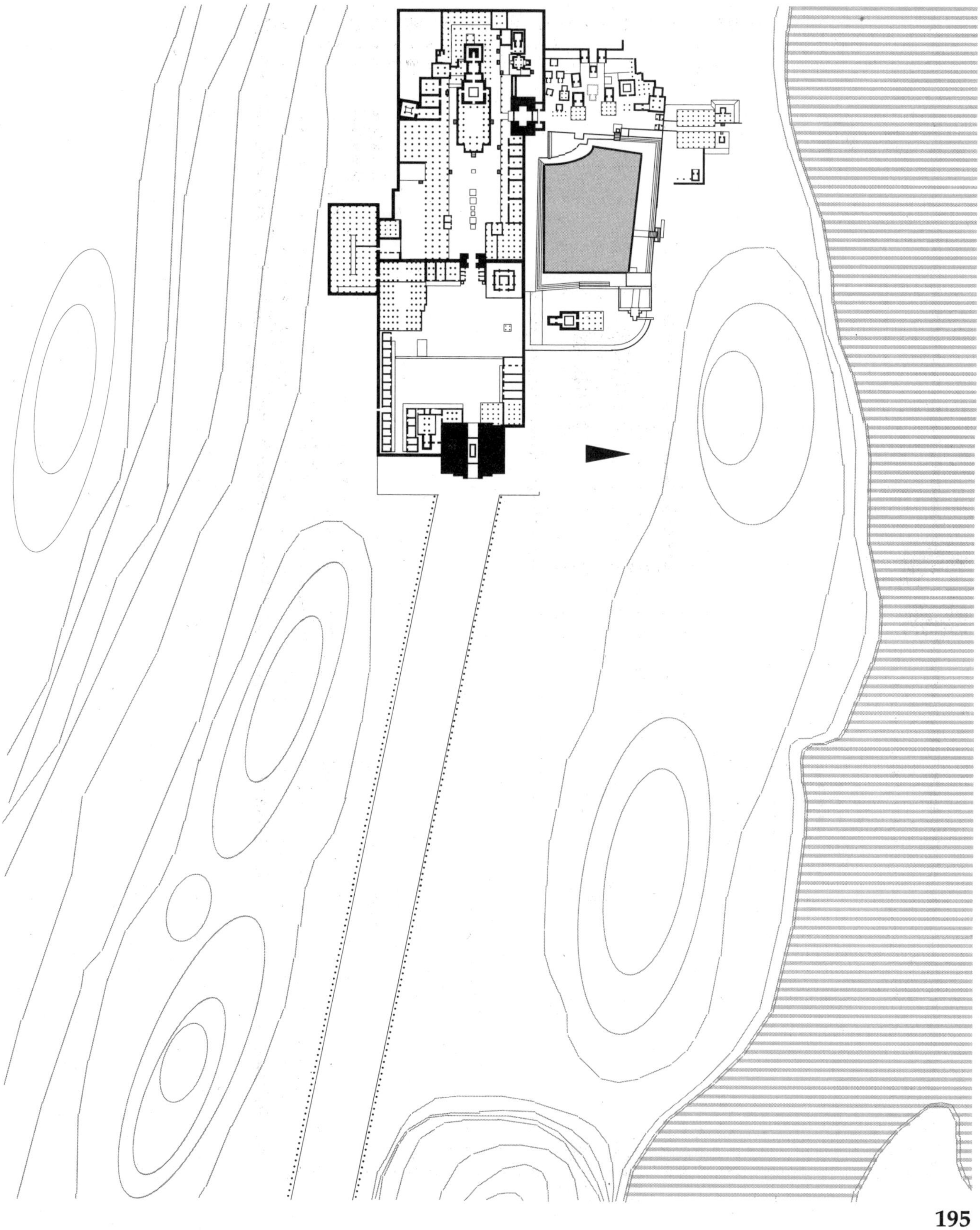

# Plate 83: Viṭṭhala Temple Complex, Hampī

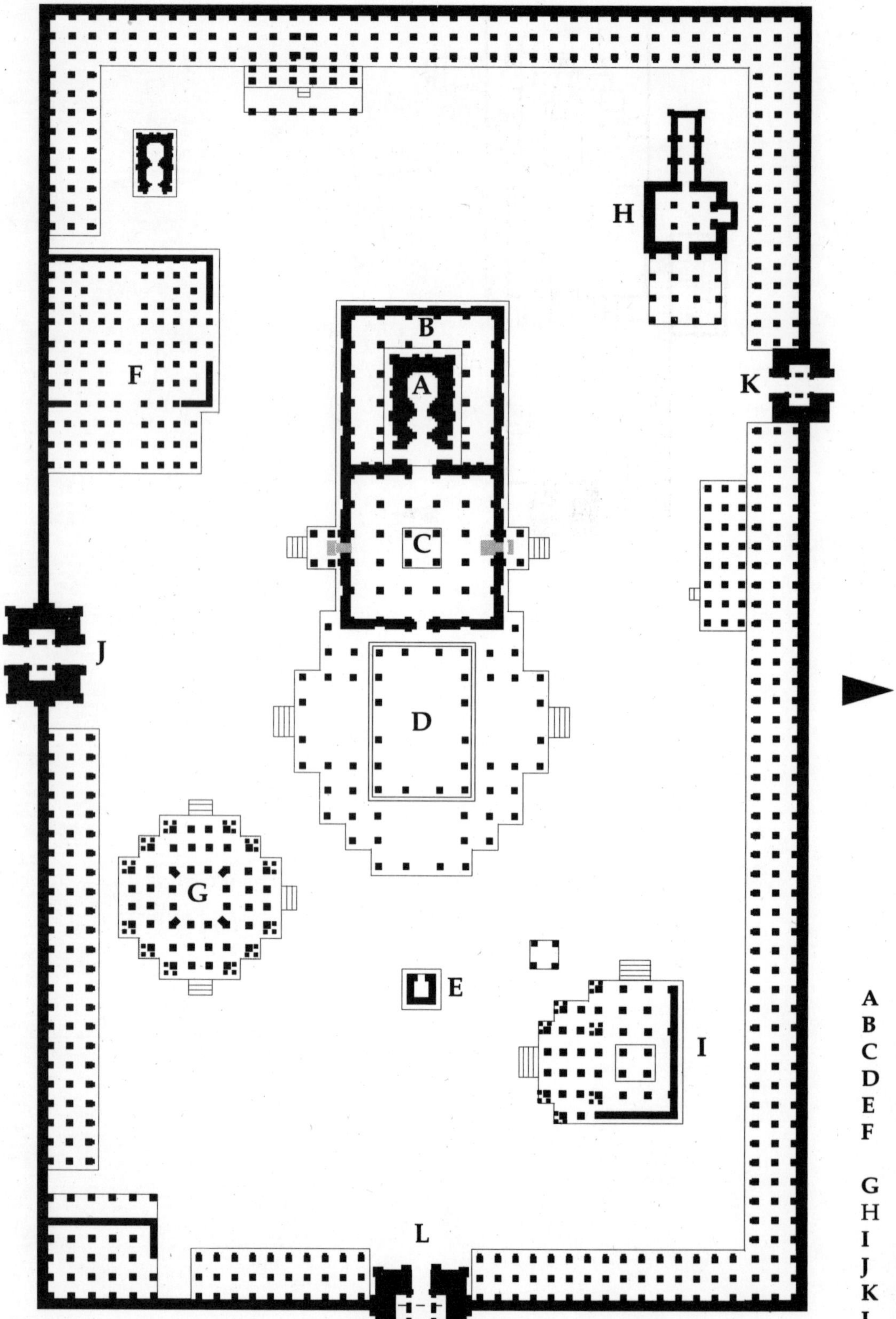


**A** *Garbhagṛha*
**B** *Pradakṣiṇāpatha*
**C** *Ardha-maṇḍapa*
**D** *Mahā-maṇḍapa*
**E** *Garuḍa-maṇḍapa*
**F** Hundred-pillared *maṇḍapa*
**G** *Kalyāṇa-maṇḍapa*
**H** Amman Temple
**I** *Utsava-maṇḍapa*
**J** South *gopuram*
**K** North *gopuram*
**L** East *gopuram*

## **Plate 84:** Kṛṣṇa Temple Complex, Hampī

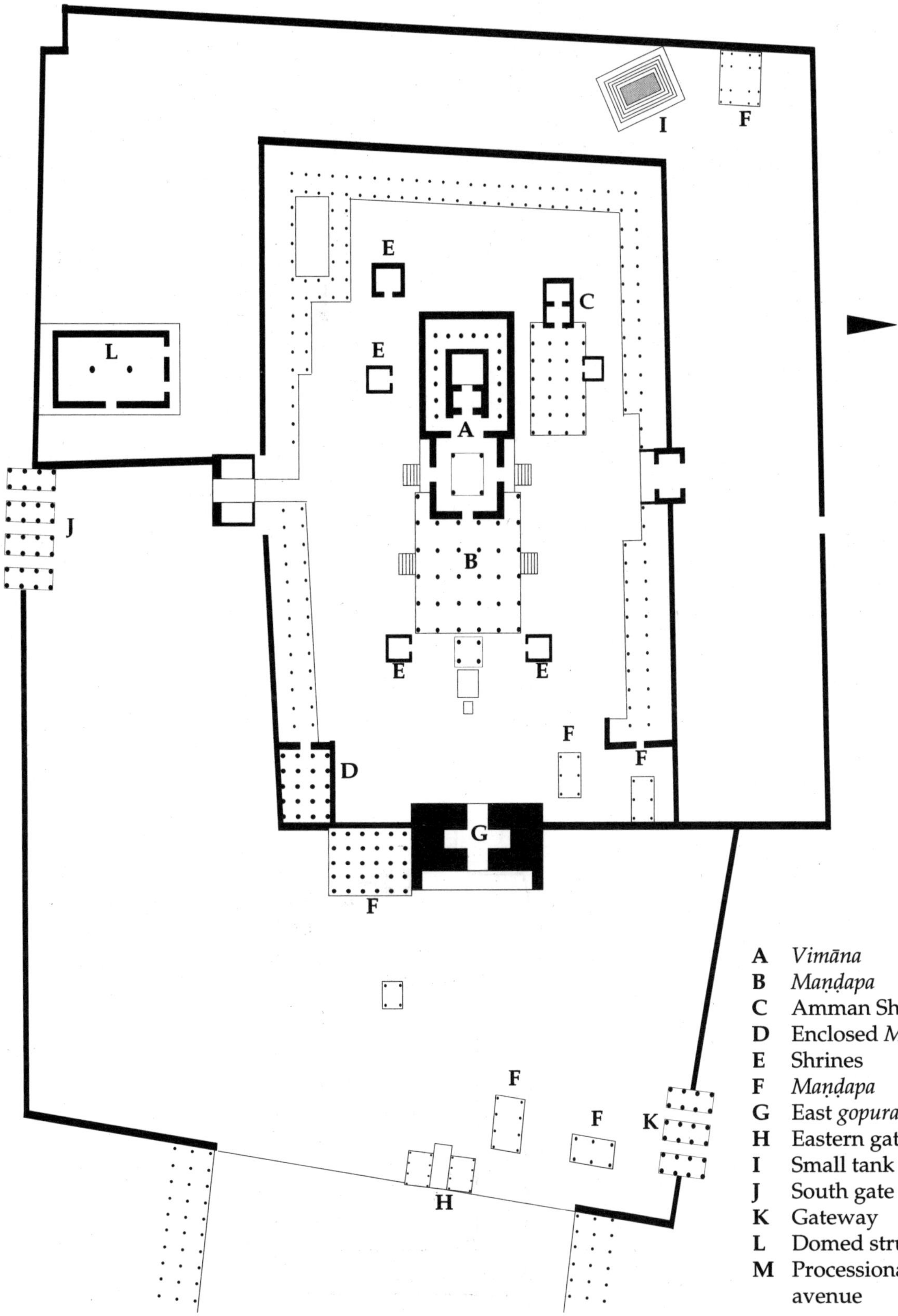


**A** *Vimāna*
**B** *Maṇḍapa*
**C** Amman Shrine
**D** Enclosed *Maṇḍapa*
**E** Shrines
**F** *Maṇḍapa*
**G** East *gopuram*
**H** Eastern gate
**I** Small tank
**J** South gate
**K** Gateway
**L** Domed structure
**M** Processional avenue

**Plate 85:** Tiruveṅgalanātha (Acyutarāyapura) Temple Complex, Hampī, indicating tank

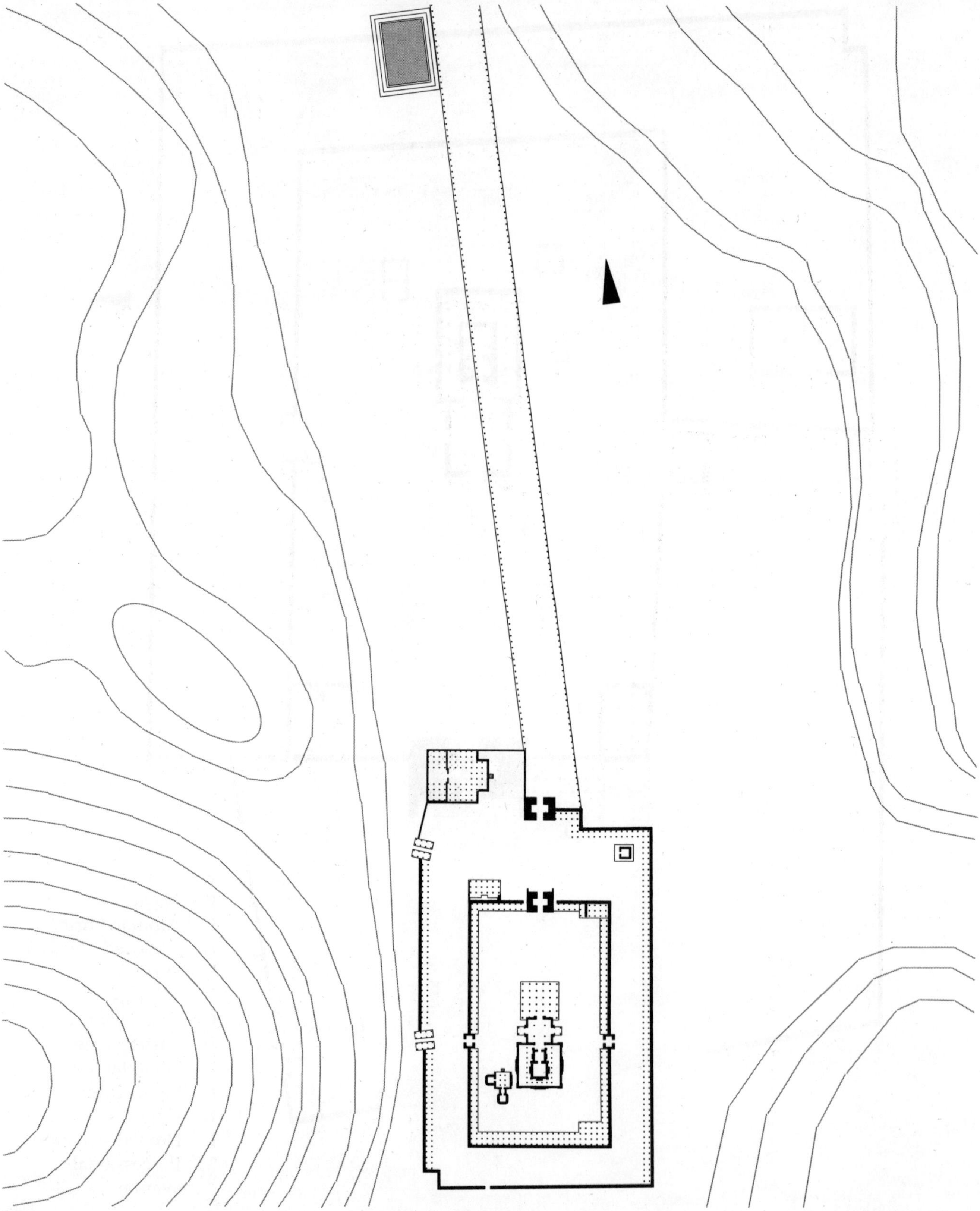

## Plate 86: Tiruveṅgalanātha (Acyutarāyapura) Temple Complex, Hampī

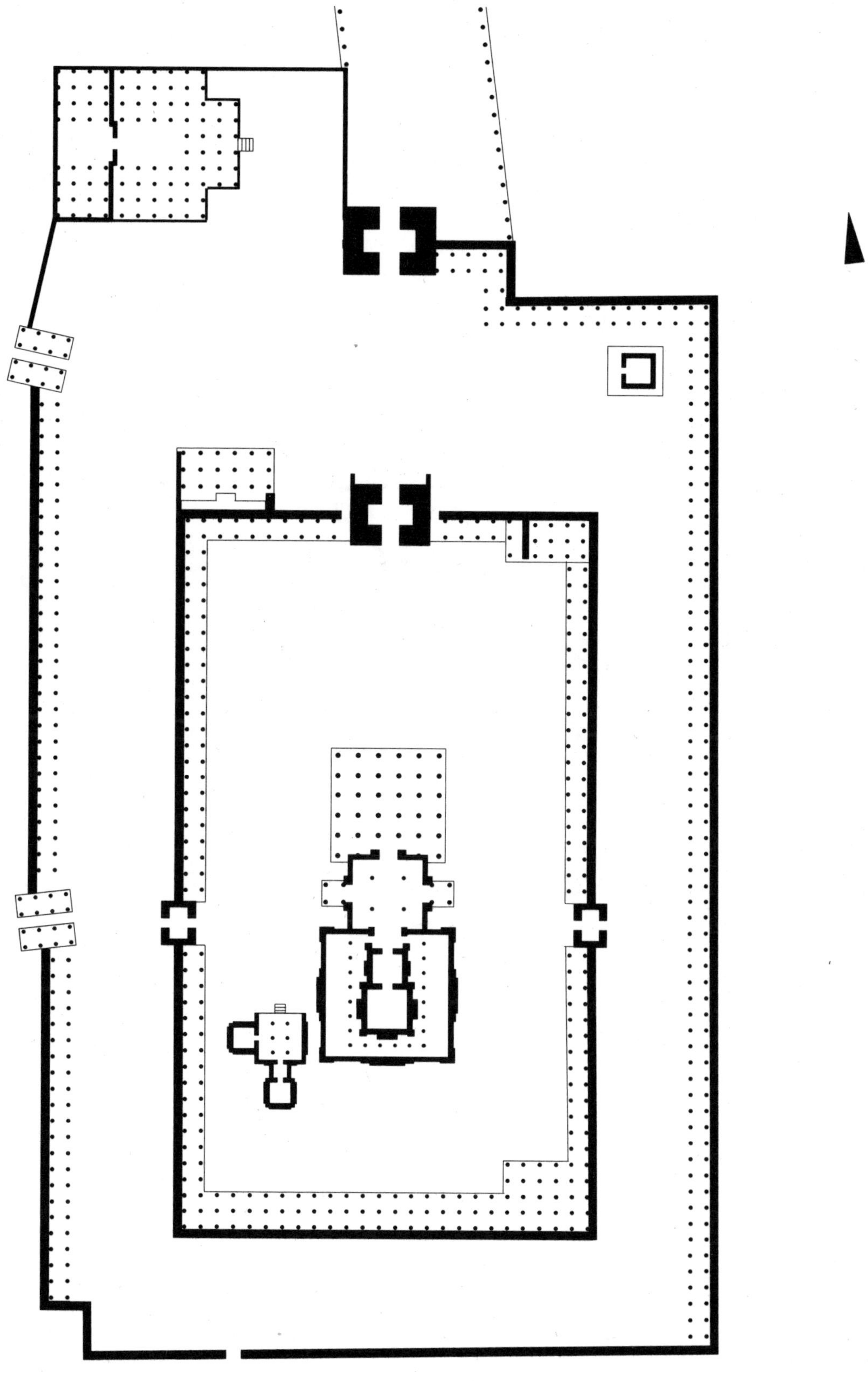

## Plate 87: Gingee Citadel, plan and topography (simplified)

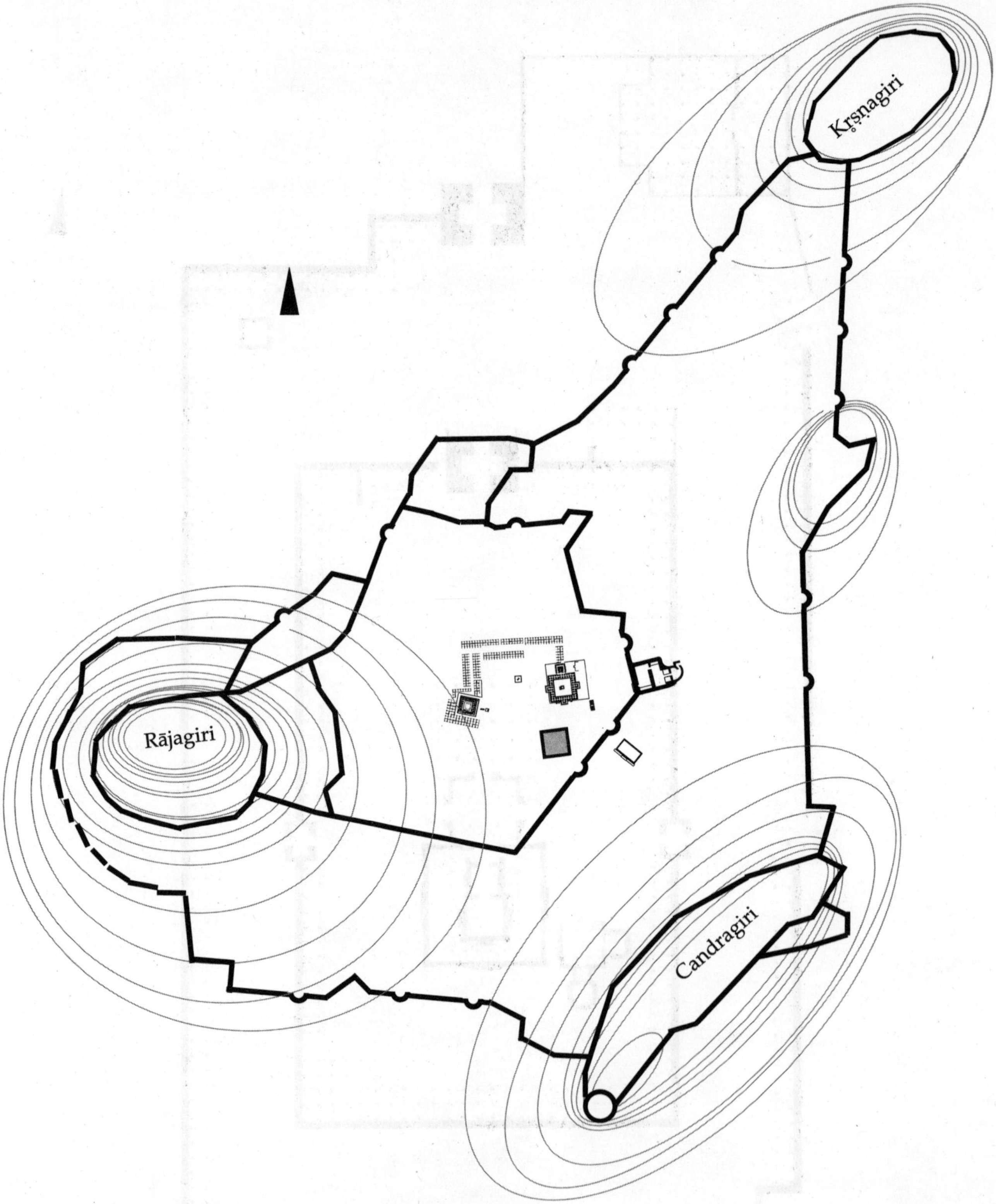

**Plate 88:** Rājagiri fortifications indicating the Kamalakanniammaṅkuḷam

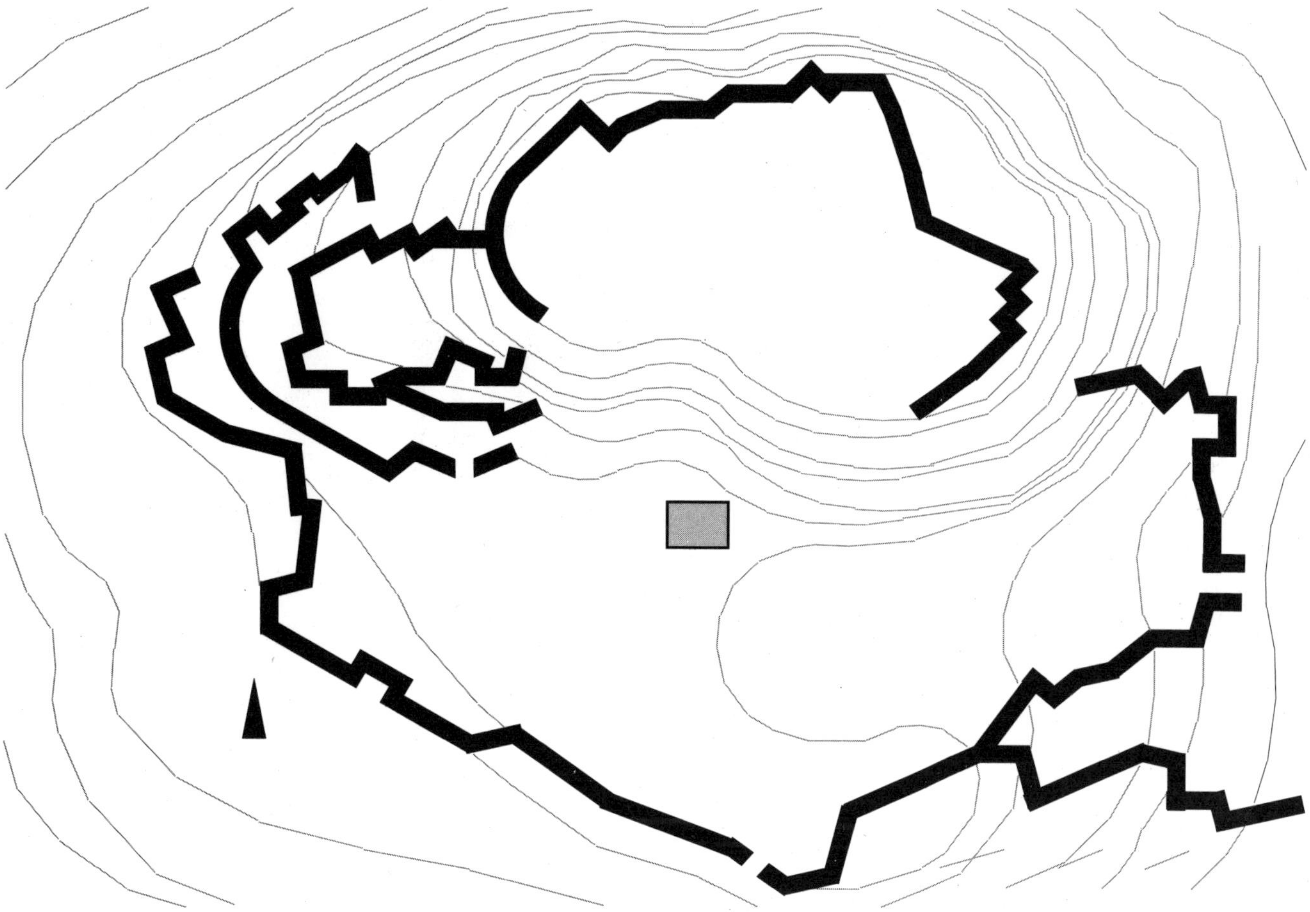

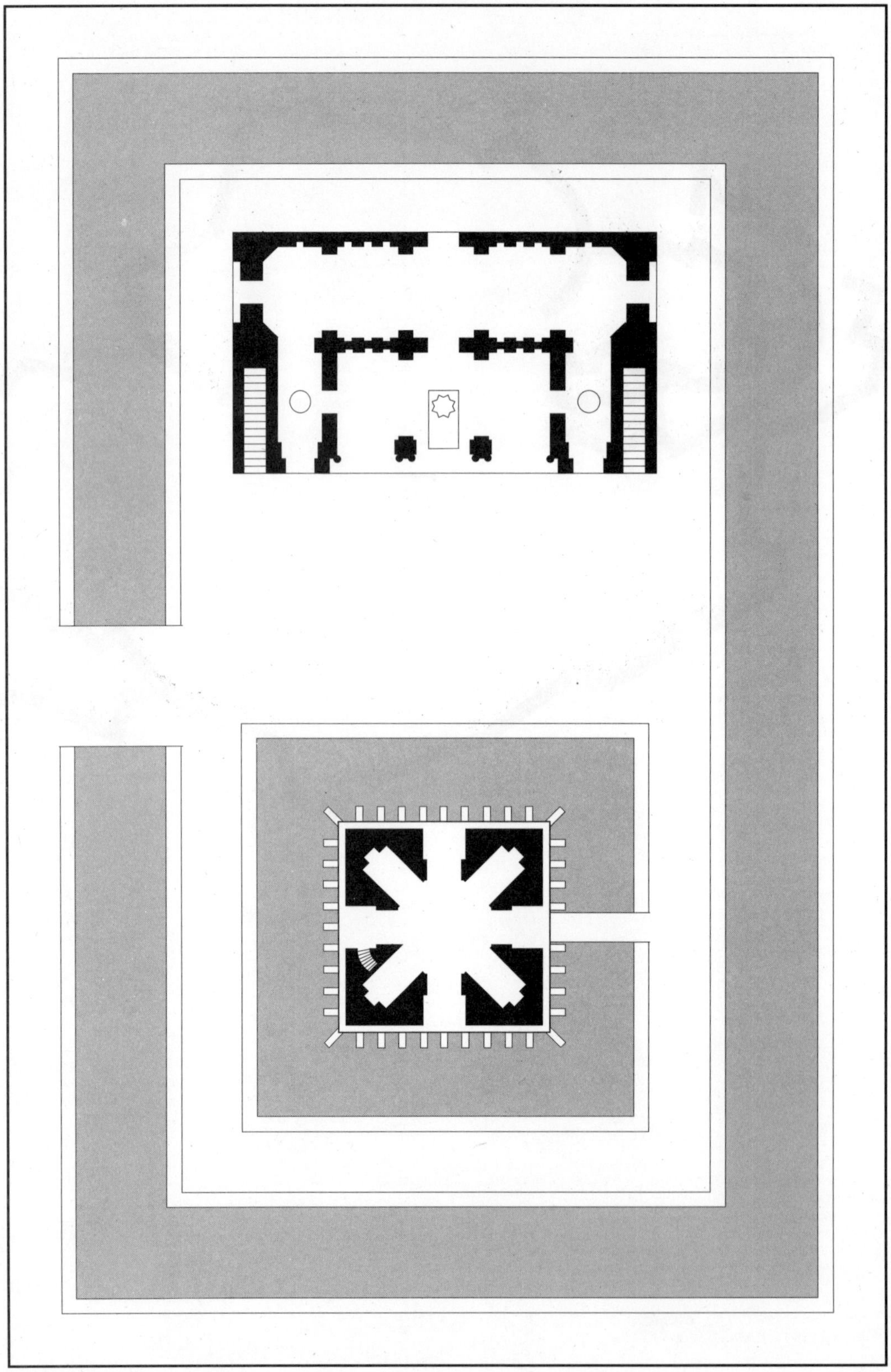

## **Plate 90:** Masjid-i Jāmī, Campāner (a. CE 1485)

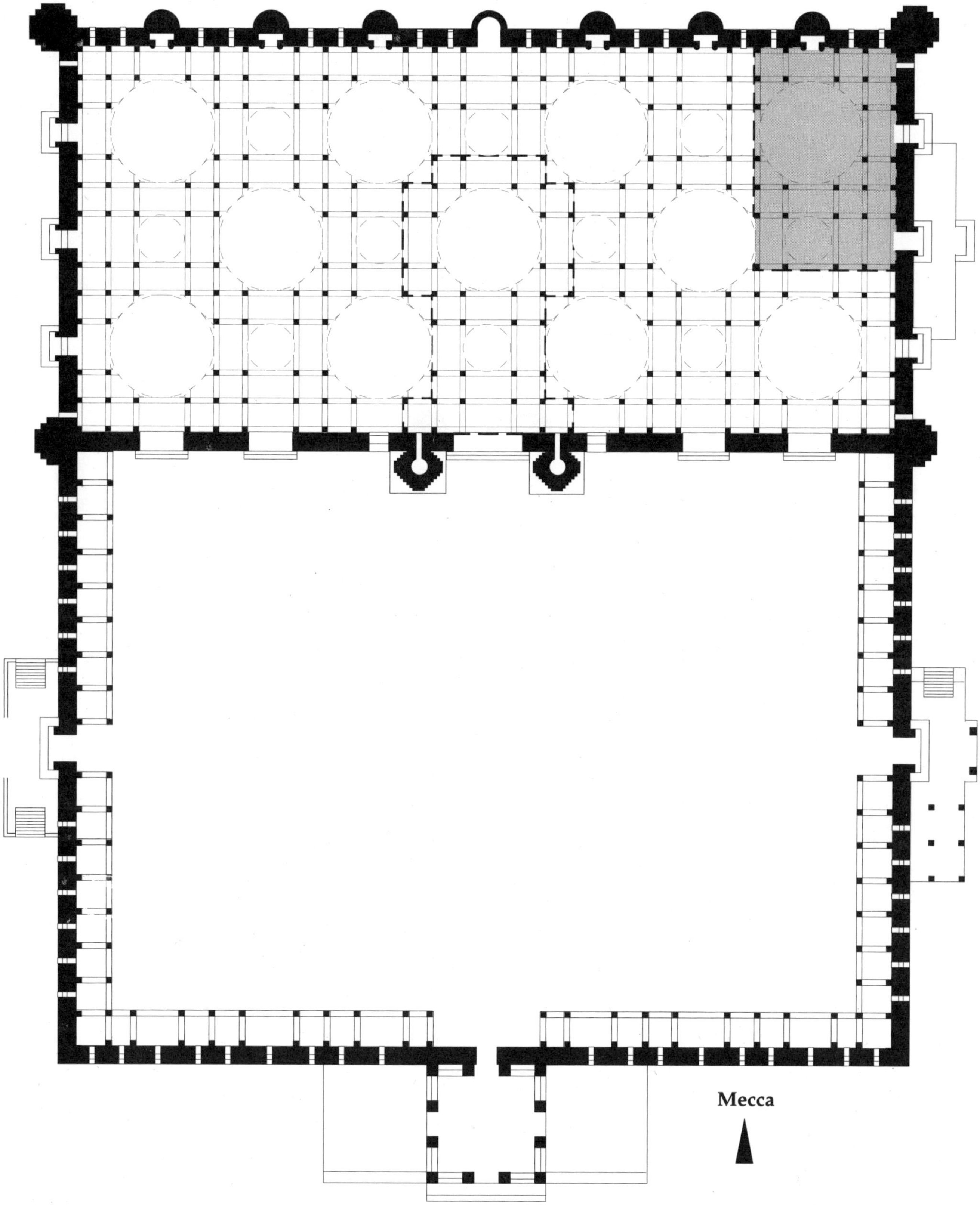

# Plate 91: Typical nirazhi maṇḍapa design

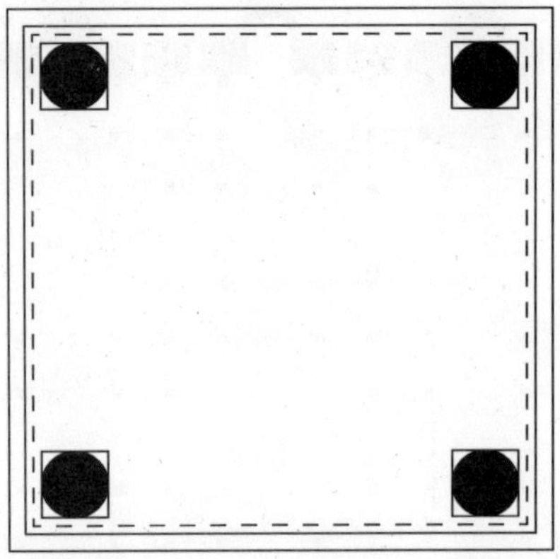

Four-column design after *nirazhi maṇḍapa*, Kailāsanāthar Temple Tank, Kāñcīpuram

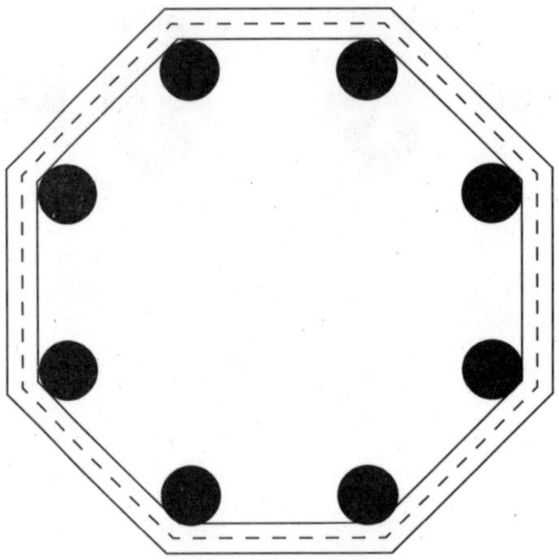

Eight-column design after *nirazhi maṇḍapa,* Fatehpur Bāwḍī, Rajasthan

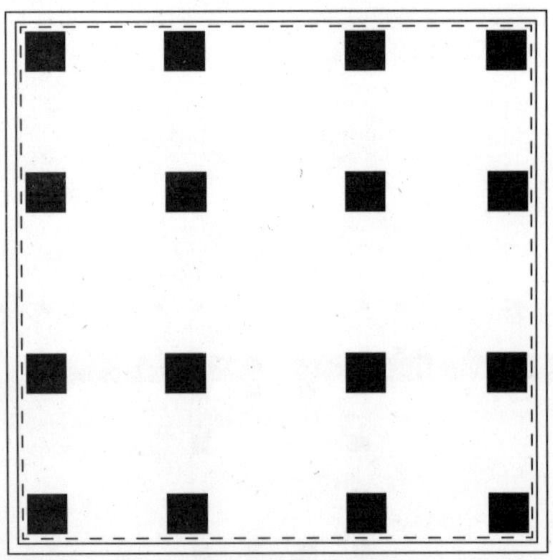

Sixteen-column design (*bārādarī*) after *nirazhi maṇḍapa*, Kāmākṣī-Amman Temple Tank, Kāñcīpuram

**Plate 92:** Bādshāhī Masjid plan indicating the bāolī (upper left)

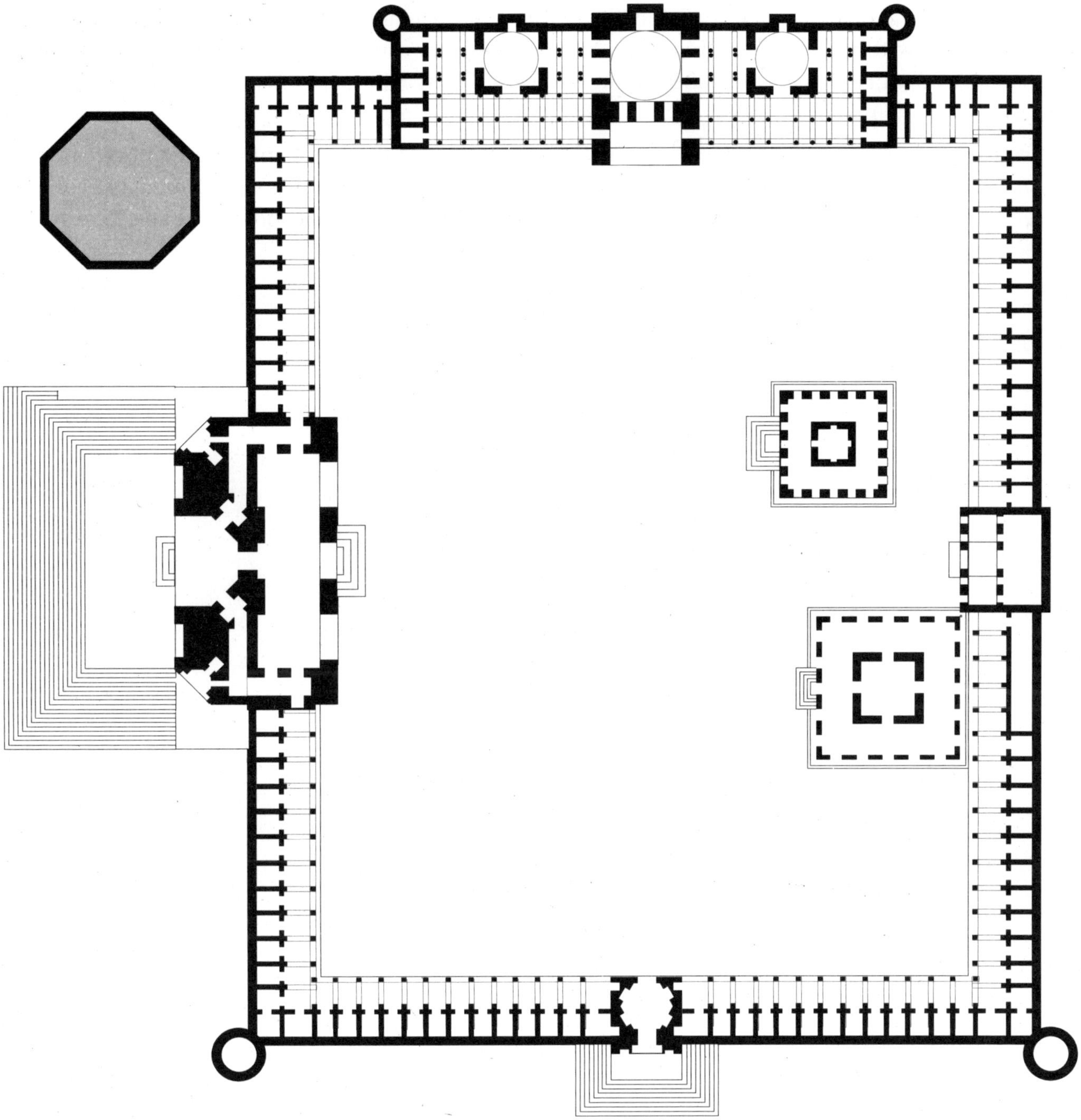

# Unverified plans

The following three plans — i.e. **Plates 91-93** — are examples of relatively important tanks and *bāolīs*. However, there is not enough information available or sufficient photographic evidence to properly consider them within the body of this work. None the less, they are here presented. The three are: the Kāmākṣī-Amman Temple Tank, Kāñcīpuram, the Swāmi Puṣkariṇī, Śrī Veṅkaṭeśvara (Bālājī) Temple, and the Lucknow Bāolī.

Following **Plates 91-93** are two conjectural possibilities for the Elephant Gate Bāolī at Fatehpur Sīkrī. These were two among the many possibilities considered for the interior placement of elements. Three were considered as most feasible — i.e. **Plates 94-95** as well as that shown in **Plate 51**, which is considered to be the most likely.

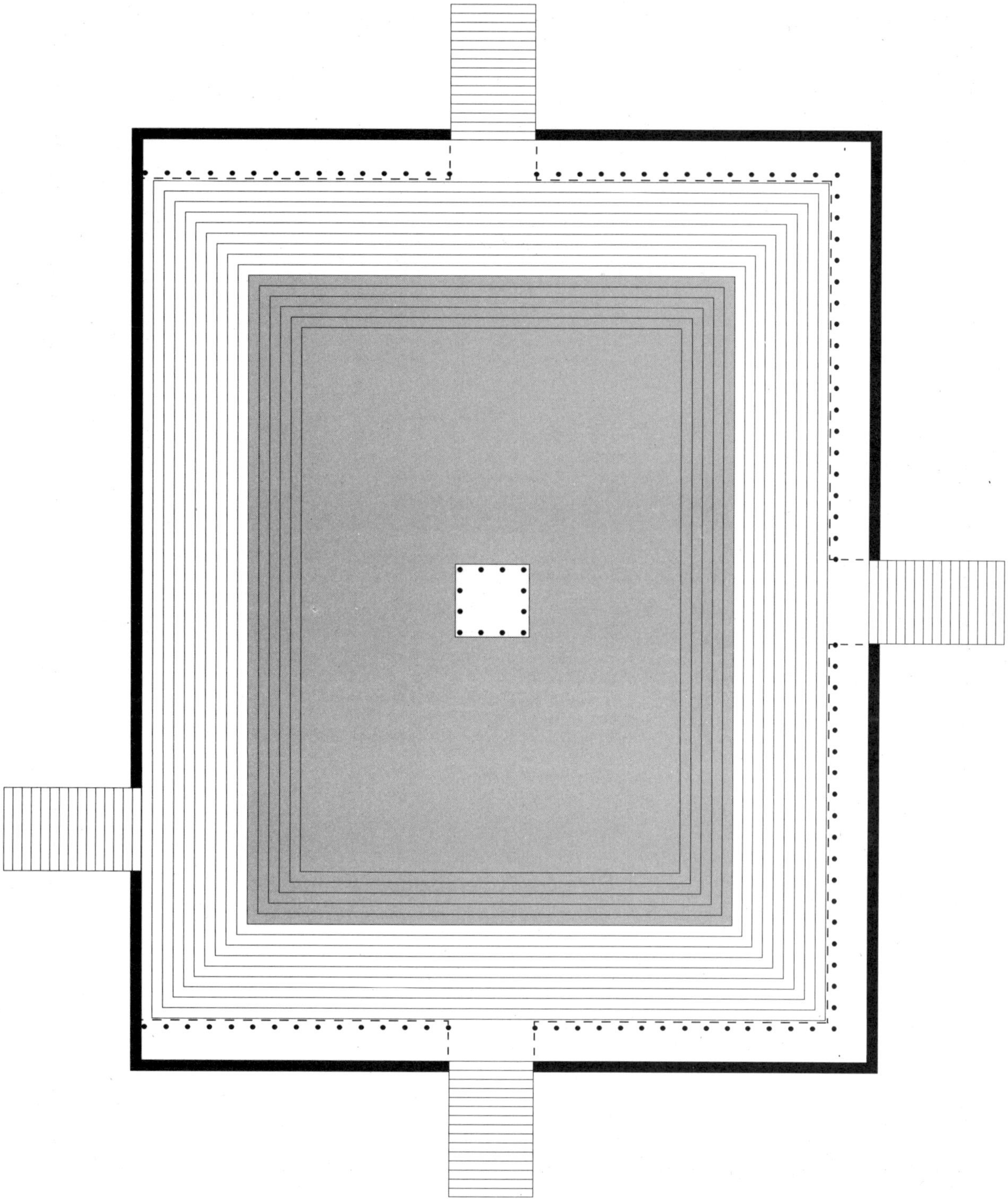

**Plate 94:** Unverified plan of Swāmi Puṣkariṇī, Śrī Veṅkaṭeśvara (Bālājī) Temple

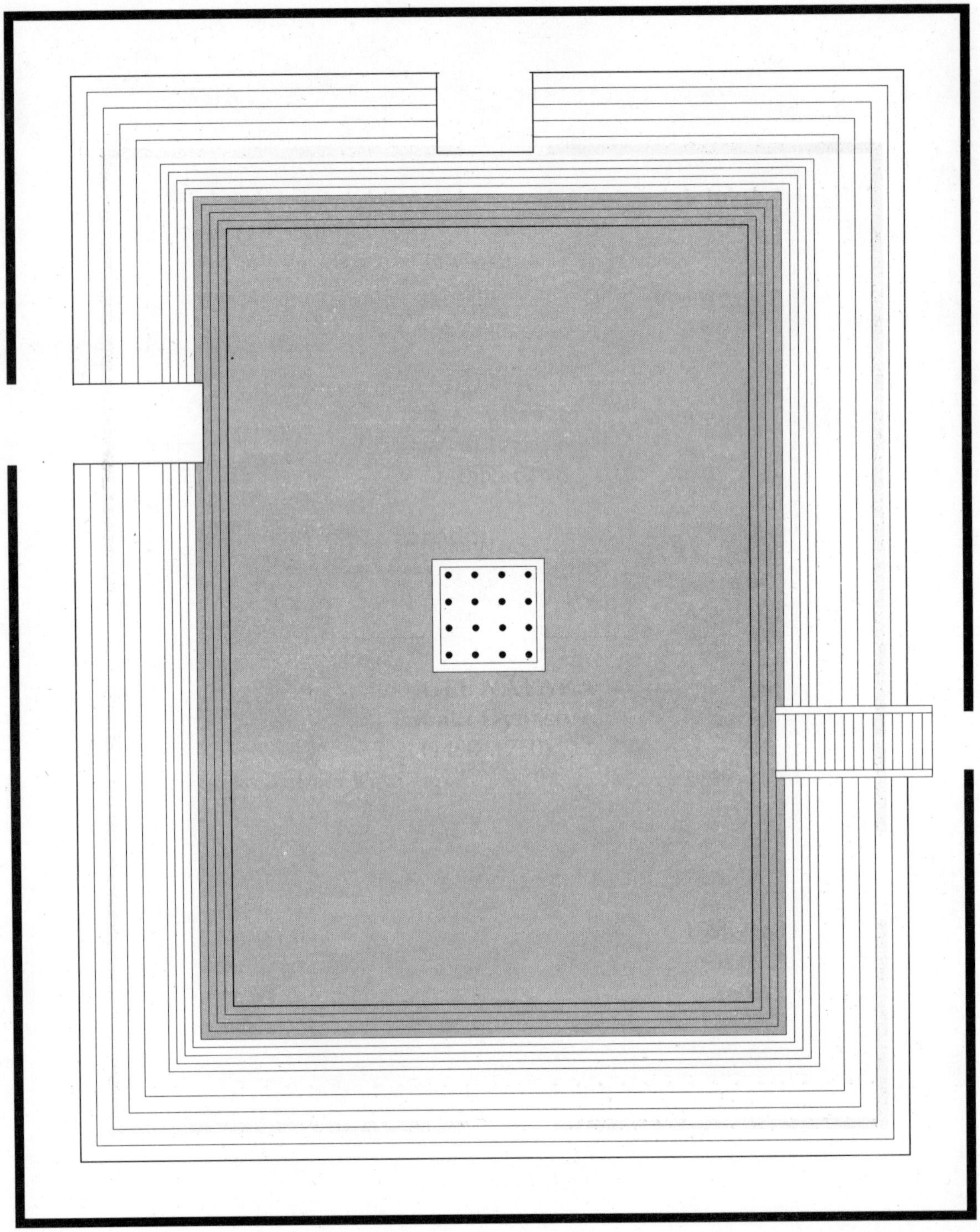

**Plate 95:** Unverified plan of Lucknow Bāolī

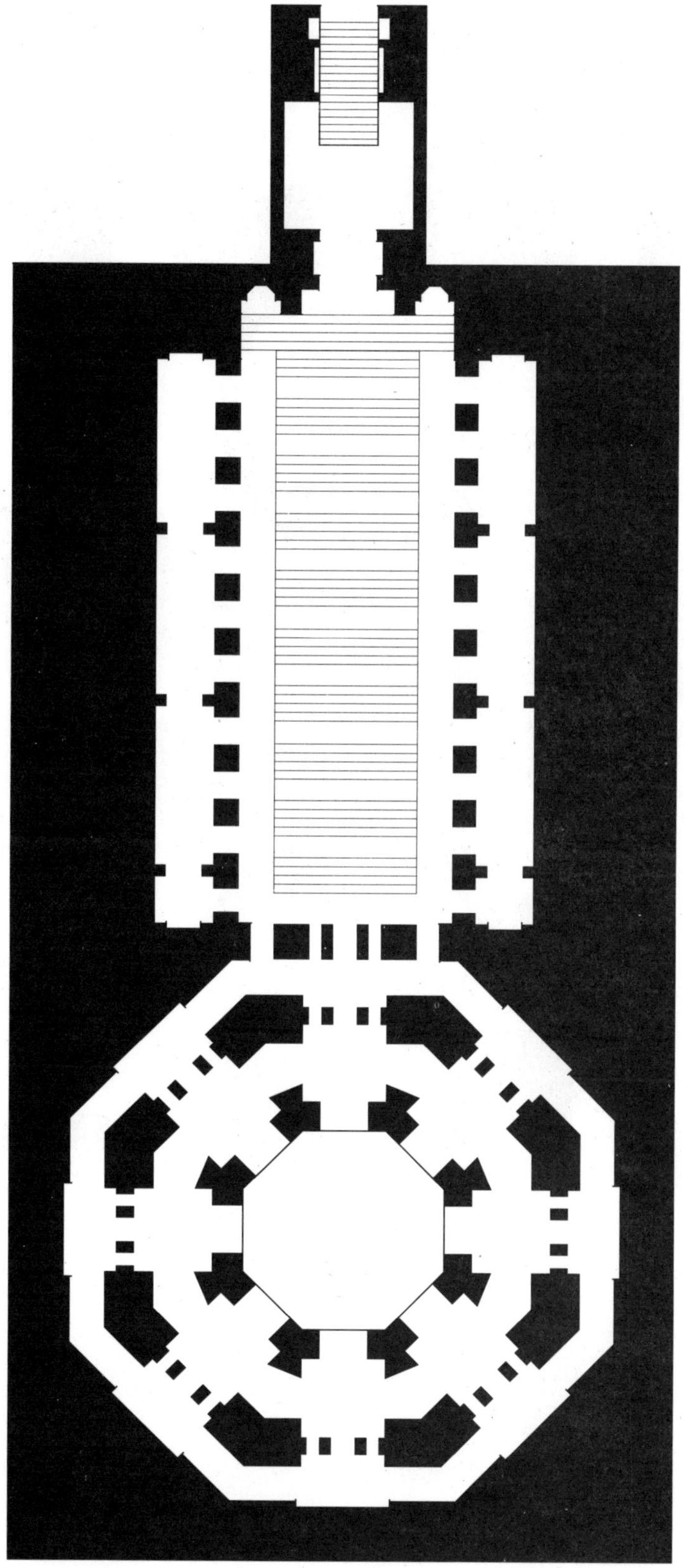

**Plate 96:** Elephant Gate Bāolī plan I (conjectural)

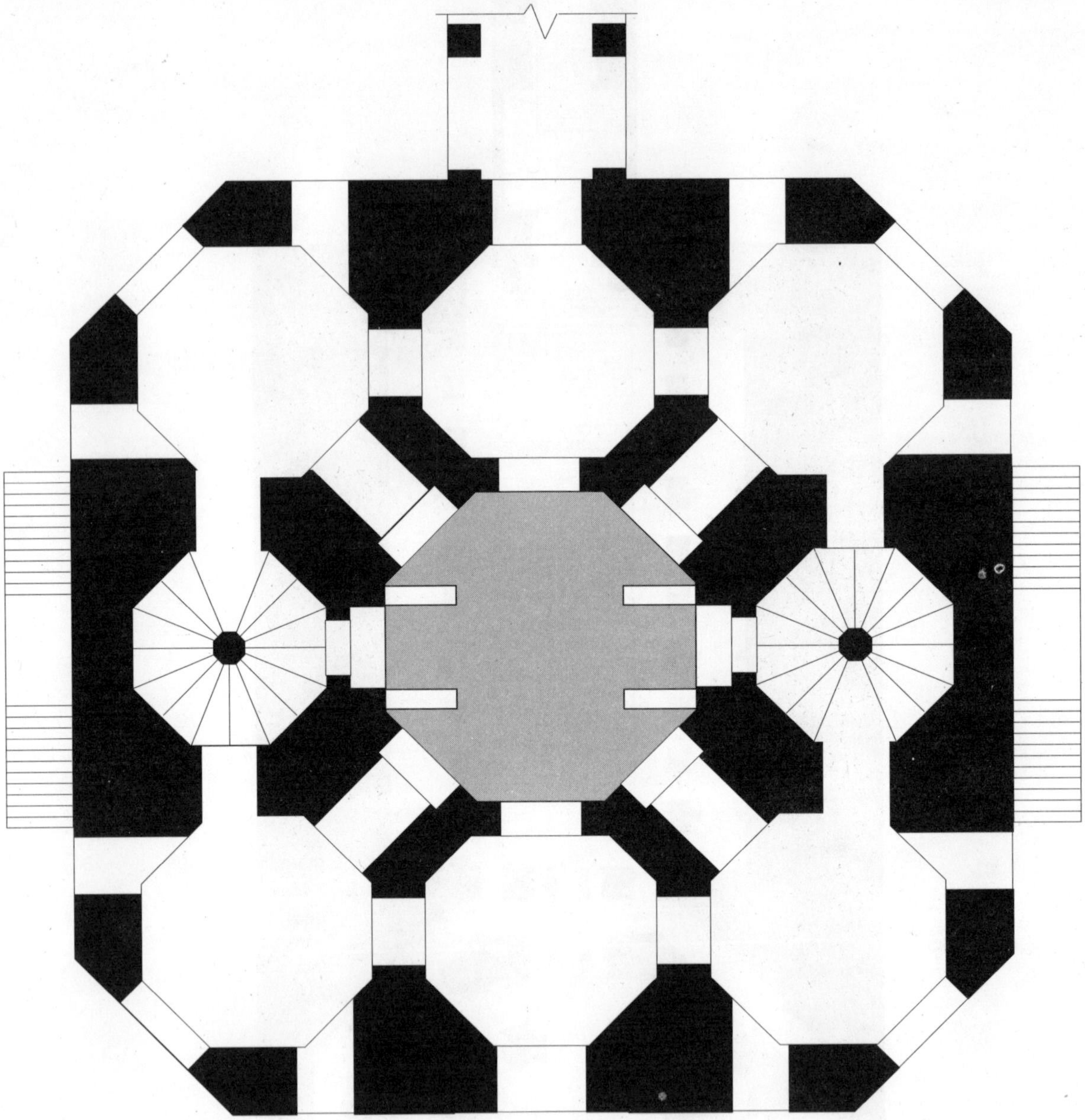

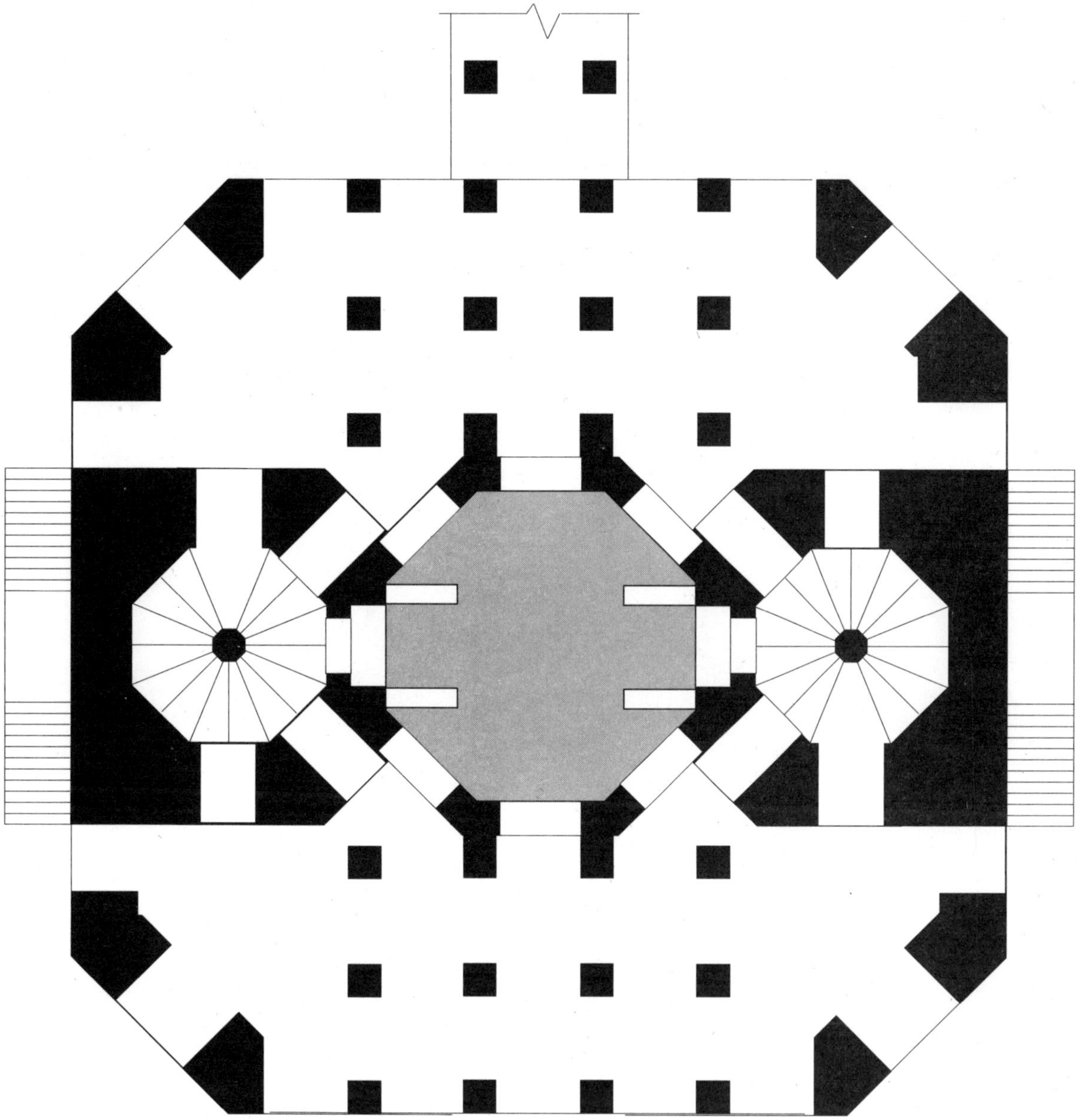

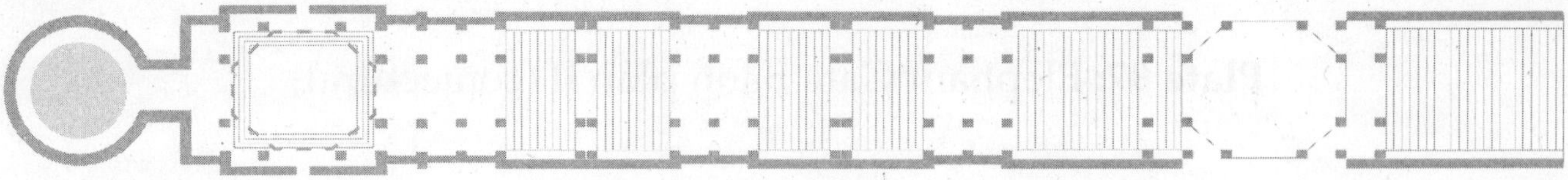

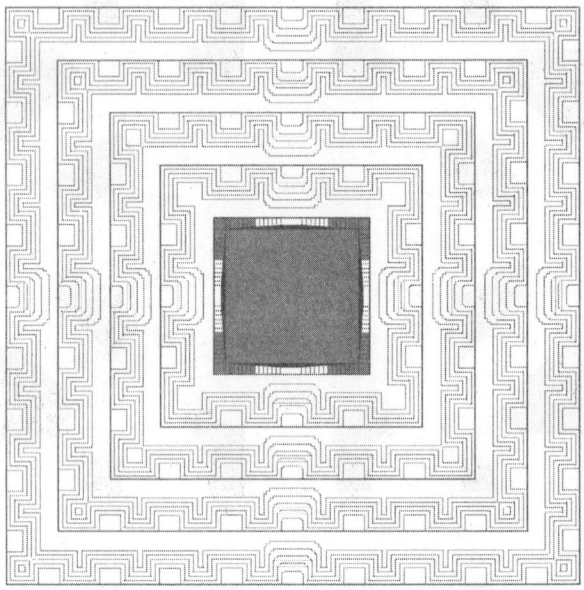

# Appendix D

# Relevant Ruling Dynasties

## GUJARAT
### Kingdom of Vallabhi
### Maitraka Dynasty
(*c*. CE 475–767)

| | |
|---|---|
| Bhatarka | *c*. 475- ? |
| Dharasena I | ?-*c*. 500 |
| Dronasiṁha | *c*. 500-25 |
| Dhruvasena I | *c*. 525-45 |
| Dharapaṭṭa | *c*. 545-56 |
| Guhasena | *c*. 556-70 |
| Dharasena II | *c*. 570–606 |
| Śilāditya I | *c*. 606-16 |
| Kharagraha I | *c*. 616-23 |
| Dharasena III | *c*. 623-40 |
| Dhruvasena II | *c*. 640-44 |
| Dharasena IV | *c*. 644-51 |
| Dhruvasena III | *c*. 651-56 |
| Kharagraha II | *c*. 656-62 |
| Śilāditya II | *c*. 662- ? |
| Śilāditya III | ? |
| Śilāditya IV | ? |
| Śilāditya V | ? |
| Śilāditya VI | *c*. 766-67 |

to Kabul

---

## ODISHA/KALIṄGA
### Gaṅga (Eastern) Dynasty
*c*. CE 496–1435

| | |
|---|---|
| Indravarman I | *c*. 496–535 |
| Hastivarman Rājasiṁha | *c*. 535-50 |
| Indravarman II | *c*. 550-90 |
| Indravarman III | *c*. 590–625 |
| Dānārṇava | *c*. 625 |
| Indravarman IV | *c*. 625-50 |
| Devendravarman I | fl. *c*. 679 |
| Anantavarman I | fl. *c*. 700 |
| Nanadavarman | fl. *c*. 717 |
| Devendravarman II | fl. *c*. 747 |
| Rājendravarman I | fl. *c*. 760 |
| Anantavarman II | *c*. 780–803 |
| Devendravarman III | *c*. 803-09 |
| Rājendravarman II | *c*. 809-40 |
| Devendravarman IV | *c*. 840 |
| Satyavarman | fl. *c*. 847/9 |
| Vajrahasta I | fl. *c*. 856 |

| | |
|---|---|
| Madhu Kāmārṇava I | ? |
| Vajrahasta II | ? |
| Devendravarman V | ? |
| Vīrasiṁha | ? |
| Kāmārṇava I | ? |
| Dānārṇava | ? |
| Guṇārṇava I | ? -869 |
| Kāmārṇava II | ? |
| Rānārṇava | ? |
| Vajrahasta III | 869–913 |
| (Kāmārṇava III) | ? |
| (Guṇārṇava II) | ? |
| (Potaṅkuśa) | *c.* 921-36 |
| (Kaligalaṅkuśa) | *c.* 936-46 |
| Guṇḍama I | *c.* 946-53 |
| Kāmārṇava IV | *c.* 953-78 |
| Vinayāditya | *c.* 978-80 |
| Vajrahasta IV | *c.* 980–1015 |
| Kāmārṇava V | *c.* 1015 |
| Guṇḍama II | *c.* 1015-? |
| Madhu Kāmārṇava II | 1019-38 |
| Vaḍjrahasta VI | 1038-70 |
| Rājrājadeva I | 1070-78 |
| Ānandavarman Choḍagaṅga | 1078–1147 |
| (Anaṅgabhīmadeva I) | *c.* 1148 |
| Kāmārṇava VI | *c.* 1148-57 |
| Rāghava | *c.* 1157-71 |
| Rājarāja II | *c.* 1171–1205 |
| Rājarāja III | *c.* 1205-16 |
| Anaṅgabhīmadeva III | *c.* 1216-38 |
| Narasiṁhadeva I | 1238-64 |
| Bhānudeva I | 1264-78 |
| Narasiṁhadeva II | 1278–1306 |
| Bhānudeva II | 1306-28 |
| Narasiṁhadeva III | 1328-52 |
| Bhānudeva III | 1352-78 |
| Narasiṁhadeva IV | 1378–1414 |
| Bhānudeva IV | 1414-35 |

## COḶA EMPIRE
### South and Eastern
### Medieval Coḷa
CE 846–1070

| | |
|---|---|
| Rāja Vijayālaya | *c.* 846-71 |
| Āditya I | *c.* 871–907 |
| Parāntaka | 907-47 |
| Rājāditya I | 947-49 |
| Gaṇḍarāditya | 949-56 |
| Ariñjaya | 956 |
| Parāntaka II | 956 |
| Āditya II | 956-69 |
| Madhurāntaka Uttama | 969-85 |
| Rājarāja I Deva the Great | 985–1012 |
| Rājendra I Coḷadeva | 1012-44 |
| Rājādhirāja I | 1044-52 |
| Rājendra II Deva | 1052-60 |
| Rāmamahendra | 1060-63 |
| Vīrarājendra | 1063-67 |
| Adhirājendra | 1067-70 |

## GUJARAT
### Kingdom of Gujarat
### Cāḷukya (Solaṅkī) Dynasty
*c.* CE 940–1244

| | |
|---|---|
| Rājā Mūlarāja I | *c.* 940-95 |
| Cāmuṇḍarāja | *c.* 995–1010 |
| Vallabharāja | *c.* 1010 |
| Ghaznavid Empire sphere | |
| Durlabharāja | *c.* 1010-22 |
| Bhīma I | *c.* 1022-64 |
| Karṇaḍeva I | *c.* 1064-94 |
| Jayasiṁha I | *c.* 1094–1125 |
| Kumārapāla | *c.* 1125-71 |
| Ajayapāla | *c.* 1171-76 |
| Mūlarāja II | *c.* 1176-78 |
| Bhīma II | *c.* 1178–*c.* 1241 with . . . |
| Jayasiṁha II | *c.* 1223 |
| Tribhuvanapāla | *c.* 1241-44 |

**AMBER (Dhundhar)**
**Kachwāhā Dynasty (Rajawat Clan)**
CE 966–1743

| | |
|---|---|
| Rājā Sodhdeva | 966–1006 |
| Duḷhā Rāo | 1006-36 |
| Kaṅkal (Kākil Dev) | 1036-38 |
| Maidal | 1038-39 |
| Hūṇadeva (Hunut Dev) | 1039-53 |
| Kantal I (Janaddeva) | 1053-70 |
| Pūjanadeva | 1070-84 |
| Malesi (Malaysi Dev) | 1084–1146 |
| Byāla (Bila Dev) | 1146-79 |
| Rājadeva | 1179–1216 |
| Kilhan (Kilhan Dev) | 1216-76 |
| Kantal II (Kuntal Dev) | 1276–1317 |
| Jansi (Junasi Dev) | 1317-66 |
| Udayakarṇa | 1366-88 |
| Nara Singh | 1388–1413 |
| Banbīr (Banvīr Dev) | 1413-24 |
| Udha Rāo (Uddhran Dev) | 1424-53 |
| Candrasena | 1453–1502 |
| Pṛthvī Singh I | 1502-27 |
| Pūraṇmal | 1527-34 |
| Bhīma | 1534-37 |
| Ratan | 1537-47 |
| Bhārmalla | 1547-74 |
| Bhagwān Dās | 1574-89 |
| Mirẓā Rājā Mān Singh I | 1589–1614 |
| Jagat Singh I | 1614 |
| Bhāo Singh | 1614-22 |
| Jaya Singh I | 1622-67 |
| Rāma Singh I | 1667-88 |
| Bishan Singh | 1688–1700 |
| Mahārājā Sawāī Jaya Singh II | 1700-43 |

moved to Jaipur

---

**MYSORE**
**Hoysaḷa Dynasty**
CE 1022–1346

| | |
|---|---|
| Nṛpakāma | 1022-47 |
| Vinayādita | 1047-98 |
| Ereyaṅga | 1063–1100 |
| Ballāla I | 1100-10 |
| Viṣṇuvardhana | 1110-52 |

| | |
|---|---|
| Narasiṁha I | 1152-73 |
| Ballāla II | 1173–1220 |
| Narasiṁha II | 1220-38 |
| Someśvara | 1233-67 |
| Narasiṁha III | 1254-91 |
| Ballāla III | 1291-42 |
| Virūpākṣa Ballāla IV | 1342-46 |

---

**KARNATAKA**
**Western Cāḷukya (Kalyāṇa) Dynasty**
CE 973–1156

| | |
|---|---|
| Taila II Āhavamalla | 973-97 |
| Satyāśraya Irivabedaṅga | 997–1008 |
| Daśavarman | 1008 |
| Vikramāditya V | 1008-14 |
| Ayyana | 1014-15 |
| Jayasiṁha II | 1015-42 |
| Someśvara I | 1042-68 |
| Someśvara II | 1068-76 |
| Vikramāditya VI | 1076–1127 |
| Someśvara III | 1127-38 |
| Jagadekamalla II | 1138-51 |
| Tailapa III | 1151-64 |
| Bijjala Kaḷacuri | 1164-68 |
| Someśvara IV | 1168-77 |
| Saṅkama II | 1177-80 |
| Āhavamalla | 1180-83 |
| Siṅghana | 1183-84 |
| (?) Jagadekamalla III | 1163-83 |
| (?) Someśvara V | 1184–1200 |

---

**COḶA EMPIRE**
**South and Eastern**
**Cāḷukya Coḷa**
CE 1070–1279

| | |
|---|---|
| Kulottuṅga Coḷa I (Rajendra Cāḷukya III) | 1070–1122 |
| Vikrama Coḷa | 1122-35 |
| Kulottuṅga Coḷa II | 1135-50 |
| Rājarāja II | 1150-73 |
| Rājādhirāja II | 1173-79 |
| Kulottuṅga III | 1179–1218 |

| | |
|---|---|
| Rājarāja III | 1218-46 |
| Rājendra IV | 1246-79 |

to Pāṇḍya

---

**DELHI (Muslim)**
**Ghulyam (Mamluk or Mu'izz) Dynasty**
CE 1206-90

| | |
|---|---|
| Sultān Qutb ad-Dīn Aibak ibn Mu'izz | 1206-10 |
| Aram Shāh | 1210-10/11 |
| Shams ad-Dīn Iltutmish al-Qutbi ibn Yalam Khan | 1210/11–1236 |
| Rukn ad-Dān Firūẓ Shāh | 1236 |
| Begum Radiyyā (queen) | 1235-40 |
| Sultān Mu'izz ad-Dīn Bahram Shāh | 1240-42 |
| Alāddīn Masūd Shāh | 1242-46 |
| Nasr ad-Dīn . . . | 1246-66 |
| Giyās ad-Dīn Balban Ulugh Khan | 1266-87/8 |
| Mu'izz ad-Dīn Kai-Kubad ibn Bugra Khan | 1287/8-90 |
| Shams ad-Dīn Kaiumarth | 1290 |

---

**DELHI (Muslim)**
**Tughluq Dynasty**
CE 1320-99

| | |
|---|---|
| Sultān Tughluq Shāh I | 1320-25 |
| Muhammad Shāh II | 1325-51 |
| Mahmūd ibn Muhammad | 1351 |
| Firūẓ Shāh | 1351-88 |
| Tughluq Shāh II | 1388-89 |
| Abu Bakr Shāh | 1389-89/90 |
| Muhammad Shāh III | 1389/90-93 |
| Sikander Shāh I | Mar.-Apr. 1393 |
| Mahmūd Shāh II | 1393-94/5 |
| Nusrat Shāh | 1394/5-99 |

Sack of Delhi by Timur
interregnum

---

**DELHI (Muslim)**
**Lodi Dynasty**
CE 1313-14

| | |
|---|---|
| Sultān Daulat Khan | 1413/14-1414/15 |

---

**UDAIPUR/MEWĀṚ (Rajasthan)**
**Sisodiā (Guhiloṭ) Dynasty**
CE 1326–1948

| | |
|---|---|
| Mahārāṇā Hamīr Singh I | 1326-64 |
| Kahetra Singh | 1364-82 |
| Lakhā Singh | 1382–1421 |
| Mokal Singh | 1421-33 |
| Kumbhakarṇa Singh | 1433-68 |
| Udai Singh | 1468-73 |
| Rāimal Singh | 1473–1509 |
| Saṅgrāma Singh I | 1509-28 |
| Ratan Singh | 1528-31 |
| Vikramāditya Singh | 1531-36 |
| Udai Singh II | 1537-72 |
| Pratāpa Singh I | 1572-97 |
| Amara Singh I | 1597–1620 |
| Karaṇa Singh | 1620-28 |
| Jagat Singh I | 1628-52 |
| Rāja Singh I | 1652-80 |
| Jai Singh | 1680-98 |
| Amara Singh II | 1698–1710 |
| Saṅgrāma Singh II | 1710-34 |
| Jagat Singh II | 1734-51 |
| Pratāpa Singh II | 1751-53 |
| Rāja Singh II | 1753-61 |
| Ari Singh | 1761-73 |
| Hamīr Singh II | 1773-78 |
| Bhīm Singh | 1778–1828 |
| Jawān Singh | 1828-38 |
| Sardār Singh | 1838-42 |
| Swarūp Singh | 1842-61 |
| Śambhu Singh | 1861-74 |
| Sajjan Singh | 1874-84 |
| Fateh Singhji | 1884–1930 |
| Bhūpāl Singh | 1930-48 |

to India

---

## GUJARAT
### Gujarat Sultanate
### Zafarid Dynasty
CE 1391–1583

| | |
|---|---|
| Sultān Zafar Khan | 1391–1403 |
| Muhammad Shāh I Tatar | 1403-07 |
| Muzaffar Shāh | 1407-11 |
| Ahmad Shāh I Shihāb-ud-Dīn | 1411-42 |
| Muhammad Shāh II Karīm | 1442-51 |
| Ahmad Shāh II Qutb-ud-Dīn | 1451-58 |
| Dāwud Khan | 1458 |
| Mahmūd Shāh I Begra Saif-ud-Dīn | 1458–1511 |
| Muzaffar Shāh II | 1511-26 |
| Sikandar | 1526 |
| Mahmūd Shāh II | 1526 |
| Bahādur Shāh | 1526-35 |
| to the Mughal Empire | |
| Bahādur Shāh (restored) | 1536-37 |
| Mahmūd Shāh III | 1537-54 |
| Ahmad Shāh III | 1554-61 |
| Muzaffar Shāh III | 1561-73 |
| to the Mughal Empire | |
| Muzaffar Shāh III (restored) | 1583 |
| to the Mughal Empire | |

---

## VIJAYANAGARA
### Saṅgama Dynasty
CE 1336–1485

| | |
|---|---|
| Harihara I | 1336-56 |
| Bukkā I | 1356-77 |
| Harihara II | 1377–1404 |
| Virūpākṣa | 1404-05 |
| Bukkā II | 1405-06 |
| Devarāya I | 1406-22 |
| Rāmacandra | 1422-30 |
| Vīra Vijaya I Bukkā Rāya | 1422-24 |
| Devarāya II | 1424-46 |
| Vijaya II | 1446-47 |
| Mallikārjuna | 1446-65 |
| Virūpākṣa II | 1465-85 |
| Prauḍha Rāya | 1485 |

---

**MYSORE**
**Waḍeyār Dynasty**
CE 1399–1796

| | |
|---|---|
| Mahārājā Vijayarāja (Yadu Rāya) | 1399–1423 |
| Hiriya Bettada Chamarāja I | 1423-59 |
| Timmarāja I | 1459-78 |
| Hiriya Chamarāja II | 1478–1513 |
| Hiriya Bettada Chamarāja III | 1513-53 |
| Timmarāja II | 1553-72 |
| Bola Chamarāja IV | 1572-76 |
| Bettada Devarāja | 1576-78 |
| Rājā Waḍeyār | 1578–1617 |
| Chamarāja V | 1617-37 |
| Immadi Rājā | 1637-38 |
| Kanthirava Narasarāja I | 1638-59 |
| Kempa Devarāja | 1659-73 |
| Cikkadevarāja | 1673–1704 |
| Kanthirava Narasarāja II | 1704-14 |
| Kṛṣṇarāja I | 1714-32 |
| Chamarāja VI | 1732-34 |
| Kṛṣṇarāja II | 1734-66 |
| Nañjarāja | 1766-70 |
| Bettada Chamarāja VII | 1770-76 |
| Khasa Chamarāja VIII | 1776-96 |

---

**VIJAYANAGARA**
**Saluva Dynasty**
CE 1485–1505

| | |
|---|---|
| Rājā Narasiṁhadevarāya | 1485-90 |
| Thimma Bhūpāḷa | 1490-91 |
| Immadi Narasimha | 1491–1505 |

---

**BĪJĀPUR**
**'Adil Shāhī Dynasty**
CE 1490–1686
from Bhamani Sultanate

| | |
|---|---|
| Sultān Yusuf 'Adil Shāh | 1490–1510 |
| Isma'il Shāh | 1510-34 |
| Mallu 'Adil Shāh | 1534 |
| Ibrahim 'Adil Shāh I | 1534-58 |
| 'Ali 'Adil Shāh I | 1558-80 |
| Ibrāhim 'Adil Shāh II | 1580–1627 |

| | |
|---|---|
| Mohammed 'Adil Shāh | 1627-57 |
| 'Ali 'Adil Shāh II | 1657-72 |
| Sikandar 'Adil Shāh | 1672-86 |

to Maharashtra
to the Mughal Empire

---

**VIJAYANAGARA**
**Tuluva Dynasty**
CE 1505-65

| | |
|---|---|
| Rājā Vīra Narasiṁha | 1505-09 |
| Kṛṣṇadevarāya | 1509-29/30 |
| Acyutadevarāya | 1529/30-42 |
| Veṅkaṭa | 1542 |
| Sadāśivarāya | 1542-65 |

---

**VIJAYANAGARA**
**Araviḍu Dynasty**
CE 1565–1660

| | |
|---|---|
| Rājā Tirumala Devarāya | 1565-72 |
| Śrīraṅga Devarāya I | 1572-86 |
| Veṅkaṭapati Devarāya I | 1586–1614 |
| Śrīraṅga Rāya II | 1614 |
| vacant - succession wars | |
| Rāmadevarāya | 1617-32 |
| Veṅkaṭapati Rāya | 1632-42 |
| Śrīraṅga Rāya III | 1642-46 |
| Veṅkaṭapati Rāya II | 1646-*c.* 1660 |

---

**MĀLWĀ**
**Mālwā Sultanate**
**Ghurid Dynasty**
CE 1390–1436

| | |
|---|---|
| Sultān Dilāvar Khan Husain | 1390–1405 |
| Alp Khan Hushaṅg | 1405-35 |
| Ghazni Khan Muhammad | 1435-36 |
| Mas'ud Khan | 1436 |

---

**MĀLWĀ**
**Mālwā Sultanate**
**Khaljī Dynasty**
CE 1436–1531

| | |
|---|---|
| Sultān Mahmūd Shāh I | 1436-69 |
| Ghiyāth Shāh | 1469–1500 |
| Nasr Shāh | 1500-11 |
| Mahmūd Shāh II | 1511-31 |

to Gujarat

---

**MĀLWĀ**
**Mālwā Sultanate**
**Shaja'tid Dynasty**
CE 1555-62

| | |
|---|---|
| Sultān Shaja'at Khan | 1555 |
| Miyān Bayezid Bāẓ Bahādur | 1555-62 |

to the Mughal Empire

---

**GINGEE NĀYAKA**
**Tubaki Dynasty**
(1490–1750)

| | |
|---|---|
| Nāyaka Tubaki Kr̥ṣṇappā | 1490–1521 |
| ? | 1521-29 |
| Viśvanātha | 1529-64 |
| Kr̥ṣṇappa I | 1564-72 |
| Vārappā | 1572-95 |
| Kr̥ṣṇappā II | 1595–1601 |
| Muttu Kr̥ṣṇappā | 1601-09 |
| Muttu Vīrappā I | 1609-23 |
| Tirumala | 1623-59 |
| Muttu Vīrappā II | 1659-82 |

to Marāṭhā Confederation 1676
to Mughals 1691

| | |
|---|---|
| Rājā Sarūp Singh | 1691–1714 |
| Desingh | 1714 |

to Nawāb of Ārcoṭ 1714
to France 1750
to England 1761

---

**DELHI**
**Sūri Dynasty**
CE 1540-55

| | |
|---|---|
| Sultān Sher Shāh | 1540-45 |
| Islam Shāh | 1545-53 |
| Mohammed V | 1553-54 |
| Firūẓ | 1554 |
| Ibrahim III | 1554-55 |
| Sikander Shāh | 1555 |

---

**MUGHAL EMPIRE (Restored)**
**Timurid Dynasty**
CE 1555–1858
from Sūri Dynasty

| | |
|---|---|
| Pādshāh Nasir-ud-dīn Muhammad Humāyūn | 1555-56 |
| Fath Jalāl-ud-dīn Muhammad Akbar I, The Great | 1556–1605 |
| Shāhb-ud-dīn Salīm Jahāṅgīr | 1605-27 |
| Khurram Shāh Jahān I | 1627-57 |
| Muhin-ud-dīn Muhammad Auraṅgzeb | 1658–1707 |
| Muhammad Āẓam Shāh | 1707 |
| Mu'azzam Bahādur Shāh I | 1707-12 |
| Muhammad Aẓimushān | 1712 |
| Muhammad Jahāndar Shāh | 1712-13 |
| Farrukhsiyar | 1713-19 |
| Rafi'-ud-Darajat | 1719 |
| Rafi'-ud-Daulā Shāh Jahān II | 1719 |
| Nekyu-siyar | 1719 |
| Roshan Akhtar Muhammad Shāh | 1719-48 |
| Mujāhid-ud-dīn Ahmad Shāh | 1748-54 |
| Muhammad Aẓīẓ-ud-dīn 'Alamgir II | 1754-59 |
| Muhiulmillat Shāh Jahān III | 1759-60 |
| Jalāl-ud-dīn Shāh Ālam II | 1759-88 |
| Baidar Bakht Kuchak Shāh | 1788-90 |
| Muhin-ud-dīn Muhammad Akbar II Shāh | 1806-37 |
| Sirāẓ-ud-dīn Bahādur Shāh II | 1837-58 |

to Great Britain

---

**GOṆḌVĀNĀ/DEVAGOṆḌA**
**Goṇḍvānā Dynasty**
(a. 1556–a. 1739)

| | |
|---|---|
| Rājā Jaṭba (Jāṭav) | ?-*c.* 1600 |
| Koka Shāh | *c.* 1600-? |
| Bakht Buland Shāh (Islam convert) | ?-1706 |
| Chānd Sultān | 1706-38 |
| Bahādur Shāh | 1738 |
| Wali Shāh (usurper) | 1738-39 |
| Rānī Ratankunwar (queen) | 1739 |
| Rājā Burhān Shāh | 1739-? |

to Nagpur

---

**BHARATPUR/DEEG**
**Sinsini Jāṭ Dynasty**
(1722–1948)

| | |
|---|---|
| Yādav Vanshi Sind Pāl | eleventh century CE |
| 12 generations | |
| Than Pāl | twelfth century CE |
| Dharam Pal | twelfth century CE |
| ? | |
| Gokula | ?-1670 |
| Rājā Ram | 1670–1888 |
| ? generations | |
| Churaman | 1695–1721 |
| Ṭhākur Badan Singh | 1722-33 |
| Mahārājā Sūraj Mal | 1733-67 |
| Jawāhar Singh | 1767-68 |
| Ratan Singh | 1768-69 |
| Kesrī Singh | 1769-71 |
| Newal Singh | 1771-76 |
| Ranjīt Singh | 1776–1806 |
| Randhīr Singh | 1805-23 |
| Baldeo Singh | 1823 -25 |
| Balwant Singh | 1825 -53 |
| Jaswant Singh | 1853 -93 |
| Rām Singh | 1893–1900 (exiled) |
| Girrāj Kaur (regent) | 1900-18 |
| Vrijendra Sawāī Kishan Singh | 1900-28 (exiled) |
| Vrijendra Singh | 1929-47 |

to India 1948

**JAIPUR**
**Kachwāhā Dynasty (Rajawat Clan)**
CE 1727–1948
from Amber

| | |
|---|---|
| Mahārājā Sawāī Jai Singh II | 1700-43 |
| Iśwarī Singh | 1743-50 |
| Madhu Singh I | 1750-68 |
| Pṛthvī Singh II | 1768-78 |
| Pratāpa Singh | 1778–1803 |
| Jagat Singh II | 1803-18 |
| Mān Singh II (Mohan Singh) | 1818-19 |
| Jai Singh III | 1819-35 |
| Rāma Singh II | 1835-81 |
| Mahārājādhirāja Sawāī Madhu Singh II | 1881–1922 |
| Sawāī Mān Singh II | 1922-49 |

to India

---

**KERALA**
**Mahārājās of Travancore**
**Cheramān-Kulaśekhara Dynasty**
CE 1729–1948

| | |
|---|---|
| Rājā Mārtaṇḍa Varmā III | 1729-58 |
| Kārtika Tirunāl Rāma Varmā II | 1758-98 |
| Great Britain sphere | |
| Balarāma Varmā I | 1798–1810 |
| Rānī Gaurī Lakṣmī Bāī (queen) | 1810-15 |
| Gaurī Pārvatī Bāī (queen) | 1815-29 |
| Mahārājā Rāma Varmā III Swāti Tirunāl | 1829-47 |
| Uttam Tirunāl Mārtaṇḍa Varmā IV | 1847-60 |
| Āyilyam Tirunāl Rāma Varmā IV | 1860-80 |
| Rāma Varmā Tirunāl Rāma Varmā V | 1880-85 |
| Śrī Mūlam Tirunāl Balarāma Varmā II | 1885–1924 |
| Mahārānī Setu Lakṣmī Bāī (queen) | 1924-31 |
| Mahārājā Śrī Cithira Tirunāl Balarāma Varmā III | 1931-49 |

to India

---

**NAGPUR**
**(Marathan)**
**Bhoslās (Bhoṅsle) Dynasty**
CE 1738–1853
from Goṇḍvānā/Devagoṇḍa

| | |
|---|---|
| Mahārājā Raghujī I | 1738-55 |
| Janojī | 1755-72 |
| Mūdhojī I | 1772-88 |
| Raghujī II | 1788–1816 |
| Mūdhojī II | 1816-18 |
| Raghujī III | 1818-53 |

to Great Britain

# Addendum

## Helical Step Well

## Helical Step Well, Campāner–Pāvāgaḍh, Gujarat (b. CE 1850)

Of the thousands of step wells scattered across the subcontinent, some may be noteworthy for their unusual depth, others may possess magnificent architectural–sculptural details, still others may have cool areas for rest and relaxation for those who come to draw water, but, none is as unique, *en plan*, or concept as the Helical Step Well at Campāner–Pāvāgaḍh.

Initially, it may appear to conform to the design of a number of other step wells which utilizes a trench and shaft (*kūpa*). A stepped trench leads down to a circular shaft with steps descending around its inner edge. However, there is a difference here! The stairs do certainly descend along the outer edge of the *kūpa*, but the shaft is not strictly cylindrical. Its radius decreases as the steps descend, creating a spiral. *En plan*, the effect is much like that encountered in the cross section of a Chambered Nautilus (**Figure 58** [right]). This spiral is related to the *Phi* principle and further related to the "golden section". Therefore, the term "helical" is not accurate as a helix is merely a spiral with the same diameter (**Figure 58** [left]) as is seen in the numerous spiral staircases encorporated such as in the Rudabāī-nī Wāv (**Plate 17**) or the Dādā Harīr Wāv (**Plate 18**).

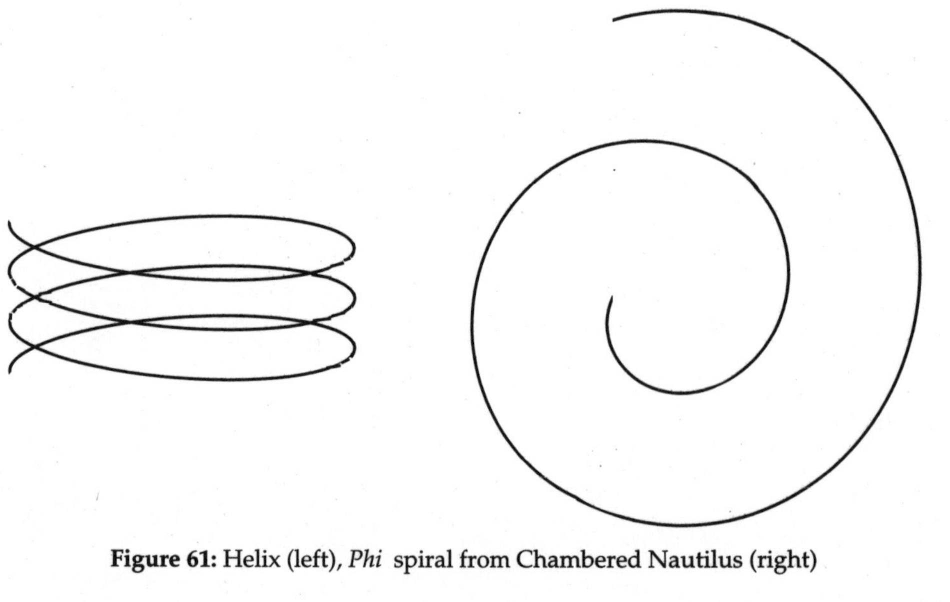

**Figure 61:** Helix (left), *Phi* spiral from Chambered Nautilus (right)

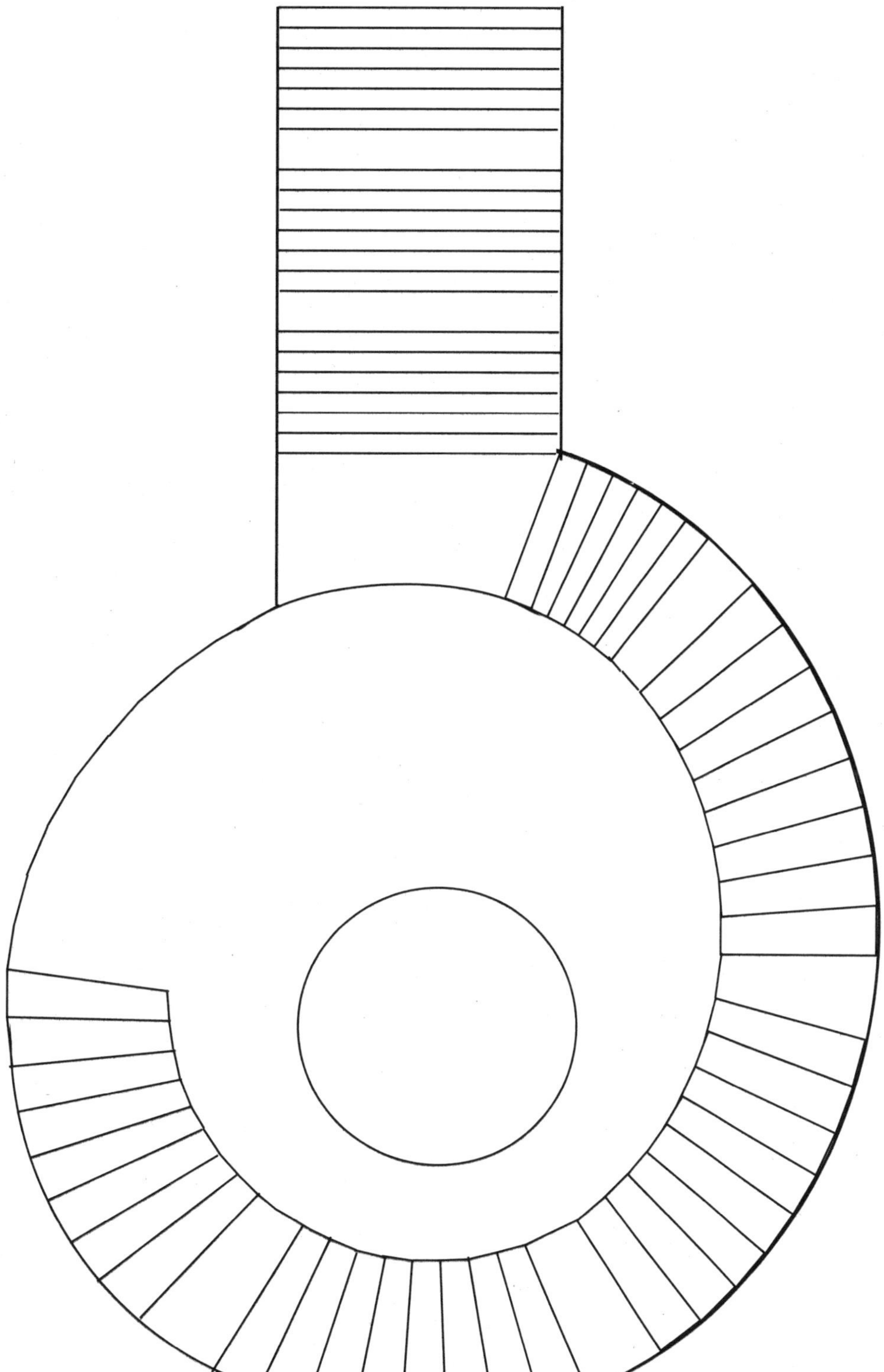

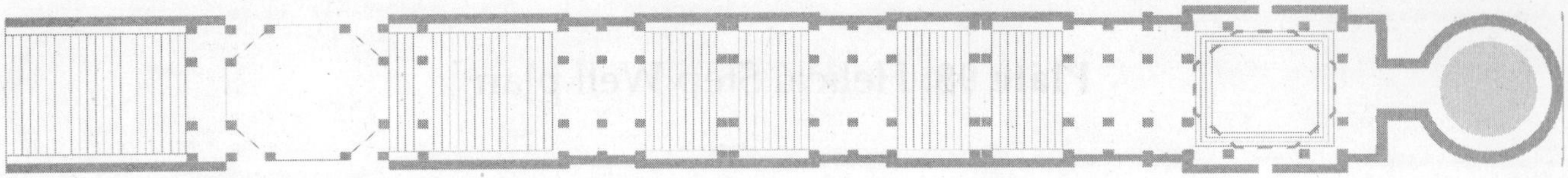

# Bibliography


Acharya, P.K. *An Encyclopaedia of Hindu Architecture*. London: 1927-46.

———, *Architecture of Manasara*, translated from the original Sanskrit. London: 1932.

Anderson, Leona M. *Vasantotsava: The Spring Festivals of India*. New Delhi: 1993.

Banerjee, Jitendranath N. *The Development of Hindu Iconography*, 2nd edn. Calcutta: 1956.

Bary, Theodore de. *Sources of Indian Tradition*, vol. I. New York: 1958.

Blurton, R.T. "Palace Structures at Vijayanagara: The Archaeological Evidence", *South Asian Archaeology 1985*, ed. K. Frifelt and P. Sorenson. London: 1989.

Brand, M. and Glenn D. Lowry. eds. *Fatehpur Sikri: A Sourcebook*. Cambridge: 1985.

Brown, Percy. *Indian Architecture (Buddhist and Hindu Periods)*. Bombay: n.d.

Bunce, Fredrick W. *The Iconography of Plans: A Study of the Influence of Buddhism and Hinduism on Architectural Plans of South and Southeast Asia*, New Delhi: 2002.

———, *Royal Palaces, Residences and Pavilions of India: 13th Through 18th Centuries — An Iconographic Consideration*, New Delhi: 2006.

Burgess, J. *Muhammadan Architecture in Gujarat*. Varanasi: 1971.

Bussagli, Mario; John Shepley, trans. *Oriental Architecture/1: India, Indonesia, Indochina*. New York: 1989.

Carr, Stephen. *Archaeology and Monumental Remains of Delhi*. Allahabad: 1967.

Chandra, Pramrod, ed. *Studies in Indian Temple Architecture*. Benares: 1975.

Chopra, P.N. , T.K. Ravindran, N. Subrahmanian. *History of South India*, 3 vols. New Delhi: 1979.

Commissariat, ManekShāh SorabShāh. *A History of Gujarat: Including a Survey of Its Chief Architectural Monuments and Inscriptions*. Bombay: 1938.

Cousens, Henry. *Bijapur and Its Architectural Remains: With an Historical Outline of the Adil Shāhi Dynasty* (Archeological Survey of India, vol. XXXVIII, Imperial Series), repr. Delhi: 1996.

———, *The Chalukyan Architecture of the Kanarese Districts* (Archeological Survey of India, vol. XLII, Imperial Series), repr. New Delhi: 1996 (1926).

Crowe, S. and S. Hawood. *The Gardens of Mughal India: A History and a Guide*. London: 1972.

Dagens, Bruno, ed. and tr. *Mayamata, Traite Sanskrit d'architecture, Premiere Partie*. Paris (?): 1970.

———, *Mayamata, Traite Sanskrit d'architecture, Deuxeme Partie*. Paris (?): 1970.

———, *Mayamatam, Treatise of Housing, Architecture & Iconography* (2 vols.). New Delhi: 1994.

Dallapiccola, Anna, et al. *The Ramachandra Temple at Vijayanagara*. New Delhi: 1992.

Danielou, Alain. *Hindu Polytheism*. New York: 1964.

Deloche, Jean. *Senji (Gingee): A Fortified City in the Tamil Country*. Pondicherry: 2005.

Deva, Krishna. *Temples of India*, 2 vols. New Delhi: 1995.

Edwardes, M. *Indian Temples and Palaces*. London: 1959.

Fass, V. *The Forts of India*. London: 1986.

Fergusson, James et al. *History of Indian and Eastern Architecture*, 2 vols. Delhi: 1967.

Foekema, Gerard. *A Complete Guide to Hoysala Temples*, New Delhi: 1996.

Fritz, J.M., G. Michell and M.S. Nagaraja Rāo. *Where Kings and Gods Met: the Royal Centre at Vijayanagara*. Tucson: 1983.

Fritz, J.M. et al. *New Light on Hampi: Recent Research at Vijayanagara*. Mumbai: 2001.

Ghurye, G.S. *Rajput Architecture*. Bombay (Mumbai): 1968.

Girhe, K.M. *Architecture of Bhoslas of Nagpur*, 2 vols. Delhi: 2004.

Gollings, J.J.M. and G. Muchell, *Vijayanagara: City of Victory, a South Indian Hindu Capital*. New York: 1991.

Goodwin, Godfrey. *A History of Ottoman Architecture*. London: 1971.

Grover, Satish. *Islamic Architecture in India*. New Delhi: 1981.

———, *Masterpieces of Traditional Indian Architecture*. New Delhi: 2004.

Harle, J.C. *Temple Gateways in South India*. Oxford: 1963.

Hardy, Adam. *Indian Temple Architecture: Form and Transformation*. New Delhi: 1995.

Hari Rāo, V.N. *The Srirangam Temple: Art and Architecture*. Tirupati: 1967.

Havell, E.B. *Encyclopaedia of Architecture in the Indian Subcontinent*. repr. New Delhi: 2000.

Husain, A.B.M. *Fatehpur-Sikri and Its Architecture*. Dacca: 1970.

Jain, J. and J. Jain-Neubauer, eds. *Rajput Art and Architecture*. Wiesbaden: 1966.

Jain-Neubauer, Jutta, *The Stepwells of Gujarat: In Art–Historical Perspective*. Bangalore: 1981.

Jung, C.G. *Mandala Symbolism*, tr. R.F.C. Hull. Princeton: 1973.

Khanna, Amar Nath. *Archaeology of India*. New Delhi: 1981.

Koch, Ebba. *Mughal Architecture: An Outline of Its History and Development 1526–1858*. London: 1991.

Kramrisch, S. *The Hindu Temple*: Calcutta: 1946.

Krishna Murthy, K. *Early Indian Secular Architecture*. Delhi: 1987.

Livingston, Morna. *Steps to Water: The Ancient Stepwells of India*. Princeton: 2002.

Mahapatra, Sthapaka Miranjana. *Silpararnakosa: A Glossary of Orissan Temple Architecture*. New Delhi: 1994.

Majumdar, R.C., ed. *The Delhi Sultanate*. Bombay: 1960.

Mehta, Rustam. *Masterpieces of Indo-Islamic Architecture*. Bombay: 1976.

Michell, G., ed. *The Royal Palaces of India*. London: 1998.

———, *The Vijayanagara Courtly Style*. New Delhi: 1992.

Misra, B.D. *The Forts and Fortresses of Gwalior and Its Hinterlands*. New Delhi: 1993.

Mukherji, Anisha Shekhar. *The Red Fort of Shahjahanabad*. New Delhi: 2003.

Nath, R. *Fatehpur Sikri and Its Monuments*. Agra: 2000.

———, *History of Mughal Architecture*. New Delhi: 1985.

———, *History of Sultanate Architecture*. New Delhi: 1978.

———, *Islamic Architecture and Style in India*. New Delhi (?): 1982.

Prabhakar Apte, June. "Vastu-Purusha Mandala: Dramatic Architectural and Artistic Study of Vastu-Purusha Mandala in Vastu-Vedic, Agamic and Puranic Tradition with Special Reference to Mayamata". *1993 Seminar on Mayonic Science & Technology*, Madras (unpublished: Library of the American Institute of Indian Studies, Gurgaon, Haryana).

Rāo, T.A. Gopinatha. *Elements of Hindu Iconography*, 4 vols. Madras: 1914-16; repr. Delhi: 1993.

Rowland, Benjamin. *The Art and Architecture of India: Hindu, Buddhist, Jain*. Baltimore: 1967.

Sachdev, Vibhuti and Giles Tillotson. *Building Jaipur: The Making of an Indian City*. New Delhi: 2002.

Safrani, Shehbaz H. *Golconda and Hyderabad*. Bombay: 1992.

Sarkar, H. *Some Contributions of India to the Ancient Civilization of Indonesia and Malaysia*. Calcutta: 1970.

———, *An Architectural Survey of Temples of Kerala*. New Delhi: 1978.

Sastri, Nilakanta. *South India Influences in the Far East*. s.l.: 1949.

Schotten Merklinger, E. *Indian Islamic Architecture: The Deccan 1347–1686*. Warminster: 1981.

Sharma, B.N. *Iconographic Parallelism in India and Nepal*. Hyderabad: 1975.

Sharma, P.K. *Mughal Architecture of Delhi*. New Delhi: 2000.

Sherring, M.A. *Banaras: The Sacred City of the Hindus* (London: 1868), repr. Delhi: 1990.

Singh, A.P. *Forts and Fortifications of India*. Delhi: 1993.

Sivaramamurti, C. *The Art of India*. New York: 1977.

Soundara Rajan, K.V. *Ahmadabad*. New Delhi: 1980.

———, *Islam Builds in India*. Delhi: 1983

Srinivasan, K.R. et al. *Sacred Tanks of South India*. Chennai: 2002.

Stierlin, Henri, ed. *Architecture of the World: India*. Lausanne: n.d.

———, *Architecture of the World: Islamic India*. Lausanne: n.d.

———, *Hindu India: From Khajuraho to the Temple City of Madurai*. Köln: 2002.

Sthapaka Miranjana Mahapatra. *Silpararnakosa: A Glossary of Orissan Temple Architecture*. New Delhi: 1994.

Stuart, C.M. Villiers. *Gardens of the Great Mughals*. New Delhi: 1983.

Sumet Jumsai. *Seen: Architectural Forms of Northern Siam and Old Siamese Fortifications*. Bangkok: 1970.

———, *Water: Origin of Ritual, Architecture and Urban Planning*. Bangkok: 1982 (in Thai).

Suresh, K.M. *Temples of Hampi and Its Environs*. Delhi: 2003

Tadgell, Christopher. *The History of Architecture in India*. London: 1990.

Tewari, S.P. *Hindu Iconography*. New Delhi: 1979.

Thapar, Bindia. *Introduction to Indian Architecture*. Singapore: 2004.

Tillotson, G.H.R. *Mughal India*. London: 1990.

———, *The Rajput Palaces*. New Haven and London: 1987.

———, *The Rajput Palaces: The Development of an Architectural Style, 1450–1750*. London: 1999.

Toy, S. *Strongholds of India*. London: 1957.

———, *The Fortified Cities of India*. London: 1965.

Vatsyayan, Kapila. *The Square and the Circle of the Indian Arts*. New Delhi: 1983.

Volwahsen, A. *Living Architecture: Islamic-Indian*. London: 1970.

Waliullah Khan, Mohammad. *Lahore and Its Important Monuments*. Karachi: 1973.

Yazdani, Ghulam. *Bidar: Its History and Monuments*. London: 1947.

———, *Māṇḍu: The City of Joy*. Oxford: 1929.

## Internet sources

http://images.google.com/imgres?imggurl=http

http://pruned.blogspot.com/2006/03/chand-Baoli.html

http://news-sv.alj.or.jp/jabs/s1/jabs0208-019.pdf

www.cheeseholiday.com/photos/photointro.php?loc=61

www.dewindia.org/images/kin1.jpg

www.ecoheritage.cpreec.org/sacred_tanks_karnataka.htm

www.ecoheritage.cpreec.org/sacred_tanks_kerala.htm

www.ecoheritage.cpreec.org/sacred_tanks_tnpondy.htm

www.gardenvisit.com/royalty-free-images/photos/index.php

www.indiantemples.com

www.indiantemples.com/tamilnadu.html
www.indiasite.com/gujarat/stepwellsofgujarat.html
www.indiawaterportal.org/tt/wbr
www.jm(Im).cahs.colostate.edu
www.joffrehorior.com/gallery
www.rainwaterharvesting.org/Rural/img/Khadins-img.jpg
www.sacredsites.com/asia/india/india/html
www.sacredsites.com/asia/India/tiruvanamalai.html
www.sacredsites.com/asia/tamilelibrary.org/teli/temples.html
www.spiritofchennai.com/rainwaterharvestingrhstp_files/images
www.srirangam.org
www.sscnet.ucla.edu/southasia/Culture/archit/stepw.html
www.tamilnation.org/culture/architecture/srirangam.html
www.templenet.com/Tamilnadu/df001.html
www.theconnection.org/photogallery/step(2water
www.the-south-asian.com/Sept2000/Baoli.htm
www.ukhap.nic.in/reports/SACTANKS.PDF
www.uq.net.au/~zzhsoszy
www.vijayanagara.org/HTML/Sacred.html
www.worldstatesmen.org/India_princes_A-J,html

# Acknowledgement

From the Great Indian Desert (Thar Desert) in the north-west of the country to the myriad tanks and pools scattered throughout India, one, in particular, caught my attention and interest — the Geometric Tank or Stepped Tank of Vijayanagara. Numerous times, I returned to the reproductions of this monument and marvelled at its beauty and elegance.

This tank led me to search out others. From the vertiginous Caṇḍī Bāolī of Ābhānerī to the simple Zenana Tank of the Padmanābhapuram Palace complex in Tamil Nadu, I became aware that many of these architectural/engineering marvels were little known to the population at large and to many scholars. Needless to say, the architects, engineers and designers of these monuments were even less known, if their names were ever recorded.

It is, therefore, to these nameless architects, engineers and designers, many of whom must be considered geniuses, that I acknowledge and dedicate this endeavour.